# Lecture Notes in Mechanical Engineering

**Series Editors**

Francisco Cavas-Martínez, Departamento de Estructuras, Universidad Politécnica de Cartagena, Cartagena, Murcia, Spain

Fakher Chaari, National School of Engineers, University of Sfax, Sfax, Tunisia

Francesco Gherardini, Dipartimento di Ingegneria, Università di Modena e Reggio Emilia, Modena, Italy

Mohamed Haddar, National School of Engineers of Sfax (ENIS), Sfax, Tunisia

Vitalii Ivanov, Department of Manufacturing Engineering Machine and Tools, Sumy State University, Sumy, Ukraine

Young W. Kwon, Department of Manufacturing Engineering and Aerospace Engineering, Graduate School of Engineering and Applied Science, Monterey, CA, USA

Justyna Trojanowska, Poznan University of Technology, Poznan, Poland

Francesca di Mare, Inst of Energy Tech, Building IC-2/63, Ruhr-Universität Bochum, Bochum, Nordrhein-Westfalen, Germany

**Lecture Notes in Mechanical Engineering (LNME)** publishes the latest developments in Mechanical Engineering—quickly, informally and with high quality. Original research reported in proceedings and post-proceedings represents the core of LNME. Volumes published in LNME embrace all aspects, subfields and new challenges of mechanical engineering. Topics in the series include:

- Engineering Design
- Machinery and Machine Elements
- Mechanical Structures and Stress Analysis
- Automotive Engineering
- Engine Technology
- Aerospace Technology and Astronautics
- Nanotechnology and Microengineering
- Control, Robotics, Mechatronics
- MEMS
- Theoretical and Applied Mechanics
- Dynamical Systems, Control
- Fluid Mechanics
- Engineering Thermodynamics, Heat and Mass Transfer
- Manufacturing
- Precision Engineering, Instrumentation, Measurement
- Materials Engineering
- Tribology and Surface Technology

To submit a proposal or request further information, please contact the Springer Editor of your location:

**China:** Ms. Ella Zhang at ella.zhang@springer.com
**India**: Priya Vyas at priya.vyas@springer.com
**Rest of Asia, Australia, New Zealand:** Swati Meherishi at
swati.meherishi@springer.com
**All other countries:** Dr. Leontina Di Cecco at Leontina.dicecco@springer.com

To submit a proposal for a monograph, please check our Springer Tracts in Mechanical Engineering at http://www.springer.com/series/11693 or contact Leontina.dicecco@springer.com

**Indexed by SCOPUS. All books published in the series are submitted for consideration in Web of Science.**

More information about this series at http://www.springer.com/series/11236

Lionel Roucoules · Manuel Paredes ·
Benoit Eynard · Paz Morer Camo ·
Caterina Rizzi
Editors

# Advances on Mechanics, Design Engineering and Manufacturing III

Proceedings of the International Joint
Conference on Mechanics, Design
Engineering & Advanced Manufacturing,
JCM 2020, June 2–4, 2020

 Springer

*Editors*
Lionel Roucoules
Arts et Métiers Institute
Aix En Provence Cedex 1, France

Benoit Eynard
Université Technologie de Compiègne
Compiegne Cedex, France

Caterina Rizzi
Department of Management Information
and Production Engineering
University of Bergamo
Bergamo, Italy

Manuel Paredes
INSA Toulouse
Toulouse, France

Paz Morer Camo
Departamento de Mecánica
Universidad de Navarra
San Sebastián - Donostia, Guipúzcoa, Spain

ISSN 2195-4356          ISSN 2195-4364   (electronic)
Lecture Notes in Mechanical Engineering
ISBN 978-3-030-70565-7          ISBN 978-3-030-70566-4   (eBook)
https://doi.org/10.1007/978-3-030-70566-4

This Springer imprint is published by the registered company Springer Nature Switzerland AG
The registered company address is: Gewerbestrasse 11, 6330 Cham, Switzerland

# Preface

It is my great pleasure to write this preface on the 10th International Joint Conference on Mechanics, Design Engineering and Advanced Manufacturing (JCM 2020) proceedings.

The pleasure is even more important regarding the current world situation that reminds us how important humans, societies and earth are. This situation unique, unexpected, unprecedented with no known solution has highlighted human behaviors and values face to the emergency. While advanced IT technologies are used as real support for collaboration and health equipment manufacturing, the world economy and society are strongly impacted.

What can we learn from that situation? What can we learn to extend and imagine the future of our teaching and research activities?

H2020 program and Industry 4.0 roadmap have been defining our current research roadmap. It is now time to assess if these expectations have become reality or not. System, engineering, PLM, robotics, virtual and augmented reality, artificial intelligence and sustainability are the current topics of scientific conferences. The JCM 2020 proceedings presents 65 papers that present the maturity level of those technologies and organizations in the academic and industrial world.

Nevertheless, we also have to think of what could be those topics in the next future and the next conferences. What could we expect for 2040? Other daily human behaviors, other industrial technologies or other industrial and world economy? What is the future world expected today by the young generation in the next 10, 20 or 30 years? From our side, how we can imagine academic activities, new curriculum, new syllabus, what could be new teaching modalities, what could be scientific research to provide a result to society?

We have, certainly, to think of new equilibriums between earth, human and technology. How frugality and technologies can be mixed and think together? How do we have to provide new engineering industrial technologies and services? How do we have to think about human values and organizations? How do we have to think about environmental and planetary concerns?

All those thoughts should be the next topic of future conferences in the academic world.

Lionel Roucoules

# Organization Committee JCM 2020 Members

## Conference Chair

Lionel Roucoules                    Arts et Métiers Institute of Technology, France

## Conference Program Chair

Manuel Paredes                      Institut National des Sciences Appliquée
                                    Toulouse, France

## Conference Advisory Chairmen

Benoît Eynard                       Université de Technologie de Compiègne, France
Paz Morer-Camo                      Universidad de Navarra Tecnun, Spain
Caterina Rizzi                      University of Bergamo, Department
                                       of Management, Informatio And Production
                                       Engineering, Italy

## Scientific Committee

Rita Ambu                           Università di Cagliari
Antonio Boccaccio                   Politecnico di Bari
Ileana Bodini                       Università di Brescia
Yuri Borgianni                      Libera Università di Bolzano
Francesco Buonamici                 Università di Firenze
Rosaria Califano                    Università degli Studi di Salerno
Marina Carulli                      Politecnico di Milano
Daniel Chindamo                     Università di Brescia
Filippo Cucinotta                   Università di Messina
Luigi De Napoli                     Università della Calabria
Claudio Favi                        Università di Parma

| | |
|---|---|
| Frizziero Leonardo | Università di Bologna |
| Marco Mandolini | Università Politecnica delle Marche |
| Giuseppe Marannano | Università di Palermo |
| Marco Marconi | Università della Tuscia |
| Domenico Marzullo | Università degli Studi di Trieste |
| Barbara Motyl | Università di Udine |
| Rocco Mozzillo | Università degli Studi della Basilicata |
| Paolo Neri | Università di Pisa |
| Pietro Piazzolla | Politecnico di Torino |
| Fabio Pini | Università di Modena e Reggio Emilia |
| Vito Ricotta | Università di Palermo |
| Gianpaolo Savio | Università degli Studi di Padova |
| Gaetano Sequenzia | Università di Catania |
| Domenico Speranza | Università degli Studi di Cassino e del Lazio Meridionale |
| Andrea Tarallo | Università degli Studi di Napoli Federico II |
| Davide Tumino | Università degli studi di Enna Kore |
| Andrea Vitali | Università di Bergamo |
| Elisabetta Zanetti | Università di Perugia |
| Fernando Aguilar Torres | Universidad de Almería |
| Francisco Cavas Martinez | Universidad Politécnica de Cartagena |
| Cristina Doñate | Univesidad de Jaén |
| Daniel Garcia Fernandez Pacheco | Universidad Politécnica de Cartagena |
| Cesar Garcia Hernandez | Universidad de Zaragoza Martín |
| Valentín Gomez Jauregui | Universidad de Cantabria |
| Rafael Hidalgo | Universidad de Cordoba |
| Cristina Manchado | Universidad de Cantabria |
| Maria Luisa Martinez Muneta | Universidad Politécnica de Madrid |
| Rikardo Minguez Gabiña | Universidad del País Vasco Aguilar |
| Ramón Miralbes Buil | Universidad de Zaragoza |
| David Ranz | Universidad de Zaragoza |
| José Ignacio Rojas Sola | Universidad de Jaén |
| Irene Sentana Gadea | Universidad de Alicante Rojas Sola |
| Egoitz Sierra | Universidad del País Vasco |
| Eneko Solaberrieta | Universidad del País Vasco |
| Nabil Anwer | Université Paris-Saclay |
| Alex Ballu | Université de Bordeaux |
| Richard Bearee | Arts et Métiers - Lille |
| Alain Bernard | Ecole Centrale Nantes |
| Jean-François Boujut | Grenoble INP |
| Sébastien Campocasso | Université de Toulon |
| Vincent Cheutet | INSA Lyon |
| Alain Daidie | INSA Toulouse |
| Jean-Yves Dantan | Arts et Métiers - Metz |

| Emmanuel DUC | SIGMA Clermont |
| Benoit Eynard | Université de technologie de Compiègne |
| Xavier Fisher | ESTIA Bidart |
| Pascal Lafon | Université de technologie de Troyes |
| Samir Lamouri | Arts et Métiers - Paris |
| Yann Landon | Université Paul Sabatier Toulouse |
| Florent Laroche | Ecole Centrale Nantes |
| Julien Le Duigou | Université de technologie de Compiègne |
| Frédéric Noël | Grenoble INP |
| Manuel Paredes | INSA Toulouse |
| Yann Quinsat | Université Paris-Saclay |
| Tatiana Reyes | Université de technologie de Troyes |
| Lionel Roucoules | Arts et Métiers - Aix-en-Provence |

## Organizing Committee

| Grazyna Cauquil | Arts et Métiers Institute of Technology, France |
| Nathalie Comte | Arts et Métiers Institute of Technology, France |
| Fakhreddine Ababsa | Arts et Métiers Institute of Technology, France |
| Richard Bearee | Arts et Métiers Institute of Technology, France |
| Aline Cauvin | Aix-Marseille University, France |
| Jean-Remy Chardonnet | Arts et Métiers Institute of Technology, France |
| Julien Gomand | Arts et Métiers Institute of Technology, France |
| Nathalie Klement | Arts et Métiers Institute of Technology, France |
| Adel Olabi | Arts et Métiers Institute of Technology, France |
| Jean-Philippe Pernot | Arts et Métiers Institute of Technology, France |
| Esma Yahia | Arts et Métiers Institute of Technology, France |
| Arnaud Polette | Arts et Métiers Institute of Technology, France |
| Pierre Garambois | Arts et Métiers Institute of Technology, France |
| Florian Huet | Arts et Métiers Institute of Technology, France |

# Contents

# Manufacturing Engineering

# Designing Powder Metallurgy Process - The Influence of High Sintering Temperature on Dimensional and Geometrical Precision

Marco Zago[✉], Ilaria Cristofolini, and Sasan Amirabdollahian

Department of Industrial Engineering, University of Trento, Via Sommarive 9, 38123 Trento, Italy
marco.zago-1@unitn.it

**Abstract.** The precision of parts produced by Powder Metallurgy (PM) strongly depends on the careful design of PM process parameters. Among them, high sintering temperature is generally considered as detrimental for dimensional and geometrical precision, and therefore neglected in industrial production. Nevertheless, high sintering temperature would strongly improve mechanical characteristics of PM parts, so that the real influence of high sintering temperature on dimensional and geometrical precision is of great interest for PM companies. This study investigates the influence of sintering temperature (up to 1350 °C) on dimensional and geometrical precision of real parts. Dimensional changes on sintering and the effect of sintering temperature have been evaluated. Geometrical characteristics have been measured both in the green and in the sintered state, and the real influence of sintering temperature has been highlighted. As a conclusion, it has been demonstrated that the larger shrinkage due to the high sintering temperature is not detrimental with respect to the dimensional precision, being it reliably predictable. Moreover, the influence on geometrical characteristics is unexpectedly low. The encouraging results of this study convinced the main PM companies in Europe to further investigate the influence of high sintering temperature, as partners in a Club Project within the European Powder Metallurgy Association (EPMA).

**Keywords:** Dimensional and geometrical precision · Product development · Precision engineering · Design for powder metallurgy · High temperature sintering process

## 1 Introduction

Dimensional and geometrical precision is one of the most interesting characteristics of Powder Metallurgy (PM) parts, strongly depending on the proper design of the PM process parameters [1–3]. In the conventional press and sinter technology, the green parts are produced by cold compaction in rigid dies and sintered in controlled atmosphere. Dimensional and geometrical precision of green parts is generally very good and may get worse during sintering. Sintering temperature is one of the most critical variables, determining the shrinkage (or swelling) of green parts [4, 5]. On increasing sintering

© The Author(s) 2021
L. Roucoules et al. (Eds.): JCM 2020, LNME, pp. 3–8, 2021.
https://doi.org/10.1007/978-3-030-70566-4_2

temperature, the larger shrinkage combined with the possible deformation of the parts due to the phenomena occurring at high temperature (creep, viscous flow, phase transformation…) may negatively influence the dimensional and geometrical precision. On the other hand, the positive effect of high sintering temperature on the mechanical properties of porous steels, resulting from the enhanced densification and from the improvement of pore morphology, is well known [6–8]. Better understanding the real effect of high sintering temperature on dimensional and geometrical precision is consequently a priority for PM companies.

This work investigates the influence of high and very high sintering temperature on the dimensional and geometrical precision of industrial Cr steel parts. Sintering temperatures are higher than those conventionally used in the industrial practice. The shrinkage of dimensions perpendicular and parallel to the compaction direction has been firstly evaluated, aiming at checking any anisotropy of dimensional changes, which is expected on the basis of previous work [1]. Geometrical characteristics in the green and sintered state have been also compared, aiming at distinguishing the influence of the different process steps. Aim of the work is demonstrating that high sintering temperature is not significantly detrimental for dimensional and geometrical precision.

## 2  Experimental Procedure

The dimensions and the geometrical characteristics of the parts studied are shown in Fig. 1. From here on, the parts will be identified as Part 1, the smaller, and Part 2, the larger.

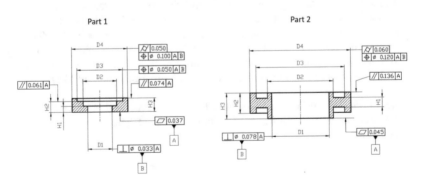

**Fig. 1.** Parts studied.

The parts were produced compacting a commercial water atomized iron powder pre-alloyed with 3%wt Cr and 1.5%wt Mo, to which 0.8%wt lubricant and 0.55%wt C (graphite) were added. Parts were compacted up to 7.0 g/cm$^3$ and sintered in vacuum furnace at four different sintering temperatures, higher than the standard ones (1200 °C, 1250 °C, 1300 °C, and 1350 °C).

Due to confidentiality reasons, neither minor details nor exact dimensions are explicitly reported. The two parts allowed investigating a broad range of dimensions (5 to

70 mm for diameters, 4 to 20 mm for heights), so that the influence of both sintering temperature and size was highlighted.

Parts were measured both in the green and in the sintered state by a Coordinate Measuring Machine (CMM), scanning mode. Acquired data were processed by MATLAB to calculate the features, from which dimensions and geometrical characteristics were derived, according to the procedure for data processing explained in depth in [9].

## 3  Results and Discussion

### 3.1  Dimensional Precision

Dimensions were grouped in diameters (orthogonal to compaction force), and heights (parallel to compaction force). Dimensional change on sintering was calculated by Eqs. (1) and (2) for diameters and heights, where subscript $g$ and subscript $s$ refer to green and sintered state, respectively.

$$\varepsilon_D = \frac{D_s - D_g}{D_g} \tag{1}$$

$$\varepsilon_H = \frac{H_s - H_g}{H_g} \tag{2}$$

Dimensional change of diameters and heights vs. size are reported in Fig. 2, as affected by sintering temperature.

Dimensional change increases on increasing sintering temperature, as expected, both for diameters and for heights [10]. Anisotropy of dimensional changes is also confirmed, in all the cases dimensional change of diameter is different than dimensional change of height.

Dimensional change is almost unaffected by size for diameters, while for height dimensional change tends to increase on increasing height, and the trend is particularly evident in Part 1 (the smallest one). Further study might prove the role of powder compaction (triaxial stress field, density gradient and ejection procedure) on the different height shrinkage, as proposed in [11]. Nevertheless, dimensional changes can be reliably predicted according to the design procedure defined in previous work [10, 12].

### 3.2  Geometrical Precision

Firstly, considering individual features, flatness and cylindricity are evaluated. Results are shown in Fig. 3, where the values measured in the green state and after sintering at the different temperatures are shown. From here on, the green bars will be related to the geometrical characteristics measured in the green state, the blue bars to the geometrical characteristics measured after sintering at the different temperatures, and the red bar to the tolerance.

Flatness slightly worsens on increasing sintering temperature, and same for cylindricity. Larger scatter bands are observed for Part 2, mainly due to the higher geometrical complexity, which makes homogeneous filling more difficult. Nevertheless, measured values are kept lower than tolerances in all the cases.

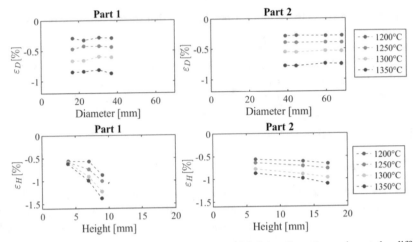

**Fig. 2.** Dimensional change of diameters (upper) and heights (lower) vs. size at the different sintering temperatures.

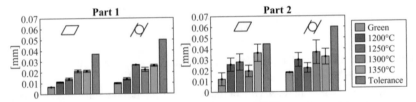

**Fig. 3.** Flatness and cylindricity in the green state and at the different sintering temperatures for the two parts studied – comparison with the tolerance.

Parallelism, perpendicularity, and co-axiality, both in the green state and at the different sintering temperatures, are shown in Fig. 4. In all the cases, bottom surface in compaction (that is the surface contacting parts holder on sintering) has been considered as primary datum feature. In part 1, according to the drawing shown in Fig. 1, the flatness is identified by the height determining the corresponding surface, and the coaxiality (as by the position tolerance) is identified by the diameter of the corresponding cylindrical surface.

In Part 1, parallelism slightly worsens on increasing the sintering temperature, while perpendicularity is almost unaffected. Coaxiality slightly worsens considering the surface defined by diameter 3, while no influence is observed for the surface defined by diameter 4. In all the cases, tolerance is never exceeded.

In part 2, measured values of parallelism are largely scattered, and this is mainly due to the uneven filling sometimes occurring in large parts, given that same scatter is observed in the green part. Tolerance, in terms of mean values, is never exceeded. Perpendicularity is almost unaffected by sintering temperature, but tolerance is exceeded. However, this comes from compaction step, given that perpendicularity measured in the green parts is only slightly worse after sintering. No significant influence of sintering temperature on coaxiality is observed, and tolerance is never exceeded.

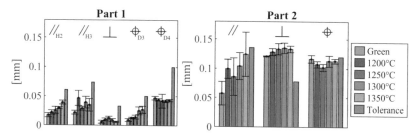

**Fig. 4.** Parallelism, perpendicularity and coaxiality in the green state and at the different sintering temperatures for the two parts studied – comparison with the tolerance.

Considering Part 2, it can be observed that effort could be required in the compaction step, aiming at improving geometrical characteristics, while the influence of sintering temperature is lower than expected.

## 4  Conclusions

The influence of high sintering temperature, up to 1350 °C, on the dimensional and geometrical precision of low alloyed steel parts has been investigated in this work. Main results are hereafter summarized.

- Dimensional change of diameters increases on increasing sintering temperature, irrespective on size, while dimensional change of height increases on increasing sintering temperature and size, and the effect is more evident for small sizes. As by the design procedure previously developed, this result does not represent an obstacle to sintering at high and very high temperature.
- The influence of high sintering temperature on geometrical characteristics is surprisingly low. The worsening, if any, never determines tolerance exceeding.
- High sintering temperature, determining noticeable improvement in the mechanical properties of sintered parts, can be used without significant detriment of dimensional and geometrical precision.

Further work is in progress within the EPMA Design for Sintering project.

## References

1. Cristofolini, I., Pilla, M., Rao, A., Libardi, S., Molinari, A.: Dimensional and geometrical precision of powder metallurgy parts sintered and sinterhardened at high temperature. Int. J. Precis. Eng. Manuf. **14**(10), 1735–1742 (2013)
2. Engstrom, U., Lindberg, C., Tengzelius, J.: Powder and process for high performance PM steels. Powder Metall. **35**(1), 67–72 (1992)
3. Griffo, A., Ko, J., German, R.M.: Critical assessment of variables affecting the dimensional behavior in sintered iron-copper-carbon alloys. Adv. Powder Metall. Part. Mater. **3**, 221–223 (1994)

4. Emanuelli, L., Menapace, C., Cristofolini, I., Molinari, A., Larsson, M.: Influence of sintering temperature on shrinkage anisotropy in Cr-Mo low alloy steel green compacts. Adv. Powder Metall. Part. Mater. **05**, 99–107 (2014)
5. Takata, J., Nawai, N.: Dimensional changes during sintering of iron based powders. Powder Metall. **38**(3), 209–213 (1995)
6. Sanderow, H., Pease, L.: Effect of high temperature sintering conditions on the mechanical properties and microstructure of FLN4-4405 P/M steel. Adv. Powder Metall. Part. Mater. 51–63 (2005)
7. Piotrowski, A., Biallas, G.: Influence of sintering temperature on pore morphology, microstructure, and fatigue behavior of MoNiCu alloyed sintered steel. Powder Metall. **41**(2), 109–114 (1998)
8. Beiss, P., Dalgic, M.: Structure property relationships in porous sintered steels. Mater. Chem. Phys. **67**(1–3), 37–42 (2001)
9. Cristofolini, I., Menapace, C., Cazzolli, M., Rao, A., Pahl, W., Molinari, A.: The effect of anisotropic dimensional change on the precision of steel parts produced by powder metallurgy. J. Mater. Process. Technol. **7**(212), 1513–1519 (2012)
10. Zago, M., Larsson, M., Cristofolini, I.: An improved design method for net-shape manufacturing in powder metallurgy. In: Lecture Notes in Mechanical Engineering, pp. 257–267 (2020).
11. Zago, M., Cristofolini, I., Molinari, A.: New interpretation for the origin of the anisotropic sintering shrinkage of AISI 316L rings based on the anisotropic stress field occurred on uniaxial cold compaction. Powder Metall. **62**(2), 115–123 (2019)
12. Cristofolini, I., Molinari, A., Zago, M., et al.: Design for powder metallurgy – predicting anisotropic dimensional change on sintering of real parts. Int. J. Precis. Eng. Manuf. **20**, 619–630 (2019)

# Process Parameters Effect on Weld Beads Geometry Deposited by Wire and Arc Additive Manufacturing (WAAM)

Supasit Manokruang(✉), Frederic Vignat, Matthieu Museau, and Maxime Limousin

Univ. Grenoble Alpes, CNRS, Grenoble INP, G-SCOP, 38000 Grenoble, France
supasit.manokruang@grenoble-inp.fr

**Abstract.** Among Additive Manufacturing technologies, Wire and Arc Additive Manufacturing process is strongly dependent of deposition conditions such as welding parameters, substrate temperature, trajectory. In this research, geometry and temperature evolutions of single beads have been investigated according to process parameters modifications. For our experiment, a heating device have been used in order to control the substrate temperature from room temperature up to 400 °C. Considering the Cold Metal Transfer technology, welding parameters, Wire Feed Speed (*WFS*) and Travel Speed (*TS*), have been modified while keeping a constant ratio λ (*WFS/TS*). Results indicate that weld bead geometry, height (*h*) and width (*w*), is influenced by substrate temperature and welding parameters. It has been shown that substrate temperature, itself influenced by process parameters, tends to produce thicker and lower weld beads while it increases.

**Keywords:** Wire and Arc Additive Manufacturing · Additive Manufacturing · Cold Metal Transfer · Single weld bead · Thermal effect

## 1 Introduction

Additive Manufacturing (AM) technologies build up materials layer by layer, allowing complex shapes to be produced. Among various processes in AM, wire-base techniques show a lower energy requirement, about 85%, than those based on powder [1, 2]. For Wire and Arc Additive Manufacturing (WAAM) technology, several modes of energy distribution are available according to suppliers.

Cold Metal Transfer (CMT) is one of them and is very popular for arc welding. In effect, CMT process requires much less energy to work than tungsten inert gas (TIG) or plasma processes. Thanks to this lower energy, thin plates and low cladding of aluminum alloys have been possible. This specificity have been exploited in WAAM as, for example, several studies have shown its interest to be able to produce parts with maximize density [3–5]. Moreover, various study have shown that temperature and thermal accumulation in WAAM strongly influence parts geometry [6, 7]. As CMT technology uses a lower amount of energy compared to other techniques, it allows to minimize these deformation induced by heat accumulation. Unfortunately, when it comes to parts with complex

© The Author(s) 2021
L. Roucoules et al. (Eds.): JCM 2020, LNME, pp. 9–14, 2021.
https://doi.org/10.1007/978-3-030-70566-4_3

shapes, it is difficult to predict nor control the temperature and its influence on the final geometry. This study places the first steps aiming at this understanding. It has been proposed to focus, at first, on weld bead as it constitutes the elementary brick of any volumic parts manufacturing.

This study investigates the interacting influences between the substrate temperature and the geometry evolution base on welding parameters of weld beads.

## 2  Experimental Setup

A YASKAWA MA1440 Robot and a Fronius TPS CMT 4000 Advanced welding station (Fronius, Pettenach, Austria) have been used in order to produce the different weld beads of 80 mm length. A 1.2 mm diameter aluminum wire, 5356 alloy, have been involved with a working distance of 15 mm. An argon shielding gas was used with a flow rate of 13 L/min. Weld beads have been deposited on a ER5356 Al alloy substrate whose dimensions were 350 mm × 350 mm × 5 mm ($w * l * h$). Then, using the 875 CMT synergic law provided by Fronius, Wire Feed Speed ($WFS$) and Travel Speed ($TS$) have been modified while their ratio $\lambda = WFS/TS$ was kept constant at a value of 8.33 because at previous of the research experiment, these things use $WFS = 5$ m/min, $TS = 0.60$ m/min and ratio = 8.33. Therefore, adding four different sets of parameter have thus been fixed as detailed in Table 1.

**Table 1.** Process parameters for experiment

| Symbol | WFS (m/min) | TS (m/min) | Ratio |
|--------|-------------|------------|-------|
| A | 3 | 0.36 | 8.33 |
| B | 4 | 0.48 | |
| C | 5 | 0.60 | |
| D | 6 | 0.72 | |

In order to control precisely the substrate temperature ($T_{sub}$) of our constructions, a specific equipment has been designed. It is composed by 3 heating resistances (6 kW) allowing temperature regulation up to 500 °C thanks to an electrical controller. For the experiment needs, several substrate temperatures have been applied to all conditions A, B, C and D. Starting at room temperature (25 °C), tests have been thus led at 100, 200, 300 and 400 °C. As shown in Fig. 1. Smaller temperature increase have been revealed insufficient in our previous experiments to involve significant geometry evolutions. During production, evolutions of substrate temperature have been recorded thanks to four thermocouples (k type) placed along with the welding direction at a distance of 10 mm from weld beads and spread by 20 mm. In the end, a 3D laser scanner (Artec Space Spider), with a resolution of 0.10 mm, was used to acquire a precise numerical model of weld beads geometry from which height and width have been extracted.

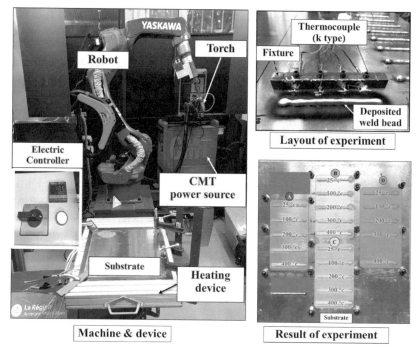

**Fig. 1.** Installation device and experiment set up

## 3   Results and Discussion

### 3.1   Temperature Evolution

The recording system previously describe is able to give various information on substrate temperature during welding and its evolution. In this way, the maximum temperature peak ($T_{max}$) reached during the process can be easily obtained for each conditioned as well as the cooling time ($t_{cool}$) needed to come back to the initial condition. All results have been compiled Table 2.

**Table 2.** Temperature evolution according to process parameters

| $T_{sub}$ (°C) | $T_{max}$ (°C) | | | | $t_{cool.}$ (s) | | | | $\Delta T = T_{max} - T_{sub}$ (°C) | | | | $(T_{max} - T_{sub})/T_{sub}$ (%) | | | |
|---|---|---|---|---|---|---|---|---|---|---|---|---|---|---|---|---|
| | A | B | C | D | A | B | C | D | A | B | C | D | A | B | C | D |
| 25 | 107 | 121 | 126 | 148 | 88 | 102 | 112 | 121 | 82 | 96 | 101 | 123 | 328 | 384 | 404 | 492 |
| 100 | 178 | 183 | 186 | 207 | 73 | 91 | 93 | 114 | 78 | 83 | 86 | 107 | 78 | 83 | 86 | 107 |
| 200 | 251 | 263 | 276 | 306 | 62 | 72 | 86 | 110 | 51 | 63 | 76 | 106 | 26 | 32 | 38 | 53 |
| 300 | 344 | 355 | 370 | 373 | 45 | 53 | 62 | 69 | 44 | 45 | 70 | 73 | 15 | 18 | 23 | 24 |
| 400 | 452 | 458 | 467 | 484 | 34 | 38 | 42 | 45 | 52 | 58 | 67 | 84 | 13 | 15 | 17 | 21 |

From these data, several ascertainments can be pulled. It appears that $T_{max}$ increase for higher *WFS* for each conditions, which seems logical as the deposited energy is driven by the amount of matter needed to be melted (*WFS*). As well, the rise of $T_{sub}$ induces obviously a rise in the recorded $T_{max}$ but in different proportions between a weld at room temperature and another one done at 100, 200, 300 and 400 °C. In this sense, even if $T_{max}$ is increasing with *WFS*, conversely, $t_{cool}$ is decreasing and the temperature is stabilizing rather faster. This phenomenon can be explained by the fact that the difference in temperature $\Delta T$ ($T_{max} - T_{sub}$) is smaller.

## 4 Geometry Evolution

Using the numerical model provided by the 3D scanner, beads dimensions have been extracted. At first, the height has been measured as the average maximum highness. For this, the full bead length have been considered without both first and final 10 mm in order to not be influenced by arc ignition and extinction. From here, the width is measured as the average largeness extracted from a cross-section placed at mid-height and parallel to the substrate. Results are shown in Fig. 2 according to welding conditions and substrate temperature.

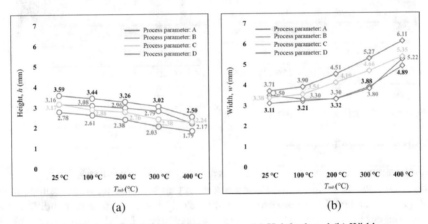

**Fig. 2.** Evolution of weld beads dimensions: (a) Height, h and (b) Width, w.

As we initially fixed a constant λ ratio, height and width are strongly linked and their product remain constant. Thus, while height is increasing or decreasing, width decreases or increases in the opposite way.

Considering this, mainly two parameters lead geometry evolutions of weld beads as can be observed in Fig. 2. In effect, as *WFS* and $T_{sub}$ rise, height decreases and width increases. This flattening of weld beads can be explained by the fact that these two factors are responsible for a general increase in temperature and leads to improve the wetting angle and fluidity of aluminum [8].

These geometrical evolutions can be important as for example, considering a constant *WFS*, there is a difference of 1 mm in height and almost 2 mm in width between a weld bead done at room temperature and another one welded on a 400 °C hot substrate.

Equally, for a constant $T_{sub}$, a WFS increase by a factor of 2 is responsible for a dimension variation about 1 mm in height and width.

These evolutions, bring back to the dimension of a weld bead, could be disastrous for parts construction if not taken into account.

## 5 Conclusion

This experiment has investigated some process parameters of Wire and Arc Additive Manufacturing process and their influences on geometry evolution of weld beads. An experiment was performed in order to track the evolution of temperature and weld beads dimensions. It has revealed that, for a constant amount of deposited material, the main parameters driving geometrical evolutions are WFS and $T_{sub}$. Whether by the fact that they directly control the input energy or by their nature, they both contribute to the general increase in temperature. In turn, this temperature rise leads to a better wetting angle and decreases the aluminum viscosity. In the end, weld beads tend to be lower and larger.

Those geometrical evolutions, potentially being driven, have to be imperatively considered as the temperature during the WAAM process is difficult to manage, especially on parts with complex shapes. Thus, a small difference on a weld bead can lead to important derives with a stack of dozens and dozens of layers.

The future work will study a single wall bead and kept constants temperature respective of deposition a new layer. It will provide important information for process planning WAAM.

## References

1. Ding, D., Pan, Z., Cuiuri, D., Li, H.: A multi-bead overlapping model for robotic wire and arc additive manufacturing (WAAM). Robot. Comput. Integr. Manuf. **31**, 101–110 (2015)
2. Jackson, M.A., Van Asten, A., Morrow, J.D., Min, S., Pfefferkorn, F.E.: A comparison of energy consumption in wire-based and powder-based additive-subtractive manufacturing. Procedia Manuf. **5**, 989–1005 (2016)
3. Xu, X., Ding, J., Ganguly, S., Diao, C., Williams, S.: Preliminary investigation of building strategies of maraging steel bulk material using wire + arc additive manufacture. J. Mater. Eng. Perform. **28**(2), 594–600 (2018)
4. Cong, B., Ding, J., Williams, S.: Effect of arc mode in cold metal transfer process on porosity of additively manufactured Al-6.3%Cu alloy. Int. J. Adv. Manuf. Technol. **76**(9–12), 1593–1606 (2014)
5. Sergio, R., Colegrove, P.A., Martina, F., Williams, S.W.: Analytical process model for wire + arc additive manufacturing. J. Addit. Manuf. **21**, 651–657 (2018)
6. Wu, B., et al.: Effects of heat accumulation on the arc characteristics and metal transfer behavior in Wire Arc Additive Manufacturing of Ti6Al4V. J. Mater. Process. Technol. **250**, 304–312 (2017)
7. Wu, B., Pan, Z., Duin, S.V., Li, H.: Thermal behavior in wire arc additive manufacturing: characteristics, effects and control. Trans. Intell. Weld. Manuf. **2**, 3–18 (2018)
8. Rodrigues, T.A., Duarte, V., Santos, T.G., Oliverira, J.P.: Current status and perspectives on wire and arc additive manufacturing (WAAM). J. Mater. **12**, 1121–1162 (2019)

# A Parametric Study of Additive Manufacturing Process: TA6V Laser Wire Metal Deposition

Valentine Cazaubon[1,2(✉)], Audrey Abi Akle[1], and Xavier Fischer[1,2]

[1] Univ. Bordeaux, ESTIA Institute of Technology, 64210 Bidart, France
v.cazaubon@estia.fr
[2] Arts et Metiers Institute of Technology, University of Bordeaux, CNRS, Bordeaux INP, INRAE, I2M Bordeaux, 33400 Talence, France

**Abstract.** Additive Manufacturing has proven to be an economically and industrially attractive process in building or repairing parts. However, the major issue of this new process is to guarantee a mechanical behavior identical to the subtractive manufacturing methodologies. The work, presented in this paper, is centered on the Laser Wire Metal Deposition (LMD-w) method with the metallic alloy TA6V. Its working principle is to fuse a coaxial wire on a substrate with a laser as a heat source. To better understand the interaction between the input parameters (Laser Power, Wire Feed Speed and Tool Speed) and the clad geometry output variables (Height, Width and Contact Angle) and the substrate displacement, we have realized an experimentation. We printed 9 clads according Taguchi's experimental design. Pearson correlation coefficient and Fisher test performed on the experimental measures showed as main result: Tool Speed is the parameter with the most significant influence on the output variables.

**Keywords:** Parameters influences · Clads geometry · Clads characterization · Taguchi's experimental design · Statistical analysis

## 1 Introduction

American Society for Testing Material (ASTM) has defined additive manufacturing (AM) as "a process of joining materials to make objects from 3D model data, usually layer upon layer, as opposed to subtractive manufacturing methodologies" [1]. AM has proven to be an economically and industrially attractive process. It achieves a buy-to-fly ratio, weight ratio between the raw material used for a part and the weight of the part, of 1:1 [2]. The wire processes allow to obtain a 99,6% deposit yield [3]. From the 7 categories of additive manufacturing processes [4] our work focuses on the Laser Wire Metal Deposition (LMD-w). Moreover, three alloys of materials are used in AM: Inconel, 316L steel and TA6V. Among them, the titanium alloy (TA6V) has low resistance to oxidation but presents the advantages of low density and high resistance to corrosion [5, 6]; that is why we have chosen to work with TA6V wire.

One major issue of AM process is to guarantee a mechanical behavior identical to the subtractive manufacturing methodologies. In this paper, we present our study

© The Author(s) 2021
L. Roucoules et al. (Eds.): JCM 2020, LNME, pp. 15–20, 2021.
https://doi.org/10.1007/978-3-030-70566-4_4

about the influences of first order input parameters on the geometric aspects and on the displacement of the substrate, metal part on which the molten wire is deposited to print the geometry [7]. To do so, we have performed an experimental campaign and have analyzed resulting data. The analysis consisted of two stages: the analysis of the correlation with the Pearson coefficient and an analysis of the variance with the Fisher test.

## 2  Protocol of Experiments

### 2.1  Model and Variables

Following the LMD-w process, the TA6V wire is deposited on the substrate, into the melting pool with a speed WFS coaxially to the laser at a power P while the robotic head is moving at a speed TS, Fig. 1.

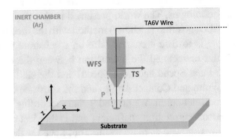

**Fig. 1.**  Schematic representation of the LMD-w process input parameters

**Table 1.**  LMD-W input parameters and measured outputs studied

| Nature | Name | Representation | Units | Domain |
|---|---|---|---|---|
| Input parameters | Laser power | P | W | [1680; 2000] |
| | Wire feed speed | WFS | m/min | [1,5; 2,1] |
| | Tool speed | TS | m/min | [0,2; 0,8] |
| Outputs | Clad height | $h_c$ | mm | Measured |
| | Dilution height | $h_d$ | mm | |
| | Clad width | $w_c$ | mm | |
| | Dilution width | $w_d$ | mm | |
| | Contact angle | $\alpha$ | ° | |
| | Substrate displacement | d | mm | |

The LMD-w process is modeled with input and output variables presented in Table 1. The objective of the experiments is to define relations between the outputs and the inputs in order to predict the clad's geometry according to a parameter combination.

## 2.2 Methods

We used, as substrates, plates of TA6V of dimensions 150 mm × 60 mm × 5 mm which have been sanded with orbital grinder (grain P60) and degreased with isopropyl alcohol (IPA). This surface treatment helps to improve surface laser absorptivity and to remove dusts. The deposit wire material is TA6V of 1.2 mm. The substrate and wire are made of titanium because of its weldability properties [8]. The substrate is bridled preventing any translational movements in the plan (x, y). TA6V is supplied by Technalloy. To protect the melting pool from oxidation, we have built the clads in an inert chamber with a protective gas: argon as recommended [6].

The inert chamber is part of a robotized cell. The cell is composed of a 6-axis robot (KUKA KR60-HA) with its controller (KUKA KRC4). The energy is bringing by a laser head (PRECITEC CoaxPrinter) fixed on the robot. As described in Fig. 1, the wire deposition is coaxial to the laser.

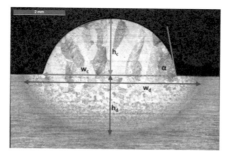

**Fig. 2.** Cross section micrography of TA6V deposit (×5). Outputs: red: clad height, green: dilution height, blue: clad width and white: dilution width.

Nine clads of 100 mm length have been deposited on each substrate. Three cuts have been made on each substrate (i.e. ¼ cord length, ½ cord length and ¾ cord length) using a silicon disc and a cutting wheel. Subsequently, the samples are coated, mirror polished and then chemically attacked with Kroll's reagent. The measurements will be carried out under a microscope (LEICA DM1750 M) and its software Leica Application Suite (see example in Fig. 2).

## 2.3 Taguchi's Experimental Design

We know according to [7] the interactions between the input parameters can be neglected. The model is a linear expression of the outputs based on the input parameters. The coefficients of the model can be calculated only after performing the experiments. In absence of these coefficients, we write the model symbolically in order to clarify the parameters inputs considered. Thus, we obtained the studied model described by the Eq. (1), with $y$ an output and $M$ the average of the output's values as the constant term. As we would like to know if there are (non-)linearity between the input and the output variables, we defined three levels for the input parameters P, WFS and TS. To determine the number of experiments, we used Taguchi's experimental design based on the use of orthogonal tables.

$$y = M + P + WFS + TS \tag{1}$$

Following the Taguchi rules, we must know the total degree of freedom (tfd) of our model (Eq. (1)) and the Least Common Multiple (LCM). With our model, we obtained tfd equals 8 and LCM equals 9. So, the chosen table contains 9 experiments ($L_9(3^3)$). The Table 2 gathers the input parameters and measured outputs.

**Table 2.** Inputs parameters and measured responses

| N | Input parameters | | | Outputs | | | | | |
|---|---|---|---|---|---|---|---|---|---|
| | *P (W)* | *WFS (m/min)* | *TS (m/min)* | *d (mm)* | *$h_c$(mm)* | *$h_d$(mm)* | *$w_c$(mm)* | *$w_d$(mm)* | *α (°)* |
| 1 | 1680 | 1.5 | 0.2 | 1.978 | 2.148 | 2.2033 | 5.151 | 6.939 | 75 |
| 2 | 1680 | 1.8 | 0.4 | 0.892 | 1.742 | 1.2847 | 3.832 | 4.873 | 76 |
| 3 | 1680 | 2.1 | 0.6 | 1.357 | 1.533 | 0.9797 | 3.493 | 4.067 | 87 |
| 4 | 1840 | 1.5 | 0.4 | 1.235 | 1.353 | 1.5443 | 4.480 | 5.633 | 55 |
| 5 | 1840 | 1.8 | 0.6 | 1.125 | 1.213 | 1.1723 | 3.878 | 4.826 | 56 |
| 6 | 1840 | 2.1 | 0.2 | 1.910 | 3.038 | 2.0567 | 4.306 | 6.320 | 103 |
| 7 | 2000 | 1.5 | 0.6 | 0.443 | 0.960 | 1.2713 | 4.345 | 5.232 | 39 |
| 8 | 2000 | 1.8 | 0.2 | 1.635 | 2.361 | 2.6157 | 5.726 | 7.592 | 72 |
| 9 | 2000 | 2.1 | 0.4 | 1.177 | 1.783 | 1.4057 | 4.493 | 5.571 | 71 |

## 3   Results and Analysis

### 3.1   Correlation Coefficient

The correlation coefficient was calculated using the Pearson's method with R software. We used it to determine whether the relationship between an input parameter and a measured output is linear or not and if it is positive or negative. If |r| > 0.6, the two studied variables tend to have a linear relationship [9]. Table 3 presents the correlation coefficients and their significance.

**Table 3.** Pearson's coefficient between the input parameters and the outputs with their significance

| Input parameters | Outputs | | | | | |
|---|---|---|---|---|---|---|
| | $h_c$ | $h_d$ | $w_c$ | $w_d$ | $d$ | $α$ |
| P | −0.05 | 0.21 | 0.42* | 0.32 | −0.27 | 0.40* |
| WFS | 0.45* | −0.11 | −0.31 | −0.20 | 0.26 | 0.69*** |
| TS | −0.86*** | −0.91*** | −0.73*** | −0.87*** | −0.77*** | −0.50** |

P-value (0; 0.001; 0.01; 0.05; 0.1) – Respective Symbols (;***; **; *;).

The analysis establishes WFS and $\alpha$ follow a positive linear law with $r = 0.69$ and p-value $< 0.001$. The results allow to conclude to negative linear relationships between TS and bead's geometries ($h_c$ and $w_c$), TS and dilution's geometries ($h_d$ and $w_d$) and TS and the substrate's displacement (d).

### 3.2 Variance Analysis

The Fisher-Snedecor test compares the variance. It has been performed between each input parameters' levels and output variables. The null hypothesis saying the variables are not independent would be rejected if the F value calculated (F) is superior to the F value read in the Snedecor table ($F_{SNEDECOR}$). $F > F_{SNEDECOR}(2, 3)$ with a p-value $< 0,05$. The Table 4 gathers the results of Fisher-Snedecor Test coefficients. If the value of $F(2, 3)$ is above 9.55 then the p-value $< 0,05$ and we can reject statistical hypothesis H0. Thus, for the model (1) the results allow concluding TS is the only input parameter to exert a significant influence on all the responses observed. However, $h_c$ also depends significantly on the WFS input parameter. Only the contact angle $\alpha$ is significantly dependent on the three input parameters.

**Table 4.** Fisher test results and their p-values

| Input | | hc | hd | wc | wd | $\alpha$ | d |
|-------|---------|--------|--------|--------|--------|--------|--------|
| P | F | 0.94 | 3.60 | 4.63 | 9.25 | 15.63 | 1.45 |
| | p-value | 0.4811 | 0.1594 | 0.1210 | 0.0521 | 0.0259 | 0.3617 |
| WFS | F | 13.27 | 2.53 | 2.57 | 5.17 | 42.14 | 0.92 |
| | p-value | 0.0324 | 0.2269 | 0.2236 | 0.1068 | 0.0064 | 0.4882 |
| TS | F | 57.21 | 67.48 | 10.93 | 68.14 | 23.50 | 8.70 |
| | p-value | 0.0041 | 0.0032 | 0.0419 | 0.0032 | 0.0147 | 0.0564 |

## 4 Conclusion

The study carried out on the LMD-W process analyses the influence of the input parameters (Power, Wire Feed Speed and Tool Speed) on the clad geometry output variables. According to Taguchi's experimental design, 9 clads were printed. The correlation coefficients and the Fisher test allow to conclude Tool Speed is the parameter with the most significant influence on the clads geometry. Five out of six measured output variables follow a linear law with Tool Speed. The clad height is also dependent on the Wire Feed Speed parameter. However, the contact angle is the only response to dependent on the three input parameters. It is therefore possible to model all these responses according to a multiple linear regression using the Least Squares method .

**Acknowledgments.** The research work reported here was made possible by the Fonds Unique Interministériel (FUI Addimafil).

# References

1. Frazier, W.E.: Metal additive manufacturing: a review. J. Mater. Eng. Perform. **23**(6), 1917–1928 (2014)
2. Lundbäck, A., Lindgren, L.E.: Finite element simulation to support sustainable production by additive manufacturing. Procedia Manuf. **7**, 127–130 (2017)
3. Javidani, M., Arreguin-Zavala, J., Danovitch, J., Tian, Y., Brochu, M.: Additive manufacturing of AlSi10Mg alloy using direct energy deposition: microstructure and hardness characterization. J. Therm. Spray Technol. **26**(4), 587–597 (2017)
4. Price, A.: Additive Manufacturing – Standards (2013). https://www.nottingham.ac.uk/research/groups/advanced-manufacturing-technology-research-group/documents/manufacturing-metrology-team/qcam-17/bsi.pdf. Accessed 6 Feb 2020
5. Peters, M., Kumpfert, J., Ward, C.H., Leyens, C.: Titanium alloys for aerospace application. Adv. Eng. Mater. **5**(6), 419–427 (2003)
6. Zwilling, V., Darque-Ceretti, E., Boutry-Forveille, A., David, D., Perrin, M.Y., Aucouturier, M.: Structure and physicochemistry of anodic oxide films on titanium and TA6V alloy. Surf. Interface Anal. **27**(7), 629–637 (1999)
7. Medrano Téllez, A.G.: Fibre laser metal deposition with wire: parameters study and temperature control. Doctoral dissertation, University of Nottingham (2010)
8. Ranatowski, E.: Weldability of titanium and its alloys-progress in joining. Adv. Mater. Sci. **8**(2), 69–76 (2008)
9. Benesty, J., Chen, J., Huang, Y., Cohen, I.: Pearson correlation coefficient. In: Noise Reduction in Speech Processing, pp. 1–4. Springer, Heidelberg (2009)

# Develop Model for Controlled Depth Milling by Abrasive Water Jet of Ti6Al4V at Jet Inclination Angle

Van-Hung Bui, Patrick Gilles[(⊠)], Guillaume Cohen, and Walter Rubio

Institut Clément Ader (ICA), CNRS-INSA-ISAE-Mines Albi-UPS, Université de Toulouse, 3 rue Caroline Aigle, 31400 Toulouse, France
patrick.gilles@insa-toulouse.fr

**Abstract.** Abrasive water jet machining (AWJM) is an interesting solution for the production of shallow pockets in metal sheets made of titanium alloys. Indeed, it produces low cutting forces and heat generation and prevents deformation of these parts after machining. In addition, it has the advantage of only using two raw materials: sand and water. It is possible to generate pocket edges with an imposed geometry using AWJM, but it is necessary to tilt the axis of the jet. The material removal mechanism is then a function of the inclination angle. The presented study propose an improved model for modelling the pocket profile in TiAl6V parts. The experimental results shows that the model is efficient as the precision is around 5%.

**Keywords:** Abrasive water jet machining (AWJM) · Titanium alloy · Alloy cutting depth model

## 1 Introduction

Titanium alloys, and the most used one Ti6Al4V, presents several advantages compared to other metallic materials, such as lightweight, excellent fatigue performance and high resistance to the aggressive environment. These advantages are the reasons of its wide employability material in industry, especially in aerospace. However, its machinability is affected by other properties as high strength, low thermal conductivity and chemical reactivity with cutting tool materials, so it is challenging to choose conventional machining processes in this case. Particularly, pylon components in the aeronautical structure have thin and large shape. Choosing conventional milling process for those parts causes dramatically reduction of the tool life due to galling tendency and deformation of parts due to heat generation caused by high cutting forces. All these restrictions led to the rapid development of machining titanium alloys by non-conventional process like laser, chemical and electrical discharge and abrasive water jet machining (AWJM). Among non-conventional machining methods, AWJM is a promising technique with outstanding characteristics in comparison with conventional machining. AWJ uses a simple tooling (pure water and abrasive particles), it imposes minimum forces on the workpiece and it

L. Roucoules et al. (Eds.): JCM 2020, LNME, pp. 21–27, 2021.
https://doi.org/10.1007/978-3-030-70566-4_5

does not produce any heat-affected zone. Also in the aspect of impact on the environment, it could be considered to be the least harmful.

Due to these advantages, AWJ process has been applied for machining with controlled depth of metals [1–3]. Some reports [4–6] has shown that the desired 3D complex geometry milled by AWJ process can be achieved by governing different slopes which are equal to the local slope of that desired 3D geometry, obtained by changing the angle between the jet axis and the target surface.

However, many studies have shown that there is a wide kind of process parameters in the AWJ process [7–9]. It makes difficult to control the amount of the material removal volume. A small variation in the characteristic of the jet plume causes a fluctuation of eroded material mass along the trajectory of the jet plume. This raises a significant issue for controlled-depth milling in AWJ machining. For those reasons, controlling the jet footprint geometry plays the most important role in generating desirable surface geometries. Thus, a predictive geometrical surface profile model is very importance to fulfil these challenges and improve the advantages of abrasive water jet technology.

The present work develops an efficient and simple geometrical milling model, taking into account the inclination jet angle $(\alpha)$ (Fig. 1-a). As shown by the study presented in [10], only the inclination angle variation is relevant to be considered. Impingement angle (Fig. 1-b) and 5-axis combination are intent to be studied in further work. It makes an effort to insight the kerf generation process with respect to a key kinematic process parameter $(\alpha)$ as a basis to develop strategies for machining controllable pocket shape and further for complex geometries in abrasive water jet technology.

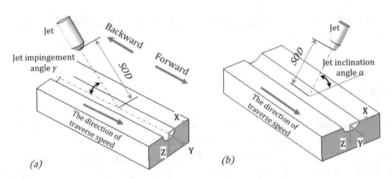

**Fig. 1.** Definition of jet angle

To enable this, several works are conducted as follows:

- (i) Introduction of an extended model [10] of the profile for the pocket machined taking into account the inclination angles of the jet $(\alpha)$.
- (ii) Experimentations with different inclination angles of the jet.

## 2 Proposed Model with the Inclination Angle of the Jet

The energy distribution in the AWJ is identified experimentally as a Gaussian distribution [9, 11], which means that at any cross-section of the jet plume (perpendicular to the jet axis), the velocity profile of water follows nearly Gaussian distribution [12]. Moving of the jet plume over a workpiece surface generates an elementary pass with bell shape as a result of Gaussian distribution. The model of the kerf profile of the elementary pass and pocket milled by abrasive water jet process have been introduced [10, 13]. Thus an open pocket can be generated by using several successive elementary passes with a suitable distance *(Pitch)* (Fig. 2).

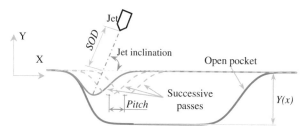

**Fig. 2.** Schematic of open pocket profile generated by successive elementary passes corresponding to the jet inclination angle at $\alpha$

In addition, in order to narrow the influence of complex operating parameters, a given configuration of the machine is defined by dividing into two categories: *(i)* the setting parameters and *(ii)* controlled parameters. The setting parameters are not easily modifiable during milling including: pressure *(P)*, grade of abrasive particle, abrasive flow rate *(ma)*, and standoff distance *(SOD)*. The controlled parameters are defined as the ones that could be modified during machining by the NC program. Considering an elementary pass, in this present study, the only controlled parameter is the traverse speed *(Vf)*.

The modelling of kerf profile of elementary pass, taking into account the influence of the jet inclination angle (Fig. 2), has been developed in the previous works [10]. The present model (Eq. 1) is developed further to predict the open pocket on abrasive water jet milling and it works with the open pocket generated at different focusing tube inclined angles corresponding to a specific machine configuration:

$$Y(x) = \sum_{i=0}^{n} [H(\alpha).Ke(\alpha).H(Vf).e^{-\left(\frac{x-i.pitch.\sin(\alpha)}{B(Vf)}\right)^2}] + cot(\alpha).x \qquad (1)$$

where $Ke(\alpha)$ is the erosion factor which depends on the jet inclination angle and it allows the governing the equation to be more suitable. $Ke(\alpha)$ is identified analytically to find the specific erosion rate on the target material. $H_e(\alpha)$ factor consider the effect of a succession of elementary passes depending on the real inclination angle of the jet. The maximum depth $H(Vf)$ and the width factor $B(Vf)$ of the corresponding elementary pass generated at the jet inclination angle of 90°, these factors have been introduced in [10, 13] and defined using Eq. 2 and Eq. 3.

$$H(Vf) = Ho \times Vf^{Hv} \qquad (2)$$

$$B(Vf) = Bo \times Bf^{Bv} \qquad (3)$$

*Ho, Hv, Bo,* and *Bv* are coefficients that are determined experimentally.

## 3  Experiment

The experiments for model validation are conducted on a five-axis AWJ machining system (FLOW MACH4C). All input parameters are selected as it was done in previous works [10, 13] which play the role of the setting parameters in a given machine configuration. In addition, the results from previous works [10, 13, 14] has demonstrated the efficiency of the model to predict the geometrical characteristics of the kerf profile of both the elementary pass and pocket for various values of traverse speed. The present work only consider the influence of the jet inclination angle ($\alpha$) on the milling process. The tests were implemented using a specific value of the traverse speed, $Vf =$ 1000 mm/min, with different jet inclination angles.

In order to demonstrate the efficiency of the model (Eq. 1), the effect of the jet inclination angles in pocket generation is performed (Fig. 3-a by varying the angle $\alpha$ in the range of 50°–90° with two pitch values (0.7 mm and 1.1 mm).

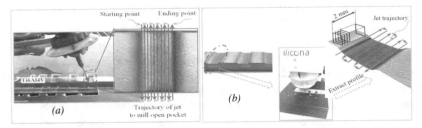

**Fig. 3.** Experimental setup and measurement employed for machining open pocket by AWJ

The identification of the dimensional characteristics of the pocket are extracted from the outcome of measuring on an ALICONA IF optical profilometer (Fig. 3-b). This apparatus is able to measure 1000 curves, distributed regularly over 2 mm, using.

## 4  Result

Based on the experimental result of milling elementary passes with three levels of the traverse speed ($Vf$) presented in Eq. 2 and Eq. 3 [10, 13], the coefficients $Ho = 336.074$, $Hv = -1.000$, $Bo = 1.224$, $Bv = -0.066$ were determined. Besides, the erosion factor $Ke(\alpha)$ was determined by the experiments and given in [10]. It should be noted that once the erosion coefficient $H_e(\alpha)$ is taken into account in the model (Eq. 1), there is a good agreement between the model predictions and the measured profiles with a mean error which is computed by Eq. 4.

$$Error = \frac{Depth_{Modeled} - Depth_{Measured}}{Depth_{Measured}} \tag{4}$$

The mean errors (Table 1) are always negative values and all the absolute values are smaller than 5%; the maximum mean value is observed at the inclination angles of 70° and also 80° for both two cases of the pitch.

**Table. 1.** Modeled and measured depth.

| $\alpha$ | Pitch = 0.7 mm | | | Pitch = 1.1 mm | | |
|---|---|---|---|---|---|---|
| | Experiment (mm) | Model (mm) | Error (%) | Experiment (mm) | Model (mm) | Error (%) |
| 90° | 0.868 | 0.844 | −1.63% | 0.497 | 0.488 | −1.84% |
| 80° | 0.789 | 0.772 | −2.04% | 0.485 | 0.466 | −4.09% |
| 70° | 0.752 | 0.740 | −1.56% | 0.466 | 0.459 | −3.35% |
| 60° | 0.709 | 0.706 | −0.36% | 0.470 | 0.453 | −2.43% |
| 50° | 0.719 | 0.710 | −1.23% | 0.496 | 0.483 | −2.57% |

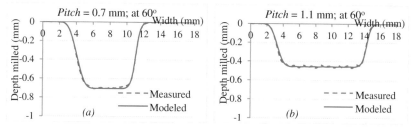

**Fig. 4.** A comparison measured and modelled profiles corresponding to the jet inclination angle at $\alpha = 60°$; *(a) - Pitch = 0.7* mm and *(b) -* Pitch = 1.1 mm.

Furthermore, Fig. 4 indicates a good fitting of measured profile with modelled profile at 60° of the jet inclination angle for the pitch = 0.7 (mm) and 1.1 (mm).

Figure 5 shows the value of $H_e(\alpha)$ factor which defined by experiment at a given jet inclination angle for two values of the pitch steps. In both cases, the $H_e(\alpha)$ factor is a function of the jet inclination angle and dependence of $H_e(\alpha)$ fully matches with the linear trendline (dot lines is the best-fit line).

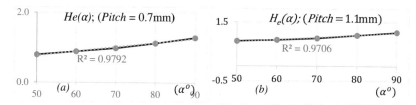

**Fig. 5.** $H_e(\alpha)$ factor defined experimentally as a function of jet inclination angle. The dot line represents the best fit line: *a) Pitch = 0.7* (mm); *b) Pitch = 1.1* (mm).

## 5 Conclusion

In this work, the generation of milled pocket at various jet inclination angles is investigated by considering the model for 2D cross-section profile milling by AWJ applications. The model has been assessed by milling open pockets in titanium alloy (Ti6Al4V). The process is controlled by the traverse speed *(Vf)* and the jet inclination angles *(α)*, other process parameters are chosen to establish a given machine configuration. The machined pocket profile is analysed to get control for the depth of the pocket by varying a jet inclination angles from 50° to 90°. The geometry of the profile of the open milled pocket significantly depends on the jet inclination angle *(α)*. The mean error is less than 5%.

## References

1. Paul, S., Hoogstrate, A.M., Van Luttervelt, C.A., Kals, H.J.J.: An experimental investigation of rectangular pocket milling with abrasive water jet. J. Mater. Process. Technol. **73**, 179–188 (1998)
2. Chakraverti, G., Kanthababu, M., Goutham, U., Hasu, B.S.: Experimental Investigation of Pocket Milling on Inconel 825 using Abrasive Water Jet Machining. FME Trans. **44**(2), 133–138 (2016)
3. Fowler, G., Shipway, P.H., Pashby, I.R.: Abrasive water-jet controlled depth milling of Ti6Al4V alloy - an investigation of the role of jet-workpiece traverse speed and abrasive grit size on the characteristics of the milled material. J. Mater. Process. Technol. **161**(3), 407–414 (2005)
4. Srinivasu, D.S., Axinte, D.: An analytical model for top width of jet footprint in abrasive waterjet milling: A case study on SiC ceramicsg. Proc. Inst. Mech. Eng. Part B J. Eng. Manuf. **225**(3), 319–335 (2011)
5. Tamannaee, N., Spelt, J.K., Papini, M.: Abrasive slurry jet micro-machining of edges, planar areas and transitional slopes in a talc-filled co-polymer. Precis. Eng. **43**, 52–62 (2016)
6. Billingham, J., Miron, C.B., Axinte, D.A., Kong, M.C.: Mathematical modelling of abrasive waterjet footprints for arbitrarily moving jets: part II - overlapped single and multiple straight paths. Int. J. Mach. Tools Manuf. **68**, 30–39 (2013)
7. Hashish, M.: Milling with abrasive-waterjets: a preliminary investigation. ASME, p. 10 (1987)
8. Fowler, G.: Abrasive water-jet - controlled depth milling titanium alloys, p. 209 (2003)
9. Alberdi, A., Rivero, A., López de Lacalle, L.N., Etxeberria, I., Suárez, A.: Effect of process parameter on the kerf geometry in abrasive water jet milling. Int. J. Adv. Manuf. Technol. **51**(5–8), 467–480 (2010)
10. Bui, V.H., Gilles, P., Cohen, G., Rubio, W.: A modeling of elementary passes taking into account the firing angle in abrasive water jet machining of titanium alloy. In: AIP Conference Proceedings, vol. 1960 (2018)
11. Ojmertz, K.M.C.: Abrasive waterjet milling: an experimental investigation (1997)
12. Sultan, T., Gilles, P., Cohen, G., Cenac, F., Rubio, W.: Modeling incision profile in AWJM of Titanium alloys Ti6Al4V. Mech. Ind. **17**(4), 403 (2016)
13. Bui, V.H., Gilles, P., Sultan, T., Cohen, G., Rubio, W.: A new cutting depth model with rapid calibration in abrasive water jet machining of titanium alloy. Int. J. Adv. Manuf. Technol. **93**(5–8), 1499–1512 (2017)
14. Bui, V.H., Gilles, P., Sultan, T., Cohen, G., Rubio, W.: Adaptive speed control for waterjet milling in pocket corners. Int. J. Adv. Manuf. Technol. **103**, 77–89 (2019)

# Simulation and Experimental Validation of Secondary Dendrite Arm Spacing for AlSi7Mg0.3 Chassis Parts in Low Pressure Die Casting

Alberto Vergnano[1(✉)], Umberto Bergamini[1,2], Daniele Bianchi[3], Paolo Veronesi[1], Roberto Spagnolo[2], and Francesco Leali[1]

[1] Department of Engineering "Enzo Ferrari", University of Modena and Reggio Emilia, Via Pietro Vivarelli 10, 41125 Modena, Italy
alberto.vergnano@unimore.it
[2] Fonderia S. Possidonio, Via Don Giovanni Minzoni 14, 41039 San Possidonio, Italy
[3] Fonderie Mario Mazzucconi, Via Kennedy 16, 24030 Ambivere, Italy

**Abstract.** The structural properties of cast aluminum parts are strongly affected by the solidification in the production process. The solidification dynamics determines the Secondary Dendrite Arm Spacing (SDAS), directly affecting the structural strength of the alloy. Simulation techniques enable the integrated design of chassis parts and their production equipment. However, in order to effectively predict the SDAS formation, the simulation models need to be investigated and calibrated. The present research investigates the SDAS formation models and identifies a robust relation to be used in Design by Simulation phases for AlSi7Mg0.3 parts.

**Keywords:** Secondary Dendrite Arm Spacing · Integrated design · Simulation · Low pressure die casting

## 1 Introduction

Structural parts for car chassis must achieve high structural strength. For cast aluminum parts, the relation between the mechanical properties and the grains size is reported in literature [1]. The grain size is measured on micrographs as Secondary Dendrite Arm Spacing (SDAS). The SDAS is strongly influenced by the solidification dynamics [2], hence the integrated design and simulation of a casting and its manufacturing equipment is critical. Design by Simulation techniques aid designers [3, 4]. However, the heat transfer and SDAS models are not easy to compute, since they must be carefully calibrated [5].

The present research investigates the SDAS formation models and identifies a robust relation to be used in the integrated design of casting and foundry equipment. The paper is organized as follows. Section 2 presents the method for SDAS analysis with simulation and experiment techniques. A case study is discussed in Sect. 3, while the concluding remarks are drawn in Sect. 4.

L. Roucoules et al. (Eds.): JCM 2020, LNME, pp. 28–33, 2021.
https://doi.org/10.1007/978-3-030-70566-4_6

## 2    Secondary Dendrite Arm Spacing Analysis

Four steps enable the complete analysis of a casting, as described hereafter.

### 2.1    Low Pressure Die Casting Simulation

The casting simulations are modeled in Magmasoft [4]. Four characteristic times are considered for describing the casting solidification. The *FStime 60%* is the time required for the solidifying alloy to reach a 60% solid fraction. This time is significant since no macroscopic feeding is possible over 60% solid fraction for the model of this alloy. *FStime 90%* is investigated to search for possible grain modifications in the last 10% solidification phase. The *Solidification time* is the time from the start of pouring to the time when the temperature locally falls below the Solidus temperature. The *Liquidus to Solidus time* is the elapsed time in the local transition from the Liquidus and Solidus temperatures [6, 7].

### 2.2    Spectrochemical Analysis with Optical Emission Spectroscopy

In Optical Emission Spectrometry (OES), a portion of the sample is vaporized through electric discharge and the optical radiation from the excited ions is transferred to the spectrometer optics. The optical beam is here broken down into individual spectral components. Each component is characteristic of an element while its intensity is proportional to its concentration. Finally, the percentage concentration of all the elements in the alloy is displayed.

The OES analysis ensures that the casting samples under investigation complies with the chemical composition in the UNI EN 1076 standard for the AlSi7Mg0.3 alloy. For these safety critical parts, further 0.012%–0.024% strontium is added in order to modify the microstructure.

### 2.3    SDAS Mapping Through Micrographs

From the image on an optical microscope, the SDAS is evaluated by averaging at least five arm spacings, as $SDAS = L/n$, $n \geq 5$, in Fig. 1. Furthermore, the SDAS is averaged over ten measurements for each different area on the casting.

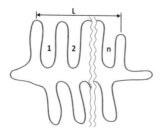

**Fig. 1.** Evaluation of secondary dendrite arm spacing.

## 2.4  Correlation Between Measured and Simulated SDAS

A correlation analysis is performed on Excel, in order to extrapolate a formula correlating the SDAS to the simulated characteristic times. From literature [6], the formula to be parameterized is:

$$SDAS = k \cdot t_S^n. \tag{1}$$

where $t_S$ is the solidification time, $k$ and $n$ are constants characteristic for the alloy.

## 3  Experiments and Simulation Analysis on Case Studies

Three different castings are analyzed on six sampling areas, for a total of 18 measurements. The experiments are averaged over ten quality-compliant samples for each one of the castings, for a total of 180 specimens. The present section reports the evaluations on a wheel hub support, while the final reported results refer to the complete evaluation on all three case studies.

The six sampling areas on the casting for the wheel hub support are shown in Fig. 2. Table 1 reports the four characteristic times for the casting solidification as simulated for the six areas. Since the simulation results depend on the model discretization with a fine mesh, sized smaller than the area investigated in the experiments, each reported value is averaged over five picked measurements close to the sampling area.

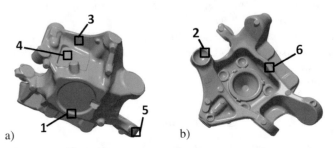

**Fig. 2.** Sampling areas on the casting for the wheel hub support in a) bottom and b) top views.

**Table 1.** Characteristic times as simulated for the six sampling areas.

| Area | Motivation | FStime 60% [s] | FStime 90% [s] | Solidification [s] | Liquidus to Solidus [s] |
|------|-----------|----------------|----------------|--------------------|--------------------------|
| 1 | High FStime | 150.01 | 174.07 | 186.75 | 122.96 |
| 2 | Thick wall | 78.51 | 85.18 | 98.69 | 49.54 |
| 3 | Thin wall | 65.91 | 76.81 | 84.38 | 39.18 |
| 4 | Short FStime | 61.14 | 73.61 | 82.40 | 37.43 |
| 5 | Zone for hardness specimens | 71.61 | 83.71 | 92.67 | 47.03 |
| 6 | Intermediate Fstime | 89.26 | 103.83 | 123.33 | 74.49 |

For each casting sample, a specimen is extracted from each one of the six sampling areas. The specimens are machined as approximately 15 × 15 × 8 mm size. They are incorporated in resin, polished with abrasive paper in order to remove any contamination from the surface and finally shortly etched in Keller's reagent. With a ARL4460 Metals Analyzer quantometer, three OES measurements are carried out on each specimen in different positions and then averaged. As example, the chemical mass compositions of the three analyzed specimens is reported in Table 2, while Fig. 3 shows the microscope image with SDAS measurements on a single specimen. The results of SDAS analysis for the six areas on the ten samples are reported in Table 3.

**Table 2.** Mass percentage composition of AlSi7MG0.3 alloy as prepared in the foundry.

| NR | Si | Fe | Cu | Mn | Mg | Zn | Ti | Sr | Others | Al |
|---|---|---|---|---|---|---|---|---|---|---|
| UNI EN 1076 | 6.50–7.50 | <0.15 | <0.03 | <0.10 | 0.25–0.45 | <0.07 | <0.18 | – | <0.10 | Balance |
| Casting1 | 7.18 | 0.103 | 0.002 | 0.002 | 0.399 | 0.001 | 0.118 | 0.028 | < 0.01 | 92.3 |
| Casting2 | 7.16 | 0.104 | 0.003 | 0.003 | 0.376 | 0.002 | 0.110 | 0.029 | <0.01 | 92.2 |
| Casting3 | 6.98 | 0.103 | 0.002 | 0.002 | 0.400 | 0.001 | 0.112 | 0.028 | <0.01 | 92.4 |

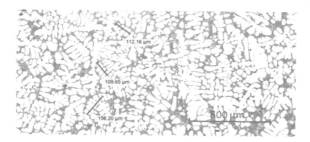

**Fig. 3.** Grain structure magnification and SDAS measurements.

**Table 3.** SDAS results on the six specimens on the wheel hub support casting.

| Nr | mean [μm] | max [μm] | min [μm] | dev [μm] |
|---|---|---|---|---|
| 1 | 41 | 47 | 33 | 14 |
| 2 | 25 | 31 | 19 | 11 |
| 3 | 23 | 27 | 20 | 7 |
| 4 | 24 | 29 | 21 | 8 |
| 5 | 27 | 32 | 20 | 12 |
| 6 | 28 | 38 | 24 | 14 |

Finally, the interpolation functions of the SDAS depending on *FStime 60%*, *FStime 90%*, *Solidification time* and *Liquidus to Solidus time* are reported in Fig. 4 over all the six sampling areas on the three analyzed castings.

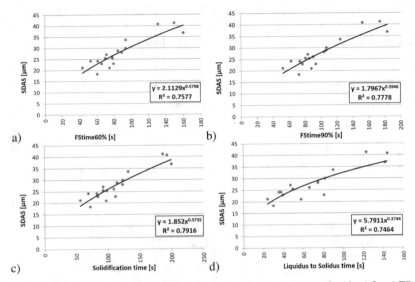

**Fig. 4.** SDAS interpolation functions (solid) depending on simulation results (dots) for a) FStime 60%, b) FStime 90%, c) Solidification time and d) Liquidus to Solidus time.

The interpolation with the *Solidification time* as variable achieves the best correlation coefficient $R^2$, in agreement with literature works. For the AlSi7Mg0.3 alloy with 0.012%–0.024% strontium addition and for the specific casting technology, the most accurate prediction of SDAS from simulations is found to be:

$$\text{SDAS} = 1.852 \cdot t_S^{0.5735}. \tag{2}$$

In the present research, (2) is considered with a reliability about $\pm 5$ μm.

## 4   Conclusions

The present research investigates the calibration of a model for the SDAS simulation in LPDC. The SDAS mapping from micrographs is studied against the maps of different characteristic solidification times from simulations. The formula for the SDAS simulation gives reliable results for the three castings, for a total of 30 analysed samples. So, the simulations are capable of investigating this parameter directly linked with the mechanical properties of the alloy. These simulations are a very important tool for designing both the product and the manufacturing equipment. In order to improve the reliability of the SDAS model, further experiments will be executed, investigating also other casting technologies.

# References

1. Zhang, L.Y., Jiang, Y.H., Ma, Z., Shan, S.F., Jia, Y.Z., Fan, C.Z., Wang, W.K.: Effect of cooling rate on solidified microstructure and mechanical properties of aluminium-A356 alloy. J. Mater. Process. Technol. **207**(1–3), 107–111 (2008)
2. Flemings, M.C.: Solidification Processing. McGraw-Hill, New York (1985)
3. Bonazzi, E., Colombini, E., Panari, D., Vergnano, A., Leali, F., Veronesi, P.: Numerical simulation and experimental validation of MIG welding of T-joints of thin aluminum plates for top class vehicles. Metall. Mater. Trans. A Phys. Mettall. Mater. Sci. **48**(1), 379–388 (2017)
4. Vergnano, A., Brambilla, E., Bonfiglioli, G.: Efficiency and reliability of gravity die casting models for simulation based design. In: Advances on Mechanics, Design Engineering and Manufacturing II, pp. 3–12 (2019)
5. Pedrazzi, S., Vergnano, A., Allesina, G., Veronesi, P., Leali, F., Tartarini, P., Muscio, A.: A simple test method for measurement of the interface thermal resistance of coated and uncoated metal surfaces. In: 37th UIT Heat Transfer Conference, Padova, June 2019 (2019)
6. Ferreira, A.F., Castro, J.A., Ferreira, L.D.O.: Predicting secondary-dendrite arm spacing of the Al-4.5 wt% Cu alloy during unidirectional solidification. Mater. Res. **20**(1), 68–75 (2017)
7. Tiryakioğlu, M.: A simple model to estimate solidification time-dendrite arm spacing relationships in cast aluminum alloys with two major alloying additions: application to the Al-Si-Cu system. Metall. Mater. Trans. A **50**(7), 3030–3032 (2019)

# Influence of the Hole Surface Integrity on the Fatigue Strength of an Aluminium Drilled Part

Alexandra Lacombe[1]([⊠]), Yann Landon[1], Manuel Paredes[1], Clément Chirol[2], and Audrey Benaben[2]

[1] Institut Clément Ader, UMR CNRS 5312, Université de Toulouse, UPS, INSA, ISAE-SUPAERO, MINES-ALBI, CNRS, 3 rue Caroline Aigle, Toulouse, France
alexandra.lacombe1@univ-tlse3.fr
[2] Airbus France, D41, 18 rue Marius Tercé, 31300 Toulouse, France

**Abstract.** Fatigue strengths of aluminium 2024-T351 open-hole specimens drilled by axial and orbital drilling processes are compared. Two drilling diameters (Ø) are studied: 6.35 mm and 9.53 mm. Surface integrity characterization tests are conducted in order to study the link between drilling processes, surface integrity and fatigue life. Fatigue test results show an increase of the fatigue life for specimens drilled by axial drilling for Ø = 9.53 mm and no significant difference in fatigue life between the two drilling processes for Ø = 6.35 mm. Surface integrity results show no impact of the roughness on the fatigue strength but a potential positive influence of the hole microhardness on the fatigue life.

**Keywords:** Drilling process · Fatigue life · Surface integrity · Aluminium alloy

## 1 Introduction

Parts of aircraft are mostly assembled using fasteners (rivets or screws) introduced into holes machined by drilling. Assembling an aircraft may require several hundred thousand to several million drilling operations depending on the aircraft size [1]. These holes for fastening are critical areas where fatigue damage can be initiated because they are areas of high stress concentration. The drilling procedure used for the machining of the fastening holes can affect the fatigue life of the drilled part [2–5]. Indeed, according to the procedure, the part undergoes different thermo-mechanical loading that can induce differences in the surface integrity of the hole (roughness, residual stress, hardness, etc.) [6].

The most common process for machining a fastening hole in the aircraft industry is axial drilling. This process involves the use of a rotating drill bit that feeds linearly into the part. Although this process is well known, it has certain drawbacks. As an example, axial drilling may result in the formation of a burr on the face of the part where the tool exits, because of the high axial forces involved [7]. So, the parts require a deburring operation after the drilling.

L. Roucoules et al. (Eds.): JCM 2020, LNME, pp. 34–40, 2021.
https://doi.org/10.1007/978-3-030-70566-4_7

Orbital drilling involves the use of a rotating end mill travelling on a helical path into the part. With this process, the cutting process is intermittent and the material is removed in the form of small chips. Orbital drilling has many advantages [8]. It avoids the formation of burrs, it allows better chip evacuation and less heating, and gives holes with high surface quality in a near dry environment and in a one-step operation [9]. However, the impact of this drilling process on the fatigue strength of the part is not well known.

Thus, the aim of this work is to compare the fatigue strength of open-hole specimens drilled by axial and orbital drilling for the aluminium alloy 2024-T351, which is commonly used in the aircraft industry for its low density and its high fatigue performance. Moreover, in order to identify the parameters that control the fatigue life, the impact of the drilling process on the surface integrity of the hole (roughness and Vickers microhardness) is also investigated.

## 2   Experimental Work

### 2.1   Fatigue Tests

Fatigue specimens were open-hole T-Type elementary specimens machined from 2024-T351 aluminium plate in such a way that their longitudinal axis was aligned with the rolling direction of the plate. The specimen width was three times the nominal diameter and its length was 200 mm (Fig. 1). The holes were obtained by drilling then deburring. Two nominal drilling diameters (Ø) were studied: 6.35 mm and 9.53 mm. For Ø = 9.53 mm, the specimen thickness was 10 mm. For Ø = 6.35 mm, three thicknesses were studied: 3.175 mm, 6.35 mm and 10 mm.

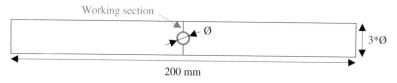

**Fig. 1.** Overall geometry of fatigue specimens

Axial and orbital drilling processes were studied. The cutting parameters and tools used for each drilling process and for each diameter are presented in Table 1.

The cutting parameters used for axial drilling were the optimum parameters provided by the tool manufacturer and those used for orbital drilling were determined through a specific Tool-Material Pair. External MQL lubrication was used for both drilling processes.

Fatigue tests were performed in a Schenk servo hydraulic machine using a sinusoidal cyclic load with a load ratio of 0.1 and a frequency of 20 Hz. Fatigue tests were carried out for various load levels in order to obtain Wöhler curves.

Fatigue test results are presented in Fig. 2 as semi-log (S-N) curves. The stress S corresponds to the maximum stress reached in the working section during a fatigue cycle. In

**Table 1.** Drilling cutting parameters and tools

|  | Ø = 6.35 mm | | Ø = 9.53 mm | |
|---|---|---|---|---|
|  | Axial drilling | Orbital drilling | Axial drilling | Orbital drilling |
| $N_{spindle}$ (rpm) | 9 000 | 40 000 | 4 000 | 40 000 |
| $N_{orb}$ (rpm) | – | 1 500 | – | 1 500 |
| $V_{fa}$ (mm/min) | 900 | 60 | 400 | 60 |
| $V_c$ (m/min) | 180 | 610 | 120 | 1 005 |
| $f_a$ (mm/rev) | 0.1 | 0.0015 | 0.1 | 0.0015 |
| Tool | Tungsten carbide drill | 4-tooth tungsten carbide end mill | Step tungsten carbide diamond coated drill | 4-tooth tungsten carbide end mill |

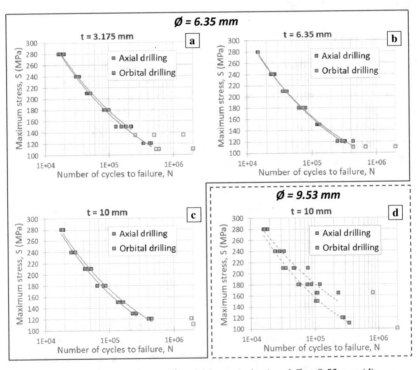

**Fig. 2.** Wohler curves for Ø = 6.35 mm (a, b, c) and Ø = 9.53 mm (d)

the aircraft industry, the fatigue performance of a structure is commonly evaluated by the fatigue quality index, which corresponds to the stress S associated with a forecast fatigue life of 100 000 cycles. Fatigue quality indexes were determined for all configurations from Wöhler curves and are shown in Fig. 3.

Fatigue test results obtained for Ø = 6.35 mm show that the specimen thickness and the drilling process have negligible influence on the fatigue strength. Except for t = 3.175 mm, for the lowest stress levels, a gain in fatigue life is observed for orbital drilling compared to axial drilling. In contrast, fatigue test results obtained for Ø = 9.53 mm show an increase in fatigue life for the specimens obtained by axial drilling for all stress levels. For this drilling configuration, a gain in the fatigue quality index of 15% is observed and an even greater gain is observed for the stress corresponding to the fatigue limit.

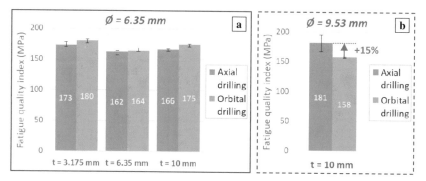

**Fig. 3.** Fatigue quality indexes for Ø = 6.35 mm (a) and Ø = 9.53 mm (b)

## 2.2  Surface Integrity Characterization

In order to identify the parameters that control the fatigue life, a test campaign was conducted to characterize the surface integrity of the holes. Roughness and microhardness were measured.

Roughness measurements were performed along the height of the hole with a profilometer with a cut-off length of 0.8 mm. The results are presented in Fig. 4. For all specimens where orbital drilling was used, the roughness average (Ra) was lower than that of specimens drilled with axial drilling, which is consistent with previous works [3]. However, no correlation could be established between these results and fatigue test results. This was probably related to the relatively low Ra values measured compared to the aeronautical specification (Ra < 1.6).

Vickers microhardness measurements were made on the hole surface with a load of 1 kgf and an indentation time of 15 s. Because of the cylindrical shape of the hole, a corrective factor was applied to the results as recommended by ASTM E92. The results are presented in Fig. 5. For Ø = 6.35 mm, holes obtained by axial and orbital drilling had similar microhardness levels whereas, for Ø = 9.53 mm, a significant difference in microhardness level was observed between the two drilling processes. For this diameter, a gain of 29% in microhardness level was observed for axial drilling. These results seem to show a correlation of fatigue test results with microhardness measurements. Indeed, the increase in microhardness for axial drilling at Ø = 9.53 mm may explain the gain in fatigue life observed for the same configuration.

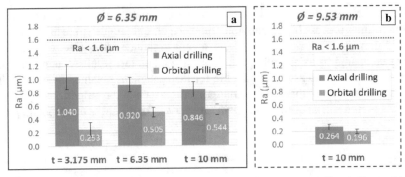

**Fig. 4.** Roughness average measurements for Ø = 6.35 mm (a) and Ø = 9.53 mm (b)

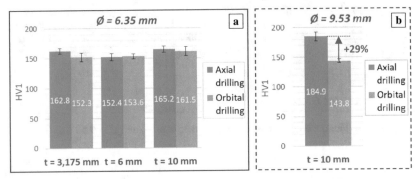

**Fig. 5.** Vickers microhardness measurements for Ø = 6.35 mm (a) and Ø = 9.53 mm (b)

## 3  Discussion and Conclusions

A fatigue test campaign was carried out in order to compare the fatigue strength of aluminium 2024-T351 open-hole specimens drilled by the axial and orbital drilling processes. The Wöhler curves obtained show a slight fatigue life difference between specimens drilled by axial and orbital drilling for Ø = 6.35 mm and a significant gain in fatigue life for the specimens drilled by the axial technique for Ø = 9.53 mm. The specimen thickness seems to have negligible influence on the fatigue strength.

Roughness measurements show no correlation between the hole roughness and the fatigue life of the drilled part. This is related to the low Ra values measured and is in accordance with the Kitagawa diagram [10], which establishes that, below a certain defect size, the fatigue limit is no longer sensitive to the defect size.

Vickers microhardness measurements seem to show a significant influence of the hole microhardness on the fatigue life of the part. The difference in microhardness level observed for 9.53 mm diameter holes may be related to the difference between the thermomechanical loads experienced by the machined surfaces in the two drilling processes. As the temperatures reached during an aluminium drilling operation are not high enough to induce a phase transformation [1], the increase in microhardness may be related to strain hardening of the hole sub-surface induced by the mechanical loading, or to precipitation hardening induced by the thermal loading [11]. In order to study these aspects, the mechanical and thermal loads involved during drilling and the material microstructure in the hole edge area will be studied.

Residual stresses present in the hole edge area are cited in the literature [12] as having an influence on fatigue strength, so this aspect of surface integrity will also be investigated. However, since the material depth affected by residual stresses after an aluminium drilling operation is relatively small (from several tens to several hundreds of microns) [13], current techniques for residual stress evaluation (X-ray diffraction, incremental hole drilling, etc.) cannot be used. So, a new strategy for residual stress evaluation will be considered.

# References

1. Girot, F., et al.: Perçage des structures aéronautiques: Les résultats du projet MEDOC, 1$^{er}$ colloque Aquitaine – Québec – Mécanique des Matériaux et des Structures pour l'Aéronautique, July 2008
2. Elajrami, M., Benguediab, M., Ronald, G.: Effect of various drilling procedures on the fatigue life of rivet holes. Synthèse: Revue des Sciences et de la Technologie **19**, 67–75 (2008)
3. Sun, D., Lemoine, P., Keys, D., Doyle, P., Malinov, S., Zhao, Q., Qin, X., Jin, Y.: Hole-making processes and their impacts on the microstructure and fatigue response of aircraft alloys. Int. J. Adv. Manuf. Technol. **94**(5–8), 1719–1726 (2018)
4. Ralph, W.C., Jonhson, W.S., Makeev, A., James, C., Newman, J.: Fatigue performance of production-quality aircraft fastener holes. Int. J. Fatigue **29**(7), 1319–1327 (2007)
5. Everett, R.A.: The effect of hole quality on the fatigue life of 2024-T3 aluminum alloy sheet (2004)
6. Ralph, W.C., Johnson, W.S., Toivonen, P., Makeev, A., Newman, J.: Effect of various aircraft production drilling procedures on hole quality. Int. J. Fatigue **28**(8), 943–950 (2006)
7. Abdelhafeez, A.M., Soo, S.L., Aspinwall, D.K., Dowson, A., Arnold, D.: Burr formation and hole quality when drilling titanium and aluminium alloys. Procedia CIRP **37**, 230–235 (2015)
8. Pereira, R.B.D., Brandao, L.C., De Paiva, A.P., Ferreira, J.R., Davim, J.P.: A review of helical milling process. Int. J. Mach. Tools Manuf. **120**, 27–48 (2017)
9. Ni, W.: Orbital drilling of aerospace materials. SAE Technical Paper (2007)
10. Kitagawa, H., Takahashi, S.: Applicability of fracture mechanics to very small cracks or the cracks in the early stage. In: International Conference on Mechanical Behaviour of Materials, Boston, pp. 627–631 (1976)
11. Choo, V.K.S., Reinhall, P.G., Ghassaei, S.: Effect of high rate deformation induced precipitation hardening on the failure of aluminium rivets. J. Mater. Sci. **24**(2), 599–608 (1989)
12. Lai, M.O., Oh, J.T., Nee, A.Y.C.: Fatigue properties of holes with residual stresses. Eng. Fract. Mech. **45**(5), 551–557 (1993)
13. Federal Aviation Administration. Assessment of residual stresses and hole quality on the fatigue behavior of aircraft structural joints (2009)

# Cold Working Process on Hard Metal Stacked Assembly

Guillaume Pichon[1]([✉]), Alain Daidie[1], Adeline Fau[1], Clément Chirol[2], and Audrey Benaben[2]

[1] Université de Toulouse, Institut Clément Ader, UMR CNRS 5312, INSA/UPS/ISAE/Mines Albi, Toulouse, France
guillaume.pichon@airbus.com
[2] Airbus Operations S.A.S., Toulouse, France

**Abstract.** Designed for aeronautical and automotive applications, the split sleeve cold expansion process is used to improve the fatigue life of bolted metallic parts. Although its application has been well tested on aluminum assemblies, hard metal applications are still being studied. This paper presents experimental results of double bolt joint assemblies under double shear fatigue tests after stacked split sleeve cold expansion. The behaviors of two sizes of assemblies with different degrees of expansion are investigated. S-N curves are the main indicators of this study but thermal aspects are also investigated to observe fretting in the specimens as bolts are preloaded. Bolt tension is a major parameter in assembly regarding fatigue life. Interference between those two phenomena is at the heart of this paper. The first results show that stacked cold expansion has a negative effect on mechanical performances, as it deteriorates the fatigue life of the assembly. However, an examination of these results provides a coherent explanation for the loss of performance that occurs.

**Keywords:** Cold expansion · Process · Fatigue · Assembly · Bolt

## 1 Introduction

Reducing the weight of structures is one of the major aims of aeronautical engineers, and specifically concerns structural parts made of hard, heavy metal. Furthermore, modern concern regarding environmental issues has strengthened the will of aircraft manufacturers to focus their innovation efforts on weight loss in order to reduce fuel consumption. For instance, the Dutch aviation industry has planned to cut its global $CO_2$ emissions by one-third by the year 2030 [1]. Some studies focus on new materials such as composites, which have created a breakthrough in the field of aeronautical structures [2]. Nevertheless, their complexity and characteristics (mechanical and thermal) make them unusable for some structural parts so far. Hard metals, and titanium alloys in particular, remain the best solutions in certain cases and are thus frequently used [3]. However, developing new and more efficient technology on ancient material is also a field of study that remains a technical challenge. Aeronautical structures are mainly made of sub-assemblies held

© The Author(s) 2021
L. Roucoules et al. (Eds.): JCM 2020, LNME, pp. 41–47, 2021.
https://doi.org/10.1007/978-3-030-70566-4_8

together by a significant number of fasteners (rivets or bolts). The cold expansion process is an innovative way to enhance the fatigue life of metal components by reinforcing the hole edges to delay their failure. It is then possible to reduce the design weight and achieve identical mechanical performances, while saving weight on the total structure. Several cold expansion process exist: expansion by conical pin, the British railways process, and a conical four-split mandrel. This paper focus on the split sleeve expansion process.

## 2   Stacked Cold Expansion Process

The Cold Expansion patented by Fatigue Technologies Inc. [4, 5] is the most widely used, mainly because of its speed of execution and portability [6]. By pulling an oversized mandrel through a fastener hole, compressive residual stresses are developed around the hole wall, fatigue crack propagation is delayed and maintenance sessions can be scheduled less frequently [7]. The process is performed with the help of a split sleeve, placed longitudinally between the mandrel and the hole. A conical taper mandrel is screwed at the extremity of a pressure cylinder. To guide the mandrel and to maintain the sleeve in position during expansion, a nosecap, symmetrically split into four sections, is inserted around the mandrel and directly mounted onto the puller unit frame. Then, the split sleeve is placed around the mandrel, and the mandrel and sleeve assembly is inserted inside the specimen hole. Finally, as the mandrel travels through the material, the hole is expanded at a ratio defined by the Degree of Cold Expansion (DCE) expressed by Eq. (1)

$$DCE = \frac{d_0^m + 2t^{ss} - d_0^{sp}}{d_0^{sp}} \times 100 \tag{1}$$

where $d_0^m$ is the mandrel major diameter, $t^{ss}$ is the split sleeve thickness and $d_0^{sp}$ is the initial hole diameter.

In this particular study, the mandrel is extracted by travelling through the 3 plates, which are clamped together. Once the first hole expansion is achieved, a temporary bolt is inserted around the sleeve, which is left in the hole to preserve correct alignment. Then the second hole is expanded with the same clamping system. The process is presented schematically in Fig. 1.

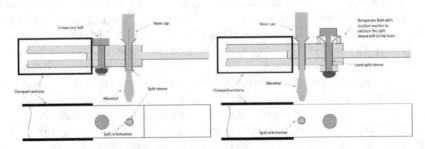

**Fig. 1.** Stacked cold working process

During the classic process, plates are expanded individually and the local deformations of the metal plates create a volcano on each face of the coupon. In his works, [8] performed cold working expansion on several similar coupons and the measured heights of such volcanoes were around a tenth of a millimeter. Although the coupons are stacked and strongly clamped in the present work, similar deformations are present, as shown by Fig. 2.

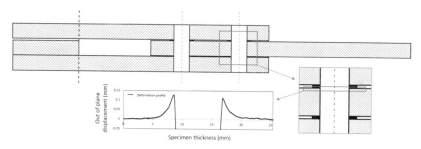

**Fig. 2.**  Volcano formation at hole edge

## 3   Experimental Campaign

Assemblies in this study were made of titanium Ti-6Al-4V. Dimensions are given in Table 1, in accordance with the design in Fig. 3. Two different sizes of fasteners were tested, code-4 (6.35 mm diameter) and code-6 (9.352 mm diameter). Non-expanded assemblies were also tested for each code as references. Hard metals are usually strongly loaded so it was interesting to study their behavior at high cold expansion rate. Therefore the chosen DCE (degree of cold expansion) was 5% for each size. The coupons were designed to favor a net section failure at the first bolt of the middle part (Fig. 3).

The last step before the test was final assembly with the dedicated fasteners. In this study, the bolts used were shear screws with self-locking nuts. Since T. Benhaddou [9, 10] has demonstrated the major impact that preload has on the fatigue life of bolted assemblies, it was decided to use a high tightening torque to compensate the clearance between plates caused by the stacked expansion. The tightening torque Co was 29 N.m for code-6 assemblies and 9.5 N.m for code-4. This tightening was executed with a numerical torque wrench having a precision range of 4%. However, strong uncertainty remained between the torque applied and the existing preload in the fasteners due to the numerous parameters involved in the process, such as torqueing speed, temperature, friction coefficients.

The fatigue tests were conducted with a frequency of 5 Hz and a load factor R = 0.1 (see Eq. 2).

$$R = \frac{\sigma_{min}}{\sigma_{max}} = 0.1 \tag{2}$$

The stress levels were chosen in order to obtain points ranging from $10^3$ to $2.10^6$ cycles. To take proper account of the dispersion effect inherent in fatigue tests, three assemblies were tested for each configuration (size and DCE).

**Table 1.** Specimen designations and dimensions

| Reference | Designation | | | | Dimensions | | | | | |
|---|---|---|---|---|---|---|---|---|---|---|
| | Code | T | TTH | DCE | Ltot | Ttot | W | Di | Dr | S |
| | [−] | [mm] | [NA] | [%] | [mm] | [mm] | [mm] | [mm] | [mm] | [mm$^2$] |
| 4-3-0-0 | 4 | 3.18 | 0 | 0 | 254 | 9.525 | 25.4 | 6.35 ± 0.02 | 6.35 | 60.48 |
| 4-3-0-5 | 4 | | | 5 | | | | 5.93 ± 0.02 | | |
| 6-4-0-0 | 6 | 4.76 | | 0 | 381 | 14.287 | 38.1 | 9.52 ± 0.02 | 9.525 | 136.09 |
| 6-4-0-5 | 6 | | | 5 | | | | 8.89 ± 0.02 | | |

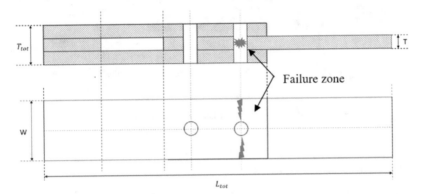

**Fig. 3.** Definition of assemblies

Specific metrology devices were installed to observe fretting behavior during the tests.

## 4　Results

The direct outputs of this experimental phase were the S-N curves or Whöler curves presented in Fig. 4. A glance at the curves shows that the fatigue lives of the expanded assemblies (green) were lower than the references (blue and red).

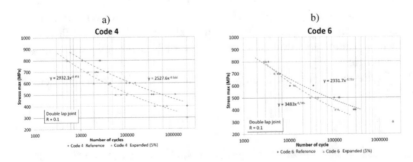

**Fig. 4.** S-N curves a) Code-4, b) Code-6

To explain such a difference, the failures of the assemblies were analyzed. For expanded assemblies, 100% of failures occurred in the net section as expected whereas, for the reference ones, only 95% of failures were located in the net section. The others were tangent to the hole. This failure mode can be explained by the strong preload of the fasteners.

Figure 5 shows the fretting zone of the assemblies tested and compares expanded and reference coupons for similar loading. On expanded assemblies, the fretting is concentrated around the holes, while the fretting zones on reference coupons are larger and scattered all over the contact surface. From these observations, it can be deduced that the load transmitted by friction was significantly lower on expanded assemblies, so the load was mainly transmitted by the fasteners and the holes. This behavior is consistent with the performance loss, assessed at about 15% (for $10^5$ cycles).

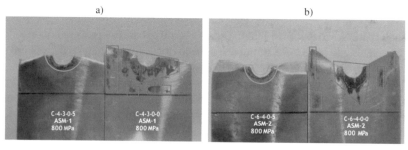

**Fig. 5.** Fretting zones of expanded (blue) and reference (red) assemblies, a) Code-4, b) Code-6

The thermal evolution of the assemblies (reference and expanded, code-6, at 500 MPa loading) is presented in Fig. 6. The reference assembly reaches a much higher temperature than the expanded one during the first phase. More friction occurs in the reference assembly, which is coherent with the previous analysis. Then fretting occurs and the heat created by friction decreases as more and more fretting occurs (appearance of black powder). The last peak of the curves shows the heat created during the plasticization of the material. Once again, it can be seen that the reference peak is much higher than that of the expanded assembly. Plastic deformation is less pronounced on expanded assemblies as the cold working process reduces their plasticizing range.

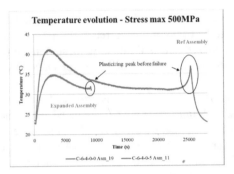

**Fig. 6.** Thermal evolution of the assemblies during the fatigue test (Code-6)

## 5   Conclusion and Perspectives

Although this study has shown that the stacked cold expansion process combined with strong tightening of bolts has a strong detrimental impact on the fatigue life of hard alloy assemblies, cold expansion alone is not to blame for such a performance decrease.

Deeper investigations should be conducted to determine the impact of cold working on titanium bolted joints, with the implementation of different parameters and configurations. In the near future, experimental tests should be carried out with interface sealant to modify the fretting behavior, while others should be performed with hand tightened bolts to remove the effects of bolt preload.

Simulations are also being developed to study the stacked cold expansion phenomenon, in particular the formation of volcanoes, which are at the origin of gaps between plates and the limited surface friction.

A scale effect was also observed in both reference cases, and expanded assemblies showed a longer life in code 4 than in code 6. This scale effect is to be confirmed by other tests with larger dimensions.

**Acknowledgments.** The research work reported here was made possible by Airbus Operations.

## References

1. Luchtvaart Nederland. Slim én duurzaam. Action plan. Ministry of Infrastructure, Amstelveen, The Netherlands (2018)
2. Huggins, G.L., West, R.R., Briscoe, R.T., Welch, J.M.: Engine pylon made from composite material (2012)
3. Boyer, R.: An overview on the use of titanium in the aerospace industry. Mater. Sci. Eng. A **213**, 103–114 (1996)
4. FTI. Cold expansion of holes using the standard split sleeve system and countersink cold expansion. Process Specification 8101C. Fatigue Technology Inc. Seattle, WA (2014)
5. Quincey, D.E., Copple, C.M., Walsh, W.B., Jarzebowicz, R.Z., Easterbrook, E.T.: Split sleeve cold expansion (1994)

6. McCLUNG, R.C.: A literature survey on the stability and significance of residual stresses during fatigue. Fatigue Fract. Eng. Mater. Struct. **30**, 173–205 (2007). https://doi.org/10.1111/j.1460-2695.2007.01102.x
7. Leon, A.: Benefits of split mandrel coldworking. Int. J. Fatigue **20**, 1–8 (1998). https://doi.org/10.1016/S0142-1123(97)00059-5
8. Achard, V., Daidié, A., Paredes, M., Chirol, C.: Optimization of the cold expansion process for titanium holes. Adv. Eng. Mater. **19** (2016). https://doi.org/10.1002/adem.201500626
9. Benhaddou, T., Stephan, P., Daidié, A., Chirol, C., Tuery, J.-B.: Effect of axial preload on double-lap bolted joints: numerical study, nantes. In: Proceedings of ASME 2012, 11th Biennial Conference on Engineering Systems Design and Analysis (2012)
10. Benhaddou, T., Daidié, A., Stephan, P., Chirol, C., Tuery. J.-B.: Optimization of fatigue behavior of metallic shear joints. In: Proceedings of 12th International Conference on the Mechanical Behavior of Materials - ICM12. Karlsruhe, Germany (2015)

# Numerical Simulation and Experimentation of Additive Manufacturing Processes with Polyurethane Foams

Elodie Paquet[1(✉)], Sébastien Le Loch[1], Benoit Furet[1], Alain Bernard[2], and Sébastien Garnier[1]

[1] University of Nantes, Laboratoire LS2N (UMR UMR6004), IUT de Nantes,
2 Avenue du Professeur Jean Rouxel, 44470 Carquefou, France
`elodie.paquet@univ-nantes.fr`
[2] Ecole Centrale Engineering School of Nantes, Laboratoire LS2N (UMR UMR6004), ECN,
1 Rue de la Noë, 44300 Nantes, France

**Abstract.** Foam Additive Manufacturing (FAM) is the additive manufacturing process allowing parts to be obtained by depositing layers of polyurethane foam using a high-pressure machine. This inexpensive technology allows large parts to be produced in a reduced time. However, the quality of the parts produced by the FAM technique is greatly affected by the various thermal phenomena present during manufacturing and by the geometrical deviations of the layers due to the expansion of the PU foam. Numerical simulation remains an effective analytical tool for studying these phenomena. The aim of this work is to build a geometric and thermal model predictive of the FAM process by the finite element method, the final objective of which is to provide temperature maps throughout the manufacturing process and also to choose the best 3D printing strategy to have a model with constant cords and the smallest possible form deviation. The proposed model and the various simulation techniques used are detailed in this article. This model is developed under the finite element code Rem3D, and validated by experimental tests carried out on a FAM machinery or a robot, an example of which is detailed in this article.

**Keywords:** Foam Additive Manufacturing · Simulation processes · 3D printing

## 1 Introduction

The FAM (Foam Additive Manufacturing) is an additive manufacturing (AM) process a process using an expanding polymer. This technology allows the production of large parts such as nautical tools, by superimposing the layers of material. It represents an interesting alternative to traditional machining processes. Indeed, it is a process which makes it possible to save significant production times in the case of the manufacture of parts with complex geometries and to limit the losses of raw material. The material used is a PU foam shear thinned reacting polymer mixture of adequate isocyanate and polyol group. The foam is deposited on a support with a free expansion where after a few seconds the material evolves from a low molecular weight emulsion (through polymerization with the evolution of heat and $CO_2$ gas) to a complex polymer [1]. Compared with other FA processes such as Fused Deposition Modeling process with plastic materials, the FAM is characterized by a high deposition rate (with, in our case, a flow of 43 g/s) because the foam used expands by 30–40 times its volume in a few seconds. The process also has the capacity to manufacture large parts as well as tools since the quantity of material required is deposited and there is no recycling to be expected. For example, the cost of manufacturing a large tool is less expensive by additive manufacturing than by traditional methods [2]. However, the FAM process also has certain non-negligible drawbacks: during successive deposits, the strong thermal gradients generated by the polyurethane foam which is an insulating material lead to the generation of internal residual stresses, and to deformations in the final part. These residual stresses have an impact both on the mechanical properties (fatigue behavior for example) and on the final geometry of the part and also on the material health of the part (appearance of cracks, burning of material, etc…). These constraints and deformations are directly linked to process parameters such as for example the speed of deposit, the height of deposit, the angle of deposition, the temperature of the components, the flow rate and the time expected between the printing of each layer. The influence of these parameters and the stacking of the cords are evaluated using a thermo mechanical simulation taking into account a Navier Stokes' behavior law for the polyurethane foaming (Rem3D software).

## 2   Foam Polyurethane Characterization

The material of interest here is polyurethane foam. The polyurethane is a polymer bi-component material obtained by mixing together isocyanate and polyol. It begins to cure in 6 s, expands to nearly 40 times its initial volume to yield a low density of 36 kg/m$^3$. Its compressive strength is 110 kPa, tensile strength is 171.5 kPa. The reactions taking place in PU foams are complex, but can be represented by only two global exothermic reactions: the first gelling reaction, where polyol and isocyanate react to form highly cross-linked polyurethane and blowing reaction where water and isocyanate react to form urea and carbon dioxide [3].

In order to model mechanical and thermal reaction, we use modelling software called Rem3D. This software computes the flow of the material by solving the Navier Stokes and the thermal equations.

The material used is very reactive and therefore cannot be characterized in the laboratory. Therefore a characterization of the material has been carried out directly on our robotic cell using a thermocouple and camera. The purpose of the experience is to recover the temperature present during the reaction as well as the material's cooling curve and the expansion heights as well as the weight losses of the material during expansion linked to the evolution of gases present during the mixing reaction.The protocol of the characterization implemented (Fig. 1) recovers the reaction temperature, the climb height, the expansion speed as well as the pressure exerted during the polymer foaming. To carry out the characterization, we programmed a casting of material with a programmed mass of 54 g over a translucent container with a diameter of 85 mm fitted with three thermocouples. A camera films the expansion and by image processing, we were able to recover the climb height as well as the speed. Another camera films the mass and the pressure difference during casting. To have a correct characterization, we did the handling and processing twice the two experiments: Experiment 1 and Experiment (Fig. 3) in order to average the results and then translate them into digital material file for the REM3D software.

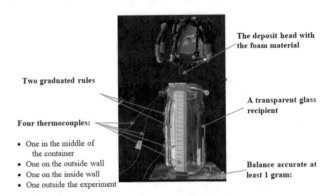

The deposit head with the foam material

Two graduated rules

A transparent glass recipient

Four thermocouples:

- One in the middle of the container
- One on the outside wall
- One on the inside wall
- One outside the experiment

Balance accurate at least 1 gram:

**Fig. 1.** Characterization protocol for foams with high reactivity in situ.

The first physical parameter measured was the reaction temperature (Fig. 3). Temperature maximum of 178° C in the heart is reached in 146 s and the cooling rate is − 2.86 °C per minute. The total cooling time for a volume of 0.0989 m$^3$ of foam poured is 4096 s. The second physical parameter measured and analyzed by image processing is the expansion rise height as well as the expansion speed measured at 32.3 mm/s. Once the material has been characterized [4], the data from the experiment (Fluid Flow rate, Height of deposition, speed....) have been translated into a digital material file and implemented in the Rem3D software. Then, the simulation was realized in order to correlate the data from simulations carried out and experiments (Fig. 2 and Fig. 3).

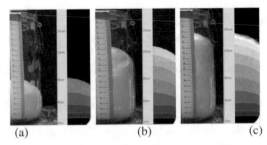

(a)                    (b)                    (c)

**Fig. 2.** Development of the flow front of the foam material in the cylinder with a 85 mm diameter after 2 s (a), 15 s (b) and 90 s (c): results from simulation (right) and practical tests (left)

The compared results are: the volume of foam and the temperature (Fig. 3). The foam volume between simulation and experiment at each time is the same (a, b, c). If we compare the reaction exothermal, the maximum temperature in simulation is reached at a time equivalent to the tests. The reaction seems to be faster when start-up.

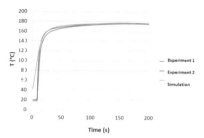

**Fig. 3.** Time evolution of the temperature in the cylinder in simulation and during experiments

## 3    Comparison Between Experiments and Numerical Simulation

The simulation of the behavior of the material during the layer deposition allows obtaining essential information such as the geometry of the manufactured layer [6], the temperature inside the material, as well as the pressures exerted.

This information is then used to determine the best strategies for the printing process to guarantee the correct geometry of the final part and correct material health. As shown in Fig. 4, it is then possible to compare the simulated image to the image taken from a printed layer section. This comparison makes it possible to highlight any differences in terms of distortion between the experiment and the simulation. Observations show a very good correlation between simulation and experience when depositing several layer.

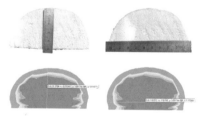

**Fig. 4.** Example of correlation between simulation and experimentation on a layer printed with a 43 g/s rate of flow and a speed deposition of 10.2 m/s.

## 4    Use of Simulation to Apply a Speed Correction on a Trajectory in Order to Minimize the Form Defect

When manufacturing a part by 3D printing, if one wants to obtain the correct geometry of the desired bead, it is essential to master certain physical parameters. In order to compensate for the geometric drifts of the printed layers, a servo-control in position of the upper layers and in speed was set up in order to achieve and maintain a desired geometry using a direct measurement from a contactless sensor and using the results from the simulation to determine the coefficients $a_{proc}$ and $b_{proc}$ to be applied in the relation of the speed of the material deposition to correct speed and smooth out flaws (Fig. 5).

The speed to be applied is given by the equation: $v = \varepsilon * a_{proc} + b_{proc}$ (per layer).

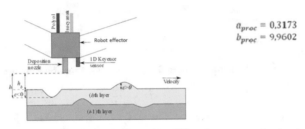

$$a_{proc} = 0.3173$$
$$b_{proc} = 9.9602$$

**Fig. 5.** The principle of the closed loop manufacturing system that has been used

The implementation of this "closed loop manufacturing system" applied to 3D printing [5] aims to better control the deposition of the layers and the final surface condition of the printed part to minimize the potential stages of completion.

Figure 6 shows on the left the surface state of printed layers with position and speed control and on the right the speed profiles associated with the surface states. The enslavement implemented makes it possible to attenuate and compensate for surface defects between the printed layers in order to get as close as possible to the theoretical geometric shape.

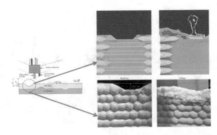

**Fig 6.** Deposition improvement from 2, 86 cm to 2, 27 cm, which corresponds to 20 % in terms of deposition accuracy

## 5   Perspectives

When printing foam of a right angle or a curve with weak radii, the robot slows down on the trajectory and this causes an addition of material in certain area thus creating inhomogeneous layers [7]. These additions of material thus create significant defects in shape deviation on the global model. The choice of strategy in terms of trajectories to be carried out at these locations is currently complex and empirical. The simulation could help us to find the optimal parameters of the process which makes it possible precisely to be able to realize these complicated printing zones and thus to optimize this set of defects (non-homogeneous layers, the recovery of layers, the end of layers which is crushed, the layers bursting…). The digital simulation of material deposition from the different operating parameters allows us to make a comparison of the best strategy to choose to ensure the geometric shape to be obtained at the end. Each time, we carry out several simulations, and we come to choose the best by making a comparative analysis of the simulations with regard to quality and productivity criteria. The knowing of thermal comportement permit us to validate a global printing in a complexe workpiece and making sure to have a correct material health of the part after printing.

## 6  Conclusion

In this paper, the potential to use foam simulation for 3D printing of a large scale part was introduced. The use of process simulation to choose the best strategy has been shown across an example of closed loop manufacturing system on a printed part. The process constraints and the development of specific strategy for printing specific area in large parts must be studied in more depth.

**Acknowledgment.** This work was carried out in the LS2N laboratory at the University of Nantes. The author acknowledges the significant contributions to this paper by the following people: Mme. Sana BAKLOUTI and Mr. Joachim MARAIS at the University of Nantes and is also grateful to TRANSVALOR society for their assistance and technical expertise.

## References

1. Keating, S.J., Leland, J.C., Cai, L., Oxman, N.: Toward site-specific and self-sufficient robotic fabrication on architectural scales. Sci. Robot. **2**(5), 15 (2017). https://doi.org/10.1126/scirob otics.aam8986
2. Poullain, P., Paquet, E., Garnier, S., Furet, B.: On site deployment of 3D printing for the building construction–The case of Yhnova TM. In: MATEC Web of Conferences, vol. 163, no. 5, p. 01001, January 2018. https://doi.org/10.1051/matecconf/201816301001
3. Barnett, E., Gosselin, C.: Large-scale 3D printing with a cable-suspended robot. Addit. Manuf. **7**, 27–44 (2015). https://doi.org/10.1007/s41693-017-0008-04
4. Rossen, W.R.: Numerical challenges in foam simulation: a review. In: Society of Petroleum Engineers (2013). https://doi.org/10.2118/166232-MS. (Delft University of Technology)
5. Barnett, E., Angeles, J., Pasini, D., Sijpkes, P.: Surface mapping feedback for robot-assisted rapid prototyping. In: 2011 IEEE International Conference on Robotics and Automation (ICRA), pp. 3739–3744. https://doi.org/10.1109/ICRA.2011.5979689
6. Talagani, F., DorMohammadi, S., Dutton, R., Godines, C., Baid, H., Abdi, V., Brett G. C., Simunovic, S.D., Chad, E., Love, L.J., Post, B.K., Blue, C.A.: Numerical simulation of big area additive manufacturing (3D printing) of a full size car. United States: N (2015)

7. Baklouti, S., Courteille, E., Lemoine, P., Caro, S.: Vibration reduction of cable-driven parallel robots through elasto-dynamic modelbased control. Mech. Mach. Theory **139**, 329–345 (2019)
8. Izard, J.-B., Dubor, A., Hervé, P.-E., Cabay, E., Culla, D., Rodriguez, M., Barrado, M.: On the improvements of a cable-driven parallel robot for achieving additive manufacturing for construction. In: Cable-Driven Parallel Robots, pp. 353–363. Springer (2018)

# Analysis of Warpage Induced by Thick Copper Metal on Semiconductor Device

Michele Calabretta[1(✉)], Alessandro Sitta[1,2], Salvatore Massimo Oliveri[2], and Gaetano Sequenzia[2]

[1] ADG R&D, STMicroelectronics, Stradale Primosole 50, Catania, Italy
michele.calabretta@st.com

[2] Dipartimento di Ingegneria Elettrica, Elettronica e Informatica (DIEEI), Università degli Studi di Catania, Viale Andrea Doria 5, Catania, Italy

**Abstract.** Electrochemical deposited (ECD) thick film copper on silicon substrate is one of the most challenging technological brick for semiconductor industry representing a relevant improvement from the state of art because of its excellent electrical and thermal conductivity compared with traditional compound such as aluminum. The main technological factor that makes challenging the industrial implementation of thick copper layer is the severe wafer warpage induced by Cu annealing process, which negatively impacts the wafer manufacturability. The aim of presented work is the understanding of warpage variation during annealing process of ECD thick (~20 μm) copper layer. Warpage has been experimental characterized at different temperature by means of Phase-Shift Moiré principle, according to different annealing profiles. A linear Finite Element Model (FEM) has been developed to predict the geometrically stress-curvature relation, comparing results with analytical models.

**Keywords:** Semiconductor · FEM · Phase-Shift Moiré · ECD · Warpage

## 1 Introduction

Thick film copper on silicon substrate is one of the major challenges in the semiconductor industry because of its excellent electrical and thermal conductivity. Compared to other materials, such as aluminum, these features allow respectively to reduce the Joule effect and help the efficient dissipation of heat generated during operation. Furthermore, Cu metallization is better than aluminum in terms of resistance to electromigration and stress voiding phenomena. The main technique to produce thick copper film on silicon substrate is electrochemical deposition (ECD), which is a highly efficient wet process for depositing a uniform layer of metal (like copper) on a semiconductor wafer. Furthermore, Cu front metal is very attractive for integrated circuit (IC) manufacturing because it enables high-reliable Cu-Cu wire bonding solution [1]. The integration of Cu into ICs is still a technical challenge due to the severe wafer warpage induced by Cu annealing, which affects the accuracy of the subsequent manufacturing processes, such as the wafer handle and the adsorption of vacuum suction cup, having also a negative impact on

L. Roucoules et al. (Eds.): JCM 2020, LNME, pp. 55–60, 2021.
https://doi.org/10.1007/978-3-030-70566-4_10

device reliability. Wafer warpage caused by thick Cu layer is mostly due to plastic deformation during the annealing [2], which is a thermal process aimed to make softer Cu compound, increasing and then stabilizing the Cu grains dimension. Stabilizing metal grain size, annealing permits to avoid electromigration issues in interconnect reliability, e.g. during reliability application-related test [3, 4]. It seems to miss in literature a detailed experimental warpage analysis for thick Cu layer: available data refer about maximum 5 μm thick Cu metal. For what concerns analytical models, Stoney equation [5] have been commonly used as reference. This formula has been developed for "membrane-like" geometries made by two materials, such as the system made by semiconductor wafer and ECD metal, in which one layer (in our case, ECD Cu) is much thinner than other. However due to complex not-linear Cu behavior during annealing, Stoney results are not reliable to predict stress and warpage during annealing. The scope of this work is to characterize the warpage induced by 20 μm thick Cu film on a rectangular wafer slice, considering two different annealing profiles. A Finite Element Model (FEM) has been developed to predict the geometrically stress-curvature relation inside the elastic regime. Numerical outcomes have been compared with the results of Stoney and Timoshenko equations.

## 2  Sample and Test Description

The considered samples for warpage analysis were $50 \times 10 \times 0.75$ mm beam, made by bulk Silicon 730 μm-thick, TiW 0.3 μm, Cu seed 0.2 μm and ECD Copper 20 μm-thick. These portions have been sliced from wafer just after copper electro-deposition at room temperature, therefore copper has not been thermally treated before samples formation. The slices have been annealed following two different temperature profile, shown in Fig. 1. The ramps have been characterized by fixed heating and cooling rate (respectively 10 and 5 °C/min), maximum temperature of 250°C. The only difference between profiles is the presence in profile "W/ plateau" of a period of 30 min in which temperature is fixed at 250 °C. Profile "W/o plateau" has been repeated three times to investigate on hysteretic behavior after the first cycle.

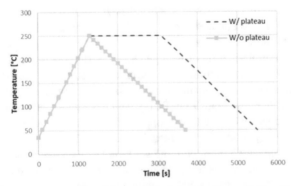

**Fig. 1.** Temperature profile considered for annealing the samples and measuring the warpage in temperature.

Warpage has been measured continuously during the annealing, according to methodology explained in following paragraph. Due to simple geometrical considerations, beam curvature has been derived from warpage and compared with the calculated by Finite Element Model in linear elastic regime.

## 3   Warpage Measurements

A method based on Moirè Phase Shift has been considered to experimentally quantify the warpage, using the commercial equipment "TDM Compact" by Insidix [6]. The sample to be analyzed is illuminated by a structured light made by stripe pattern. The stripe pattern is deformed by the sample's surface structure and the resulting image is captured by a CCD camera, that correlates the out-of-plane deformation with the xy coordinates. The desired temperature profiles have been reproduced during warpage measurement, heating sample by the infrared heater and cooling with compressed air. More detailed method explanation has been proposed in [7, 9]. Warpage has been monitored in temperature, according to thermal profiles described in Sect. 2. It has been plotted in Fig. 2, following the sign convention declared in Fig. 3.

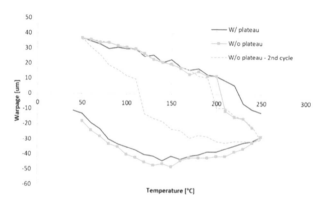

**Fig. 2.** Measured warpage during temperature profiles. Negative values are reached during the heating, while positive ones during the cooling phase. Dotted line represents the warpage vs temperature profile occurred during a second temperature cycle.

Analysis has highlighted that annealing process modifies warpage at 50 °C in both temperature profiles. Warpage is ~ −10 μm at the start, whereas it becomes ~ + 35 μm at the end of first cycle. Looking to the second cycle, warpage behaves as a closed loop because there is no warpage modification between the start and the end of cycle. It has been observed the first part of warpage-temperature diagram in the second cycle (between 50 and 90 °C) has a linear trend. The incremental ratio between warpage variation and temperature in this range has been considered to benchmark the different warpage calculation methods presented in following paragraph. The warpage trend in temperature during the heating phase shows a change in slope that occurs at 150 °C, at which warpage reaches its maximum. This observed behavior could indicate an irreversible in copper layer: dedicated physical analysis will help to better understand the phenomenology.

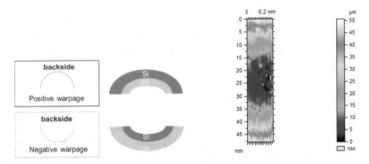

**Fig. 3.** Sign convention for warpage and warpage map at 150 °C.

## 4  Finite Element Model

A Finite Element Model has been developed with Comsol Multiphysics to calculate the warpage variation, curvature and mechanical stress due to temperature variation. Material properties, reported in Table 1, have been taken from on-line archive Matweb [10], excepting for Young Modulus of copper layer, which has been derivate multiplying the Indentation Modulus $E_{IT}$, measured by nanoindentation, for factor $1-\nu^2$. It has been assumed material behavior as elastic, neglecting the properties temperature dependence.

**Table 1.** Considered material data for simulation.

| Material | Young Modulus [GPa] | CTE [ppm/K] | Poisson ratio |
|---|---|---|---|
| Silicon [10] | 120 | 2.5 | 0.28 |
| Copper [10] | 77.5 | 16.4 | 0.364 |

Linear elastic warpage and stress behavior has been calculated with the customized FE model and two different literature approaches. One is derived from the Timoshenko theory for bi-metal thermostats [8] and the other is the Stoney equation [5], to calculate thermal warpage in thin film. Warpage results have been benchmarked with experimental warpage observed in the second cycle during the first heating phase (50–90 °C). Stress along x direction and warpage variation in temperature range 50–90 °C have been calculated with FEM, considering the system stress and strain free at 50 °C. The results, shown in Fig. 4, have highlighted stress is uniformed distributed on the strip, compressive on copper side due to positive temperature variation. Warpage is correlated by means of a parabolic law with strip length.

The comparison between warpage and stress variation coming from considered methods has been resumed in Table 2. FEM warpage is the closest with measurements and underestimates the experimental behavior of about 6%. Comparing between them the calculation methods, FEM predicts the highest warpage and highest stress.

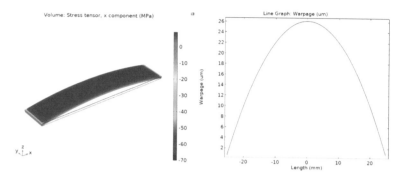

**Fig. 4.** Direction x-stress component and warpage on z-direction computed by FEM

**Table 2.** Benchmark between experiments, FE model and literature analytical approach.

| Method | Experimental | FEM | Timoshenko | Stoney |
|---|---|---|---|---|
| Warpage [$\mu$m/°C] | 0.700 | 0.660 | 0.605 | 0.632 |
| Stress [MPa/°C] | n.a. | 1.683 | 1.003 | 1.075 |

## 5  Conclusions

In this work, it has been studied the evolution of the deformation during the annealing process of ECD thick copper film (20 $\mu$m), measuring at different temperature the warpage of a wafer slice by means of dedicated optical technique. Analyses have shown an irreversible strain at a reference temperature (50 °C) induced by the first heating cycle. Plateau has not affected the residual warpage. The deformation in elastic regime has been reproduced with a preliminary linear FE model, which is a starting point to develop a more complete model to predict ECD warpage during entire annealing profile, considering the involved not-linearities and correlating it with further physical analysis dedicated to address the involved phenomenology.

## References

1. Mancaleoni, A., et al.: Copper wire bonding process characterization and simulation. In: CIPS 2020; 11th International Conference on Integrated Power Electronics Systems. VDE, 2020
2. Cheng, G., Xu, G., Gai, W., Luo, L.: Influence of observed anelasticity of Cu on the wafer warpage evolution during thermal processes. In: 2018 19th International Conference on Electronic Packaging Technology (ICEPT), pp. 1119–1122. IEEE, August 2018
3. Russo, S., Bazzano, G., Cavallaro, D., Sitta, A., Calabretta, M.: Thermal analysis approach for predicting power device lifetime. IEEE Trans. Device Mater. Reliab. **19**(1), 159–163 (2019)
4. Sitta, A., Russo, S., Bazzano, G., Cavallaro, D., Greco, G., Calabretta, M.: Numerical approach to predict power device reliability. In: 2018 13th International Conference on Design & Technology of Integrated Systems in Nanoscale Era (DTIS), pp. 1–5. IEEE, April 2018

5. Stoney, G.G.: The tension of metallic films deposited by electrolysis. In: Proceedings of the Royal Society of London. Series A, Containing Papers of a Mathematical and Physical Character, vol. 82, no. 553, pp. 172–175 (1909)
6. Weidmann, D., Dubois, G., Hertl, M., Chauffleur, X.: Determination of strength of interface in packages based on an approach using coupling of experimental and modeling results. In: 2011 12th International Conference on Thermal, Mechanical & Multi-Physics Simulation and Experiments in Microelectronics and Microsystems, pp. 1–6. IEEE, April 2011
7. Calabretta, M., Sitta, A., Oliveri, S.M., Sequenzia, G.: An integrated approach to optimize power device performances by means of stress engineering. In: International Conference on Design, Simulation, Manufacturing: The Innovation Exchange, pp. 481–491. Springer, Cham, September, 2019
8. Timoshenko, S.: Analysis of bi-metal thermostats. Josa **11**(3), 233–255 (1925)
9. Mirone, G., Sitta, A., D'Arrigo, G., Calabretta, M.: Material characterization and warpage modeling for power devices active metal brazed substrates. IEEE Trans. Device Mater. Reliab. **19**(3), 537–542 (2019)
10. Matweb, Material Property Data. www.matweb.com. Accessed 21 April 2020

# Finite Element Modeling and Validation of Metal Deposition in Wire Arc Additive Manufacturing

Akram Chergui[✉], Nicolas Beraud, Frédéric Vignat, and François Villeneuve

Univ. Grenoble Alpes, Grenoble INP, G-SCOP, 46 Avenue Felix Viallet, 38031 Grenoble, France
Mohammed-akram.chergui@grenoble-inp.fr

**Abstract.** Wire arc additive manufacturing allows the production of metallic parts by depositing beads of weld metal using arc-welding technologies. This low-cost additive manufacturing technology has the ability to manufacture large-scale parts at a high deposition rate. However, the quality of the obtained parts is greatly affected by the various thermal phenomena present during the manufacturing process. Numerical simulation remains an effective tool for studying such phenomena. In this work, a new finite element technique is proposed in order to model metal deposition in WAAM process. This technique allows to gradually construct the mesh representing the deposited regions along the deposition path. The heat source model proposed by Goldak is adapted and combined with the proposed metal deposition technique taking into account the energy distribution between filler material and the molten pool. The effectiveness of the proposed method is validated by series of experiments, of which an example is detailed in this paper.

**Keywords:** Wire-arc-additive-manufacturing · Metal deposition · Finite element · Thermal behavior

## 1   Introduction

Wire Arc Additive Manufacturing (WAAM) is one of the most promising additive manufacturing (AM) processes. It combines an electrical arc as a heat source and a wire as feedstock, and parts are obtained by depositing weld beads, layer-by-layer, using arc welding technologies. This technology is inexpensive, offers a large workspace, and allows high deposition rates. However, the quality of parts manufactured using WAAM is greatly affected by the various thermal phenomena present during the manufacturing process. Béraud et al. [1] showed that process simulation remains an efficient tool for building a more thorough understanding of the different thermal phenomena involved in metal additive manufacturing. This method allows not only to reduce experimentation, but also to better optimize the process at a lower cost. Most recent work in additive manufacturing literature addressed process simulation using finite element method. Xiong et al. [4, 5] studied the thermal behavior of cylindrical parts made by WAAM through a finite element thermal simulation of the process. Montevecchi et al. [6, 7] developed a finite element model based on a mesh coarsening technique in order to reduce the computational cost of the process simulation. In the same perspective, Ding et al. [10]

© The Author(s) 2021
L. Roucoules et al. (Eds.): JCM 2020, LNME, pp. 61–66, 2021.
https://doi.org/10.1007/978-3-030-70566-4_11

proposed a finite element approach based on two models (transient and stationary) in order to investigate the thermomechanical behavior of parts manufactured in WAAM. Michaleris [9] reviewed the existing techniques for metal deposition modeling, and proposed a new hybrid algorithm in order to reduce the computational time.

From the perspective of a simulation, WAAM process is very similar to multi-pass welding process [6]. However, the physics of welding process put forward some complex physical phenomena, involving thermodynamics, heat and mass transfer, electricity and magnetism [11]. Modeling such phenomena at a part scale level can be costly due to the unreasonable computational time requirements. For this reason, most studies in literature recommend to model the heat input using a heat source model, taking into consideration the energy contribution of the different physical phenomena occurring in the molten pool. In this article, the double-ellipsoid heat source model proposed by Goldak [3] is adapted and combined with a new finite-element metal deposition technique in order to model material and heat input in WAAM process simulation. The proposed method allows not only to take into account the energy distribution between filler material and the molten pool, but also to consider the changing in the boundary conditions during the deposition process. The overall model is validated with an experimental test case, and the results are presented.

## 2  Proposed FE Metal Deposition Technique

### 2.1  Metal Deposition Modelling

There are two finite element techniques for modeling material deposition in additive manufacturing process simulation: inactive element method and quiet element method, as reviewed in literature by Michaleris [9]. In the inactive element method, the elements representing the deposited regions are initially inactive, and activated gradually according to the deposition path. In the quiet element method, all elements are present from the start of the analysis, but low values are assigned to their material properties (conductivity and specific heat), so they do not affect the analysis. These material properties are then switched to the real values according to the deposition path. Both inactive and quiet methods can be used to model metal deposition in WAAM process, and each presents some advantages over the other. But also some disadvantages such as introducing errors into the finite element analysis, and taking a longer time to run. Furthermore, the interface between inactive (or quiet) and active elements is continuously evolving during the deposition phase, making it difficult to compute this internal interface and consider its surface convection and radiation. For this reason, convection and radiation on this interface are neglected on both inactive and quiet element techniques. Leading to additional errors in the finite element analysis [9] (Figs. 1 and 2).

In the present model, metal deposition is taken into account using a new element deposition technique. In this technique, each deposited droplet is modeled by a set of elements, representing together a numerical droplet. Every numerical droplet is created at its corresponding time-step in the deposition phase simulation. The new created elements are then added to the elements already created in the previous time-steps. Thus, the mesh representing the deposited regions is constructed gradually along the deposition path.

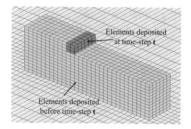

Fig. 1. Element deposition technique

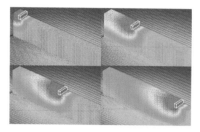

Fig. 2. Mesh evolution using the element deposition technique.

## 2.2 Heat Input Modelling

Goldak et al. [3] proposed a double-ellipsoid heat source model in order to model the heat input in welding process. The model is expressed as follows:

$$q_{f,r}(x, y, z) = \frac{(6\sqrt{3}f_{f,r}Q)}{(\pi abc_{f,r}\sqrt{\pi})} exp\left(-3\frac{x^2}{a^2} - 3\frac{y^2}{b^2} - 3\frac{z^2}{c_{f,r}^2}\right) \quad (1)$$

Where $q_f$ and $q_r$ are the heat flux densities of the front and rear ellipsoids respectively. $a$, $b$ $c_f$ and $c_r$ are the semi-axes of the two ellipsoids. $Q$ is the arc energy input, and $Q = \eta UI$. $I$, $U$ and $\eta$ are the welding current, voltage and efficiency. $f_f$ and $f_r$ are the distribution factors of the heat flux in the front and rear ellipsoids, and $f_f + f_r = 2$.

In WAAM, the arc power is not fully delivered to the part directly. According to previous works, about 50% of the total arc power is used to melt the feed wire. This energy is subsequently transmitted to the part through the enthalpy of the deposited droplet [6] (Figs. 3 and 4)

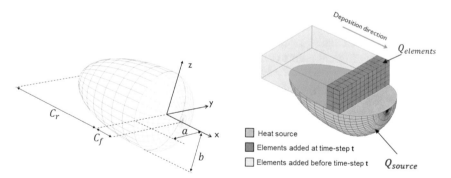

Fig. 3. Goldak heat source

Fig. 4. Adapted Goldak heat source

In the present model, the Goldak heat source is adapted and combined with the proposed element deposition technique to model the heat input in WAAM process, taking into consideration this energy distribution between the wire and the substrate. In fact, the direct energy transfer from the arc to the substrate is considered using the inferior half of the double-ellipsoid Goldak model. The remaining 50% of the total energy is delivered by means of the deposited elements. These latter are charged with an amount of energy equivalent to the energy delivered by the adapted heat source, and expressed as follows:

$$Q_{source} = Q_{elements} = \frac{\eta UI}{2} \tag{4}$$

## 2.3 Model Update

As discussed earlier, the internal interface between inactive (or quiet) and active elements is difficult to compute, because it changes continuously during the deposition phase. Therefore, surface convection and radiation are often neglected on this interface using inactive and quiet element methods, thus introducing errors to the finite element analysis. It is easier to overcome such issue using the proposed element deposition technique, as the internal active/inactive interface becomes an external free surface. In the proposed simulation, a model-update procedure is developed in order to update the boundary conditions and the material properties after each elements deposition step. This procedure allows to compute the external surface of the total mesh, to apply surface convection and radiation adequately, and to assign the relevant material properties to the new-deposited elements. The evolution of material properties such as conductivity, specific heat and density as a function of temperature is also taken into consideration, and the latent heat is considered as a triangular variation of the specific heat between solidus and liquids temperatures.

## 3  Validation

In order to check the effectiveness of the proposed technique, an experimental validation was carried out by comparing simulated and experimental thermal curves of a test case. The test case consists of fabricating two thin-walled parts of eight layers according to two different strategies: raster and zigzag. Temperatures as function of time were measured at six different points (P1, P2 and P6) on the substrate plate using six k-type thermocouples.

The test case was carried out using a WAAM cell composed of a Yaskawa DX200 6-axis robot, a Fronuis CMT welding technology, and a 2-axis positioner. Two different materials have been considered in this experiment. The substrate plate material was a 5083 welding aluminum alloy, and the thin-walls structures were manufactured using a 5356 aluminum alloy filler wire with a diameter of 1.2 mm. The beads were deposited under a 100% argon gas protection with a flow rate of 13 L/min (Fig. 5).

### 3.1  Results

Figure 6 presents the experimental temperature curves as a function of time at points P1 P2 and P3, compared to the results obtained with the proposed simulation. Comparison for both Raster and Zigzag strategies are shown in Fig. 6(a) and Fig. 6b, respectively.

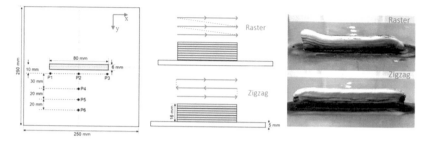

**Fig. 5.** Experimental design

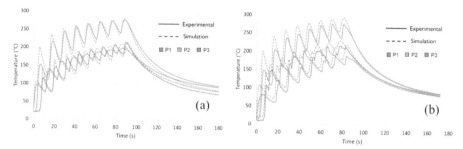

**Fig. 6.** Temperature curves comparison between the experimental measurements and simulation at P1 P2 and P3 for Raster strategy (a) and Zigzag strategy (b).

The results demonstrates that the temperature curves obtained from the finite element model correspond well to those measured in the experiment. The average error between the simulated and the experimental values is about 5%. However, the gap is higher in the first peaks. Possible sources of such error could be the thermal inertia of the thermocouples.

## 4   Conclusions

In this study, a new metal deposition technique is proposed in finite element modeling of WAAM process. This element deposition technique enables, for each deposition time-step, to add new elements to the previous mesh, following the passage of the heat source. Thus, allowing to model both material and heat input, and to better consider heat exchange in the external surfaces. The proposed modeling technique is validated through an experimental test case, where simulated results are shown to be in agreement with the experimental curves. In the future, the proposed simulation is to be used to better understand and predict the thermal related defects in parts fabricated in WAAM.

## References

1. Beraud, N.: Fabrication Assistée par Ordinateur pour le procédé EBM. Thèse, Université Grenoble Alpes (2016)

2. Vayre, B., Vignat, F., Villeneuve, F.: Metallic additive manufacturing: state-of-the-art review and prospects. Mech. Ind. **13**(2), 89–96 (2012)
3. Goldak, J., Chakravarti, A., Bibby, M.: A new element model for welding heat source. Metall. Trans. B. **15B**, 299–305 (1984)
4. Xiong, J., Lei, Y., Li, R.: Finite element analysis and experimental validation of thermal behavior for thin-walled parts in GMAW-based additive manufacturing with various substrate preheating temperatures. Appl. Therm. Eng. **126**, 43–52 (2017)
5. Xiong, J., Li, R., Lei, Y., Chen, H.: Heat propagation of circular thin-walled parts fabricated in additive manufacturing using gas metal arc welding. J. Mater. Process. Technol. **251**, 12–19 (2018)
6. Montevecchi, F., Venturini, G., Scippa, A., Campatelli, G.: Finite element modelling of wire-arc-additive-manufacturing process. Procedia CIRP **55**, 109–114 (2016)
7. Montevecchi, F., Venturini, G., Grossi, N., Scippa, A., Campatelli, G.: Finite element mesh coarsening for effective distortion prediction in wire arc additive manufacturing. Addit. Manuf. **18**, 145–155 (2017)
8. Michaleris, P.: Modeling metal deposition in heat transfer analyses of additive manufacturing processes. Finite Elem. Anal. Des. **86**, 51–60 (2014)
9. Ding, J., et al.: Thermo-mechanical analysis of wire and arc additive layer manufacturing process on large multi-layer parts. Comput. Mater. Sci. **50**, 3315–3322 (2011)
10. Hu, J., Tsai, H.L.: Heat and mass transfer in gas metal arc welding. Part I: the arc. Int. J. Heat Mass Trans. **50**(5–6), 833–846 (2007)

# Study of the Use of Sawdust and Mycelium Composite as a Substitute of EPS

R. Miralbes[1($\boxtimes$)], D. Ranz[1], and D. Zouzias[2]

[1] University of Zaragoza, Zaragoza, Spain
miralbes@unizar.es
[2] University of Leuven, Leuven, Belgium

**Abstract.** Expanded polystyrene foams are a petroleum-origin material that is usually used in some applications such as motorcyclist helmets. Despite it notably mechanical properties, it low density and its capability to absorb energy during an impact, it is necessary to find a renewable-origin substitute material. Thus, it has been studied the use of a sawdust and mycelium composite material under quasi-static and dynamic efforts. Sawdust is a waste material that has very small grains that are totally disaggregated so it has very low material properties. The use of oyster mushroom mycelium generates an internal structure that joins grains and, consequently, the resultant material has notably high mechanical properties. Then it has been compared the resultant properties (stress-strain curve, absorbed energy, decelerations, etc.) with the different densities EPS ones and it has been concluded that this composite material, despite it high density, it could be a suitable substitute material and in some cases it has better properties.

**Keywords:** Mycelium · Sawdust · Renewable · Mechanical · Helmet

## 1 Introduction

Expanded polystyrene foams (EPS) are frequently used in some applications in which it is necessary to absorb energy such as helmets and protector packaging for golds; however, this material has a non-renewable origin and high environment foot-print. As a result, some studies seek for substitutive materials such as cork agglomerates [1] that should have an internal structure similar to the EPS one. The internal structure of this material and, thus, its mechanical properties, depends on this manufacturing process in which it is possible to control the expansion of the EPS to obtain the desired density that is usually around 10–40 kg/m$^3$ for packaging applications and between 60 to 120 kg/m$^3$ for helmets and similar [2]. This structure is composed of closed air cells surrounded by EPS; depending on the desired density and the expansion process, there is more or less enclosed air. This internal structure is reflected in the mechanical stress-strain curve of the EPS under compression efforts that is the most common load that must support. This curve has three different zones [3] named elastic region, collapse region and densification region. In the elastic region the efforts are supported by the cell walls that are under elastic bending efforts and so, the material can recover it initial shape;

L. Roucoules et al. (Eds.): JCM 2020, LNME, pp. 67–72, 2021.
https://doi.org/10.1007/978-3-030-70566-4_12

this zone presents a linear behavior that usually goes until a strain around a 3–5%. Additionally, the air enclosed inside compress, but after a certain point, the pressure of this enclosed air cannot be supported by the cell walls and, as a result, they collapse so the material has a plastic behavior; consequently, the stress-strain curve presents a plane zone that is usually called the plateau or the collapse region that ends with a strain between a 40 and 60%. In this point all the cells have totally collapsed and the material behavior is similar to the non-foamed material so, it presents an exponential behavior; this zone is called densification. Due to this stress-strain shape, especially the plateau zone, EPS is adequate to absorb energy and generate a constant deceleration because high deceleration peaks can damage easily goods and, in helmets, generate more brain damages.

On the other hand, EPS has a non-renewable origin because it is generated using styrene that comes from petroleum or from natural gas, so it is necessary to find a material that have similar capability to absorb energy than the EPS but with a renewable origin. In this point, there some authors are focused on cork and their products [1] that are a renewable-origin material that can be easily recycled; however, the use of cork is limited to the quantity that can be obtained from the cork oak that only grown in a few places around the world and cannot supply the world necessities. This is the main reason why this article is going to explore another renewable-origin material, the sawdust that is a sub product and sometimes a waste.

Sawdust, like its name point, has a dust material state that is composed of very-low size grains surrounded by air; this internal structure is similar to the EPS foam so it mechanical behavior should be similar. However, sawdust grains are not join and, as a result, it mechanical properties are insignificant and without any structural application. To solve this problem, the article studies the use of mycelium to join sawdust grains and generate a material similar to the EPS. Mycelium is a set of hyphae that come from the vegetative part of a fungus or lichen and are multicellular filaments that can grow very quickly. For this reason and for their frequent ramifications, a tangle of hyphae with a huge surface appears in the substrate: the mycelium. The mycelium has two different main functions, obtain nutrients and tie the fungus to the ground. Due to these functions, mycelium generate a vast extension of interconnected ramifications and this is the reason why, combined with it high-speed growing and it renewable origin, mycelium has been selected to join sawdust to obtain a compacted material that could have a similar behavior to the EPS. There are some studies [4] that use these materials to generate furniture and complex structures but there are not studies about their mechanical properties under quasi static and dynamic compression efforts.

## 2   Materials and Methods

It has been used the grow.bio [5] material that consist in a prepared mixture of sawdust and oyster mushroom seeds that germinate with water and flour and needs to grow along ten days to generate the hyphae to compact the sawdust. After this period, the hyphae growing is stopped using a heating oven that is used too to dry the material. The material obtained can have any desired geometry because during the growing process, after four days, it is possible to put the material inside a mold and the final material will have the mold geometry; this could be perfect for some applications such as helmets and protective envelopments for goods (Fig. 1).

**Fig. 1.** Left: initial mycelium and sawdust material. Right: material after the drop tower test

According to grow.bio [5] this material must have at least a density of 121 kg/m$^3$, (measured: 121.98 kg/m$^3$) that it is similar to EPS120, a compressive stress at a 15% of strain of 0.128 MPa and a flexure strength of 0.234 MPa. However, there is not too much information about the stress-strain curve, the energy that can absorb, etc. There are only a few articles that are reviewed by Girometta [6] that study as a substrate EPS, cellulosic fibers and cotton, but there are not studies about the use of sawdust.

The resultant composite material has been studied under quasi-static compression test using an 8032 INSTRON universal test machine with a 0.2 mm/s velocity until reaching a 90% of strain. It has been tested 40 mm cubic specimens to obtain the forces-displacements curve and, afterwards, the stress-strain curve and the absorbed energy-strain curve. With the density, using these results, it has been obtained the specific stress-strain and the specific absorbed energy-strain curve.

In the case of the dynamic test, using a 75 J drop tower machine, 40 mm cubic specimens have been tested to absorb 75 J. With an internal accelerometer it has been obtained the vertical deceleration-time curve.

## 3  Results and Discussion

Analyzing the results of the quasi-static compression test, it can be pointed that the mycelium has a totally different stress-strain curve shape (Fig. 2 left) and has initially a constant slope until around a 50% of strain. After this point it has an exponential shape. It can be pointed that initially the stiffness of the mycelium is significantly lower than the EPS ones and increase constantly; on the other hand, the EPS has initially a high stiffness in the elastic zone and, afterwards a constant plateau; both are directly depending on the density of the EPS. After a certain point the EPS has an exponential behavior. Consequently, initially, during an impact, the mycelium will have a lower stiffness and will generate lower initial decelerations but, after a certain point, it stiffness will be notably higher than the EPS ones and will generate higher decelerations. As a result, the mycelium will have initially a lower energy absorption capability (Fig. 2 right) that will increase throughout it crash. In this point it can be observed that, under high deformation, it can absorb more energy than lowest density EPS. Consequently, with the same dimensions, the mycelium can absorb more energy than the EPS of densities of 60, 75 and 80 kg/m³ in case of a high energy impact and will deform less to absorb this energy. On the other hand, for lower energy impacts mycelium will deform much more but would generate lower decelerations.

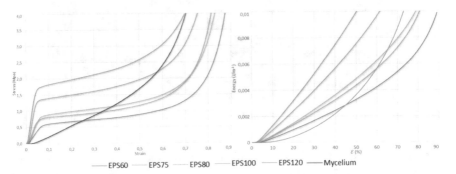

**Fig. 2.** Up: stress-strain curve. Down: absorbed energy-strain curve.

If the density is included in the analysis, despite the influence of the high density of the composite material, it can be pointed that the shape of some EPS specific stress-strain curves (Fig. 3 left) are similar and the difference between other curves is lower than for the stress-strain curve; however, due to the mycelium high density, it curve is hardly lower initially and reach the specific stiffness with higher strains; consequently, an element with the same weight but with different materials, if the EPS is used, it will be initially significantly stiffness and will deform less but, in case of high strains the mycelium will deform less. Consequently, EPS can absorb initially notably more energy (Fig. 3 right); this difference will decrease and finally with high strains (80–90%), mycelium could absorb the same energy of the EPS. As a result, with the same weight, EPS can absorb more energy but it will have a higher stiffness.

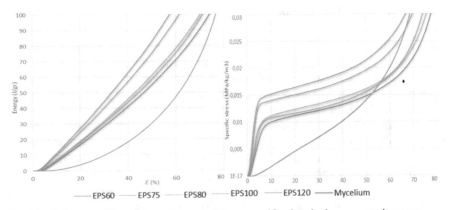

**Fig. 3.** Up: specific stress-strain curve. Down: specific absorbed energy-strain curve.

In this point it must be pointed that the deceleration depends on the stiffness of the material and the stress-strain curve shape but the relationship is not straight. Consequently, despite the mycelium has initially a lower stiffness, the deceleration depends on the stiffness along the entire stress-strain curve; consequently, a drop tower test has been carried out (Fig. 4). It can be pointed that the mycelium and the EPS 120 show lowest decelerations peaks. Additionally, due to it initially lowest stiffness, the deceleration peak of the mycelium appears later. On the other hand, the deceleration will appear during more time (15 ms vs around 10 ms).

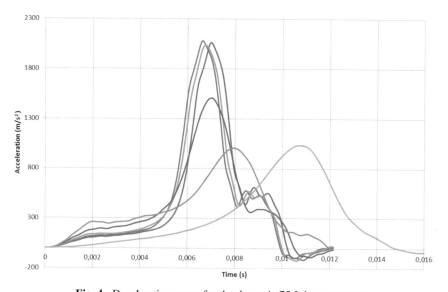

**Fig. 4.** Deceleration curve for the dynamic 75 J drop tower test

Speaking about the deformation (Table 1), all the materials have a similar deformation due to the high energy impact test but the mycelium will have a lower permanent deformation.

**Table 1.** Maximum deceleration (m/s$^2$), average deceleration (m/s$^2$), maximum deformation (%) and permanent deformation (%) for the dynamic 75 J drop tower test.

|  | EPS 60 | EPS 75 | EPS 80 | EPS 100 | EPS 120 | Mycelium |
|---|---|---|---|---|---|---|
| Max. Decel | 416.8 | 414.7 | 420.2 | 392,1 | 372,1 | 252,3 |
| Av. Decel | 207.8 | 206.2 | 2037 | 1508 | 1006 | 1035 |
| Max def | 0.81 | 0.87 | 0.85 | 0.87 | 0.86 | 0.83 |
| Per def | 0.58 | 0.56 | 0.56 | 0.55 | 0.55 | 0.42 |

## 4 Conclusions

The results show that mycelium could be a suitable material to substitute EPS in some applications such as helmets because it can absorb similar energy and would generate lower peak decelerations. Despite its high density, this material presents initially a notably lower stiffness that increase linearly until reaching and passing the stiffness of the EPS for high deformation. As a result, this material would generate lower decelerations in heads in case of low energy impact and in the initial moments of the impact; it is in this moment when the highest peak decelerations appears but after a certain point the maintained deceleration will be higher that the EPS. Consequently, the subsequent point is the design and test of a helmet built using this material to verify it behavior.

## References

1. Fernandes, F.A.O., Alves de Sousa, R.J., Ptak, M., et al.: Helmet design based on the optimization of biocomposite energy-absorbing liners under multi-impact loading. Appl. Sci. Basel **9**(3), 735 (2019)
2. Ling, C., Cardiff, P., Gilchrist, M.D.: Mechanical behaviour of EPS foam under combined compression-shear loading. Mater. Today Commun. **16**, 339–352 (2018)
3. Di Landro, L., Sala, G., Olivieri, D.: Deformation mechanisms and energy absorption of polystyrene foams for protective helmets. Polym. Testing **21**(2), 217–228 (2002)
4. Holt, G.A., McIntyre, G., Flagg, D., et al.: Fungal mycelium and cotton plant materials in the manufacture of biodegradable molded packaging material: evaluation study of select blends of cotton byproducts. J. Biobased Mater. Bioenergy **6**(4), 431–439 (2012)
5. https://grow.bio/collections/shop/products/grow-it-yourself-material
6. Girometta, C., Picco, A.M., Baiguera, R.M., et al.: Physico-mechanical and thermodynamic properties of mycelium-based biocomposites: a review. Sustainability **11**(1), 281 (2019)

# Topology Optimization Design of Internal Reinforcements in a Sailing Dinghy

Antonio Mancuso[1], Antonio Saporito[1], and Davide Tumino[2(✉)]

[1] Dipartimento di Ingegneria, Università degli Studi di Palermo, Viale delle Scienze, 90127 Palermo, Italy
[2] Facoltà di Ingegneria e Architettura, Università degli Studi di Enna Kore, Cittadella Universitaria, 94100 Enna, Italy
davide.tumino@unikore.it

**Abstract.** In this paper, a novel approach has been followed based on FEM simulation and Topology Optimization tools to locate and model the reinforcements inside the hull of a sailing dinghy. This process assumes that the inner volume included between the hull and the deck is, at the beginning of the simulation, filled with material; then a portion of this inner volume is eroded until a final free form shape of the reinforcements is obtained. A key point of this procedure is the definition of the optimization constrains because the final shape of the reinforcements must fulfill several requirements such as weight, stiffness and stress. At the end of the optimization procedure, the final shape of internal reinforcements consists of a truss-like web frame with a final weight equal to the 18% of the initial full body.

**Keywords:** Yacht design · Reinforcement · Topology optimization

## 1 Introduction

The traditional design approach of a sailing yacht is standardized by the international Rules for Classification and Construction [1]. Scantling of the internal structure derives by parametric equations that return the number and the dimensions of longitudinal and transverse frames to be applied to the hull in order to improve its local and global stiffness. Once the operating category and the material of the boat have been decided, governing mathematic relations are mainly related to the waterline length of the hull.

The possibility of a scantling procedure different from [1] has not been systematically researched yet. This also happens because, in the naval field, safety reasons suggest to be conservative in terms of reinforcement element design and that traditional manufacturing techniques based on panels and beams bonding are still preferred. Topology Optimization (TO) techniques [2] allow more freedom to the designer because the optimal structure is obtained from an initial bounding box that is gradually eroded in order to retain the minimum volume, usually with the maximum stiffness. Few attempts can be found in literature where TO is applied in marine applications. In [3], fixed shape stiffeners have been internally distributed for the composite submarine sail structure using TO

© The Author(s) 2021
L. Roucoules et al. (Eds.): JCM 2020, LNME, pp. 73–79, 2021.
https://doi.org/10.1007/978-3-030-70566-4_13

techniques. In [4] TO has been used to determine only primary reinforcements of a 46 m sailing yacht: only quasi-static rig and weight loads are considered and the TO procedure is conditioned by the interior layout design.

In the present paper, a procedure implementing FEM simulations and TO has been defined in order to place the reinforcements inside the hull of a sailing dinghy. Structural and harmonic analysis results are input data to search the optimal mass distribution inside the boat. A set of optimized shapes has been compared to a traditionally designed reinforcement layout. This innovative approach, supported by modern additive manufacturing technologies and by related process parameters optimization [5], could represent a more efficient naval design method.

## 2  Methods

TO consists in finding the optimal solid-void pattern of a material layout inside a bounding box representing the design domain once boundary conditions are defined. In typical TO software, the design domain is meshed in Finite Elements which density can vary between 0 and 1. In this paper the TO package by ANSYS has been adopted. It is a density-based topology optimization tool that uses a Solid Isotropic Material with Penalty (SIMP) method to define pseudo-density of each finite element of the mesh during optimization [2]. The optimization of a general non convex problem is solved with a Sequential Convex Programming (SCP) strategy. With this approach the convex portions of the problem are handled exactly and efficiently, while non convex portions of the problem are modelled by convex functions that are (at least locally) accurate.

## 3  Case Study

The boat studied is the 4,6 m length sailing dinghy represented in Fig. 1. It is an R3 class boat that has participated to the inter-university 1001velaCup regattas since 2012 and has been refitted in 2015 to reduce the maximum beam and the total weight. The lightening process has involved the reduction of the internal reinforcements that, at a first stage of construction, were over dimensioned. The internal layout of longitudinal frames and transverse sections was suggested by a traditional hull panel subdivision approach given in [1]. In Fig. 1 a unique longitudinal frame element can be easily distinguished, while four transverse sections are located at variable intervals. Furthermore, one of these sections (under the mast) is not properly transverse but is symmetrically rotated. The material assumed for the boat manufacturing is marine plywood, with a density $\rho = 500$ kg/m$^3$ and a Young modulus $E = 8$ GPa. Hull and deck are obtained with 16 mm thick panels, while reinforcements have a thickness that varies from a minimum of 12 mm to a maximum of 20 mm. The boat weight is 87 kg.

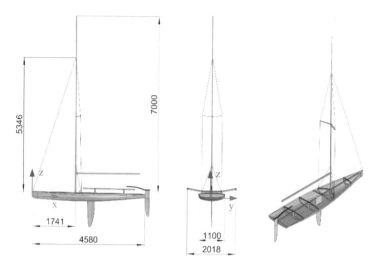

**Fig. 1.** View of the boat.

The original version of the boat has been numerically simulated with the Finite Element Method in order to obtain the deformed shape of the hull under navigation loads. The CAD model has been meshed into thick shell elements and a static structural analysis has been set up assuming that the boat is sailing beam reach with an Apparent Wind Speed of 6.4 m/s and an Apparent Wind Angle of 45°. By using open source Velocity Prediction Programs, a drive force of 349 N and a heel force of 654 N have been calculated. The simplified method used in [6] has been here adopted to impose the equilibrium between the heeling moment given by the heel force and the maximum righting moment given by the crew members on trapeze in order to keep the boat flat on the sea.

Compressive preload is applied on the mast and traction preloads are applied on forestay and shrouds to return a self-equilibrated force system. External loads are super-imposed to preloads when the boat navigates. Resulting force components are reported in Table 1. Preloads is a self-equilibrated force system, while the sum of external loads on the rig and on terrace equilibrates the heel and drive forces. The resultant force in the $z$ axis must be null. Loads have been applied on the FEM model of the original boat and displacement constraints have been added in the $z$ direction at the right ahead and at the stern to avoid undesired rigid body motion. At the trunk surfaces, a frictionless support is applied in the $x$ and the $y$ direction to simulate drag and lift actions given by the centerboard.

Results of a static structural analysis are then reported. In Fig. 2 it is evident that the critical portion of the hull is the one under the mast foot: compressive actions due to preload and to the crew weight on trapeze cause a maximum $z$ displacement of about $-1.2$ mm and a distributed stress pattern over the rotated section with some local concentrations at the beam intersections.

TO of the boat internal reinforcement has been setup starting from a preliminary structural analysis of the full boat with loads in Table 1 applied on it and the same

**Table 1.** Loads on the boat [N].

| Load | X component | Y component | Z component |
|------|-------------|-------------|-------------|
| Preload port shroud | −107.5 | 177.35 | 2238.3 |
| Preload starboard shroud | −107.5 | −177.35 | 2238.3 |
| Preload mast | 0 | 0 | −5100 |
| Preload forestay | 214.9 | 0 | 623.34 |
| External port shroud | −19.346 | 31.927 | 402.94 |
| External starboard shroud | 39.748 | 65.596 | −827.87 |
| External mast | −260.62 | −410.61 | −1630.6 |
| External forestay | 191.9 | 0 | 555.5 |
| Terrace | −300.7 | −340.9 | 0 |
| Displacement | – | – | 2364 |
| Target weight | – | – | −873.5 |
| Total | −349 | −654 | ~0 |

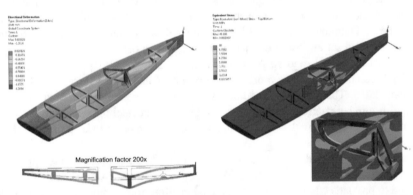

**Fig. 2.** FEM results on the original boat: displacement along $z$ and particular on the keel (left) and Von Mises stress on hull and reinforcements (right).

external constraints as the ones applied on the original boat. The only difference is that, in this case, the weight of the full boat is 445.4 kg because the volume between the deck and the hull is, at the beginning of the process, full of material. Volume is now meshed in solid hexagonal elements.

To perform the optimization procedure it has been considered that during navigation the boat can be also subjected to loads repeated at variable frequency, for example fluttering of sails or slamming caused by wave trains. These actions could interfere with natural frequency of thin hull panels and cause undesired resonance phenomena. The proposed TO procedure considers also the deformation caused by a modal analysis, then the optimization goal of the objective function is the research of this maximum value:

$$max(Z) = max(w_1 Obj_1 + w_2 Obj_2) \tag{1}$$

where $Obj_1$ is the structure stiffness, $Obj_2$ is the first harmonic frequency and $w_1$ and $w_2$ are relative weights varying between 0 and 1, stepped by 0.25.

Some response constraints have been added to all the analyzed cases: $z$ displacement of the keel line has been limited to a minimum of -0.5 mm, maximum Von Mises stress has been limited to 10 MPa and the final mass must equal the original one, that is 87 kg. Furthermore, some manufacturing response constraints have also been added: member mean dimension should be included between 15 and 300 mm and the retained volume must be symmetric with respect to the $x$-$z$ plane. Hull surfaces have been excluded from the optimization process.

## 3.1  Discussion

Results of the optimization process are reported in Fig. 3 for half boat (on the left) and for the transverse section under the mast foot (right).

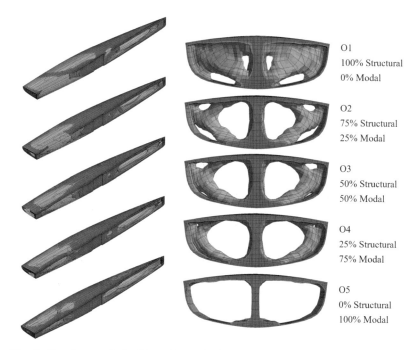

**Fig. 3.** Topology optimized boat shapes: half volume (left) and central section (right)

The position of the center of gravity has been calculated and plotted in Fig. 4 left, together with the one calculated for the original boat. The coordinate $x_{CG}$ affects the hydrostatic trim angle of the boat. The original boat has $x_{CG} = 2630$ mm while for the optimized cases it varies between 2204 mm for O1 and 2770 mm for O5. As far as the structural component is reduced in favour of the modal component (from O1 to O5), $x_{CG}$ moves towards the stern. To obtain the target original value of $x_{CG}$, a correction mass should be properly located. The position of $z_{CG}$ varies from a minimum of 67.75 mm to a maximum of 73.83 mm. This parameter should be as lower as possible in order to improve the hydrostatic heel stability of the boat.

In Fig. 4 right is reported the relative reduction of each moment of inertia of the optimized boats with respect to the original one. The highest reduction can be observed for $I_x$ (best is 32.69% saving for O2) while lower values are calculated for $I_y$ (best is 14.16% saving for O5) and for $I_z$ (best is 15.16% saving for O5).

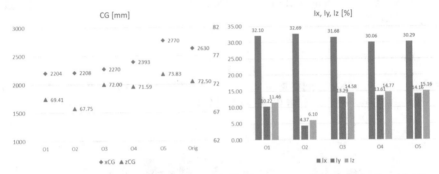

**Fig. 4.** Center of gravity position (left) and relative moment of inertia reduction (right).

## 4 Conclusions

A numerical procedure implementing FEM and TO has been defined in order to define a reinforcement layout inside the hull of a sailing dinghy. This must fulfill two concurrent objectives: maximum stiffness and highest natural frequency of the reinforced boat. Optimized shapes have been compared with traditionally designed reinforcement layout, keeping the same total weight. A consistent reduction in the moment of inertia has been obtained around the longitudinal axis (between 30% and 33%) and around the transverse axis (between 4% and 14%), meaning an improved maneuverability and sea-keeping attitude. Moreover, the optimized shapes exhibit more regular vertical deflections of the keel under static loads. In the authors' intentions, the obtained TO CAD shape should be step-wise approximated and realized by overlapping constant thickness layers of marine plywood inside the hull.

## References

1. ISO 12215-5n Small craft — Hull construction and scantlings — Part 5
2. Zhang, W., Zhu, J., Gao, T.: Topology Optimization in Engineering Structure Design. ISTE Press Ltd. (2016)

3. Rais-Rohani, M., Lokits, J.: Reinforcement layout and sizing optimization of composite submarine sail structures. Struct. Multidisc. Optim. **34**, 75–90 (2007)
4. Leidenfrost, D.: Development of a Nature Inspired Hull Structure for a 46m Sailing Yacht, Master thesis, 2015. Hochschule Bremen, Dykstra Naval Architects, Alfred-Wegener-Institut.
5. Ingrassia, T., Nigrelli, V., Ricotta, V., Tartamella, C.: Process parameters influence in additive manufacturing. In: Lecture Notes in Mechanical Engineering, pp. 261–270 (2017)
6. Mancuso, A., Pitarresi, G., Tumino, D.: Using FEM simulation to predict structural performances of a sailing dinghy. Int. J. Interact. Des. Manuf. **12**, 811–822 (2018)

# State of the Art on Robust Design Methods for Additive Manufacturing

Youssef Malyani[✉], Myriam Orquéra, and Dominique Millet

University of Toulon, COSMER, 83000 Toulon, France
`youssef-malyani@etud.univ-tln.fr`

**Abstract.** Additive Manufacturing (AM) technologies allow to produce functional parts with complex geometries that cannot be manufactured by conventional processes. However, the complexity of the product is increased and causes new constraints in the manufacturing process. Therefore, these new processes lead particularly to new needs in design methods. The objective of this paper is to explore and form an overall view of design methods, especially, robust design (RD) methods. Robust design is defined here as a methodology that enables to design a product with optimal performances and insensitivity to small variations of the inputs of the manufacturing process. In this contribution a state of the art of robust design methods applied to AM will be carried out.

**Keywords:** Robust design · Additive manufacturing · Design for additive manufacturing (DfAM) · Taguchi method

## 1 Introduction

AM groups a set of technologies that allows parts to be manufactured by adding material, usually layer upon layer [1]. In the last few years, AM processes have evolved quickly from rapid production of prototypes into manufacturing processes enable to produce end-use products [2]. This could significantly reduce the high funding in injection moulding tooling of small series production and thus, decrease cost and time to market within the product production [3]. Therefore, these processes are fully mature for industrial production. Research has indicated that the industries have begun strongly using the AM technologies such as selective laser melting (SLM) for example to directly produce end-use parts [4]. However, all AM technologies still require significant developments, especially at the design level. In addition, the skills of designers need to be completely developed in this field and guidelines must be set up. Hence, there is a necessity to develop and implement RD methodologies for AM processes to assist designers. The aim of these methodologies is to optimize products performances, i.e. guarantee robustness of the product against all the unwanted performance variations that could be generated due to the part geometry or process variables such as machine technology or material…etc. This article will present the existing RD methodologies applied in several fields, in particular AM field. In order to develop a robust design for AM, a state of the art will be conducted. Thus, various works related to this framework will be cited. First of all, the

© The Author(s) 2021
L. Roucoules et al. (Eds.): JCM 2020, LNME, pp. 80–85, 2021.
https://doi.org/10.1007/978-3-030-70566-4_14

research methodology consisted in selecting the articles that best explained the robust design of the mechanical parts. Then, an analysis of the articles will highlight their strengths. Afterwards, the authors will propose a vision of robust design for AM.

# 2  State of the Art

This section presents two main subjects. The first one is about RD and its existing methods. The second deals with RD methods with adaptation to AM processes.

## 2.1  Robust Design

The first English version of the Taguchi's robust design concept has been published in 1979 [5]. In fact, Taguchi has developed a statistical method for carrying out experimental designs. It is mainly used for the product quality improvement. On the one hand, [6] and [7] have applied Taguchi's method based on design of experiment (DOE) to simplify and limit the experimental approaches of their studies and therefore, reduce time and resources. On the other hand, RD means that a system's design is done in a way that sharply reduces variations, i.e. getting the best solution that will behave as expected by the designers [8]. In addition, [9] have identified that the main principles of RD methodology are insensitivity to noise factors, awareness of variation and finally continuous applicability. Indeed, these principles need to be achieved in order to guarantee a RD methodology. Otherwise, a manufactured product will present unwanted performance variations, as a result, it shows that the robustness of a design depends entirely on its performance variations. To avoid or at least decrease these variations, design engineers must consider them in the early stages of the design process in order to predict their effects on the product's performance and therefore, be able to control and limit them as much as possible.

In this context, [8] have presented a set of methods whose objective is to predict the effects of variations of the system to be designed. Firstly, they have identified independently a deterministic analysis of variations based on the conditioning of a matrix and afterwards, a statistical approach to analyze dimensional and geometric tolerances. According to the authors these methods are used for both technical and economical raisons. However, they are based on very precise assumptions such as the linearity of the objective functions.

Furthermore, particularly in structural dynamics, [10] proposed a RD methodology based on the coupling of stochastic multi-objective optimization with a robust condensation method with respect to structural modifications and uncertainties. The idea of this methodology is to integrate robustness as an additional objective function to be maximized. In addition, [10] have demonstrated the efficiency of their methodology in two different simulations. Firstly, the simulation of a bracket made up of an assembly of two reinforced plates and then, of a curved plate reinforced by five stiffeners. The results of this methodology present very slight performance variations caused by the uncertainties in the design parameters and a significant gain in terms of calculation time. However, this methodology remains limited only to structures. Besides that, [11] confirmed that the main intention of RD is to design a system with less performance variations and even insensitive to the process inputs variations such as material properties and operational conditions. Currently, the robustness is applied to simple forms, such as thin-walled shells like cylinders, cones and spheres [12]. However, the forms with non-uniform geometry such as the optimized parts for AM remains non-studied. Furthermore, the integration of robustness in the iterations of topological optimization might be a strong point for AM parts, however, it still applied only for conventional designs [13].

## 2.2 Robust Design for Additive Manufacturing

In this section, RD methods with adaptation to AM which mainly focus on functional improvements and the limitation of geometrical defects are presented.

In the recent years, AM processes have offered enormous manufacturing capabilities and reduced manufacturing costs. However, it is often difficult to have high levels of geometric and dimensional precision with some of these processes. For instance, the quality of parts manufactured by AM depends on several variables, such as the material used, process parameters and the machine technology. Due to the numerous additive processes and their rapid proliferation, designers suffer from a lack of design guidelines and standardization of best practices [14]. Numerous works have been done before in this framework to develop RD methodologies for AM in order to produce high quality parts.

[7] have conducted a detailed research to optimize the 3D printing (3DP) manufacturing processes. This research focused particularly on numerous process parameters, such as the layer thickness and binder setting saturation value...etc. According to the authors, these parameters mainly influence the final part quality and accuracy. In order to limit the number of experiments of their study, the authors have used the Taguchi's method based on DOE. The main idea of this work was to improve the quality of the produced 3DP parts and reduce their building time while keeping optimal manufacturing costs. The results were relevant whether at the level of building time and accuracy error which have been reduced, beside the increasing of the flexural stress at the level of strength. Despite these improvements, this study still only adapted for rapid prototyping parts, an extra effort is therefore needed to extend it for end-use parts. Another RD methodology was proposed by [3] which combines three concepts; Taguchi design of experiments (DOE), multi objective optimization and statistical process control (SPC). The authors have demonstrated the effectiveness of their combined methodology to assess AM feasibility and robustness for direct component manufacturing of a typical ABS injection moulded plastic part. This work has better encompassed almost all the contributions of RD for AM by providing relevant case studies results about three different AM processes. However, this work needs to be completed by projecting it on multi-components products, while considering the noise factors and interaction between the process variables.

AM technology and structural optimization technology have been recently well integrated in the works of [15]. Thus, they have demonstrated the effectiveness of their structural optimization design approach based on SLM for thin-walled antenna bracket. The proposed design approach for the antenna bracket might be further improved in the future by exploiting Wang's work [13]. Besides that, [16] have proposed the concept of topological optimization loops for multi-components product. Based on a case study, several topological optimization paths are compared. Thus, certain optimization principles are suggested to choose the most suitable path according to the designer's objectives (e.g. minimize the calculation time, obtain the best mechanical behavior or obtain the lowest mass). The results obtained during this work were relevant for mechanisms with an open loop kinematic chain and to minimize mass as an objective and minimize displacement as a constraint. To complete this research, it would be more interesting to conduct a study of complex systems with closed loop kinematic chains with other objectives and constraints of topological optimization.

To sum up, the Table 1 gives an overview about the works previously cited.

**Table 1.** Summary of the state of the art.

| Reference | Methods type | Proven for AM? | For which AM process? | Advantages | Drawbacks | Applicable to an assembly? |
|---|---|---|---|---|---|---|
| [3] | Robust design & Taguchi method | Yes | Stereolithography; Polyjet and laser sintering | Give guidelines for machines Automation of process parameters selection | Does not include noise factors Limited for simple parts | No |
| [7] | Taguchi method | Yes | 3D printing | Optimize process parameters Consider noise factors | Limited for simple parts Limited for RP technologies | No |
| [8] | Robust design method | No | – | Predict the effects of variations of a system | Based on very precise assumptions | Yes |
| [10] | Robust design method | No | – | Integrate robustness as an objective function to maximize | Need to be adapted for AM products | Yes |
| [11] | Robust design method | No | – | Provide guidelines for the designers | Need to be proven for AM technologies | Yes |
| [12] | Numerical design approach | No | – | Analyze the buckling behavior of critical shell structures | Need to be proven for AM's optimized parts | No |
| [15] | Structural optimization design | Yes | SLM | Improve the natural frequency and dynamic characteristics | Need to be applied for multi-body components | No |

# 3   Conclusions and Perspectives

This paper presents a state of the art of robust design methods for simple parts and products and gives some definitions and principles of RD and the Taguchi method. Firstly, this contribution has presented a set of robust design methods applied in different domains such as dimensional and geometric tolerances domain, structural dynamics and tolerance compensation. The second part of this state of art dealt with RD methods with adaptation to additive manufacturing. Generally, it is clearly noticed that some of the exiting RD methods are not adapted to multi-body systems (e.g. [3, 7, 12] and [15]), and some of them are not proven yet for AM technologies (e.g. [8, 10–12] and [13]). In addition, the majority of these methods have not been combined yet. Table 1 visualizes the characteristics of the different methods presented in this work and gives their advantages and drawbacks. Obviously, this article has highlighted the gaps in the literature, this will offer new perspectives of research. That is why the next work will be focused on the development of a robust additive design methodology allowing to achieve all the advantages of most AM processes. This methodology will be studied at the first time on a simple part, in which an industrial case study will be carried out and then, this methodology will be improved in order to perform multi-components systems design.

# References

1. I. ASTM International: Standard terminology for additive manufacturing–coordinate systems and test methodologies (2013)
2. Yang, S., Zhao, Y.F.: Additive manufacturing-enabled design theory and methodology: a critical review. Int. J. Adv. Manuf. Technol. **80**(1–4), 327–342 (2015)
3. Ituarte, I.F., Coatanéa, E., Salmi, J.T.M.: Robust design principles to evaluate additive manufacturing capabilities. In: 1st International Symposium on Robust Design, pp. 135–145 (2014)
4. Klahn, C., Leutenecker, B., Meboldt, M.: Design strategies for the process of additive manufacturing. Procedia CIRP **36**, 230–235 (2015)
5. Taguchi, G., Wu, Y.: Introduction to off-line quality control. Central Japan Quality Control Association (1979)
6. Rahmati, S., Akbari, J., Barati, E.: Dimensional accuracy analysis of wax patterns created by RTV silicone rubber molding using the Taguchi approach. Rapid Prototyp. J. **12**(2), 115–122 (2007)
7. Hsu, T.J., Lai, W.H.: Manufacturing parts optimization in the three - dimensional printing process by the Taguchi method. J. Chinese Inst. Eng. **33**(1), 121–130 (2010)
8. Caro, S., Bennis, F., Wenger, P.: Etat de l'art de la conception robuste de mécanismes, p. 10 (2010)
9. Hasenkamp, T., Arvidsson, M., Gremyr, I.: A review of practices for robust design methodology. J. Eng. Des. **20**(6), 645–657 (2009)
10. Ait Brik, B., Bouhaddi, N., Cogan, S. : Une méthodologie de conception robuste en dynamique des structures. Eur. J. Comput. Mech. **15**(1–3), 15–27 (2006)
11. Aschenbrenner, A., Schleich, B., Wartzack, S.: An overview and classification of tolerance compensation methods. In: International Conference on Engineering Design, ICED19, pp. 3471–3480 (2019)
12. Wagner, H.N.R., Sosa, E.M., Ludwig, T., Croll, J.G.A., Hühne, C.: Robust design of imperfection sensitive thin-walled shells under axial compression, bending or external pressure. Int. J. Mech. Sci. **156**, 205–220 (2019)
13. Wang, E., Sell, D., PHan, T., Fan, J.A.: Robust design of topology-optimized metasurfaces. Opt. Mater. Express **9**(2), 469–482 (2019)
14. Gao, W., et al.: The status, challenges, and future of additive manufacturing in engineering. Comput. Aided Des. CAD **69**, 65–89 (2015)

15. Zefeng Xiao, Y.B., Yang, Y., Wang, D., Song, C.: Structural optimization design for antenna bracket manufactured by selective laser melting. Rapid Prototyp. J. **24**(3), 539–547 (2018)
16. Orquéra, M. : Conception pour la fabrication additive: approche méthodologique pour les systèmes mécaniques multi-corps. Thèse de doctorat en mécanique, Université de Toulon, Toulon, France, 204 p. (2019)

# Topological Optimization of a Mechanical System with Adaptive Convergence Criterion

Myriam Orquera[✉], Sébastien Campocasso, and Dominique Millet

Université de Toulon, COSMER, Toulon, France
orquera@univ-tln.fr

**Abstract.** Topological optimization (TO) is commonly used to design a part for additive manufacturing (AM), but rarely for entire systems including several parts. How can be optimized a mechanical system in which each optimized part changes the boundary conditions? A Design method called TOMS (Topological Optimization of a Mechanical System) has been developed to take into account the variation of the boundary conditions when optimizing parts. When the using TOMS method the loops are performed until the optimization converges. The object of this article is to propose a discussion on the quantification of this convergence based on a practical case study.

**Keywords:** DfAM · Design for additive manufacturing · Product design · Additive manufacturing · Topological optimization

## 1 Introduction

Additive manufacturing processes allow to obtain parts or mechanical assemblies by adding material layer by layer, line by line or point by point [1]. This relatively recent manufacturing processes require new design methods due to new constraints and possibilities. Most of the methods proposed for these processes are applicable on a single part [2–4]. All these methods have in common a design framework with four steps: specification, topological optimization, volume reconstruction, validation. To perform the optimization of a part - during the second step - the CAD design and non-design spaces, the material, the boundary conditions, as well as the constraints and the optimization objective must be defined.

The articles dealing with the design of a product or a part manufactured by AM are carried out with a specific objective. For the mechanical product design proposed in [5], the objective was for example to apply AM capabilities to functionally improve the system. To achieve this, topological optimization (TO) and functional improvements were both applied on each component. Likewise Jilich et al. [6] have developed a gripper for garment handling with the objective of the diminution of part number and also functional improvement. A non-assembly additively manufactured vice was the purpose of the study presented in [7]. The necessary clearance and the most adapted orientation were chosen before applying a TO on each part.

L. Roucoules et al. (Eds.): JCM 2020, LNME, pp. 86–91, 2021.
https://doi.org/10.1007/978-3-030-70566-4_15

In all these articles, the parts have low inertia and the optimization of one of them has few influences on the mechanical behavior of the system. That is probably why the variation of the boundary conditions due to the optimization has not been taken into account.

However when the mass has an important influence on the mechanical behavior, new boundary conditions have to be taken into account. Our method called TOMS (Topological Optimization of a Mechanical System) and presented in detail in [8] is one of the firsts for designing a mechanical product taking into account this aspect. For this, optimization loops have been developed. These loops allow a product design taking into account the variations of the boundary conditions until the optimizations converge.

In this article, the TOMS method will be first briefly presented with a focus on the TO convergence. Then, the main goal of this paper is to analyze the improvements due to an adaptive convergence criterion for each part.

## 2   The TOMS Method

The Topological Optimization of a Mechanical System (TOMS) corresponds to the topological optimization of each rigid body of a mechanical product by taking into

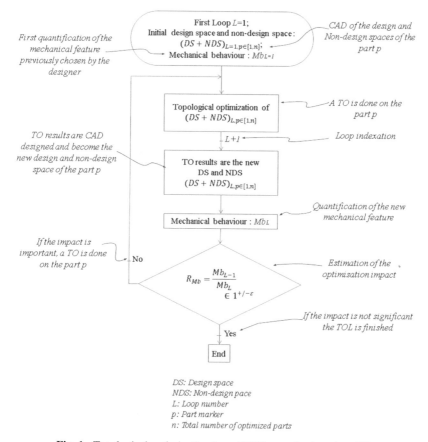

**Fig. 1.** Topological optimization loop (TOL) organization chart [8].

account the impacts of optimizations between the rigid bodies. This method was recently proposed and is more detailed in [8, 9].

Depending on the designer's objective, an optimization path must be chosen as well as an indicator of the mechanical behavior Mb. From the CAD of the product with its design and non-design spaces, the optimization loops denoted TOL (Topological Optimization Loops) will be applied to each of the rigid bodies until the ratio used as the variation indicator $R_{Mb}$ belongs to $[1 - \varepsilon, 1 + \varepsilon]$; $\varepsilon$ is called the coefficient of convergence. The TOL and the ratio $R_{Mb}$ are detailed in Fig. 1.

## 3  Adaptive Coefficient of Convergence E

The application of the TOMS method was performed on a robot arm following different optimization paths as shown the Fig. 2. The first path was carried out by optimizing first the part which has the greatest impact on the mechanical behavior Mb; while the second path began the optimization with the component which impacts the highest quantity of equations.

The initial mass of the robot is 16.2 kg, dispatched as followed: mass of the part 1 = 6.970 kg; mass of the part 2 = 8 kg; mass of the part 3 = 1.264 kg. The $R_{Mb}$ ratio is determined for the inter torque required to turn the part 1. The initial inter torque value is equal to 634 N.mm.

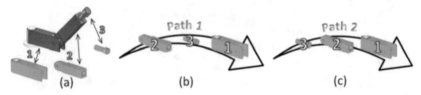

**Fig. 2.** (a) Robot arm with the three CAD parts, space design in green, non-design space in orange; (b) The first optimization path; (c) The second optimization path.

In this section, the value of $\varepsilon$ will be discussed for each path. $\varepsilon$ is defined as a constant that must be imposed at the beginning of the study. The lower the convergence coefficient is, the higher the impact of the optimization will be. In this article, another way to achieve optimization convergence is suggested. It consists in modifying the value of the convergence coefficient for each optimized part according to its impact on the mechanical behavior.

In the study carried out in [8], a unique value of $\varepsilon$ was set at 0.1, which leads to carry out a certain number of optimization loops on the parts as shown in Table 1 for path 1 and 2.

For the first path, in order to identified the part which has the greatest impact on the mechanical behavior Mb, a reduced sensitivity coefficient $S^*$ (see Table 2) of the chosen mechanical behavior was determined for each part $i$. The convergence coefficient could vary depending on this reduced sensitivity coefficient $S^*$. Indeed, the higher the coefficient $S^*$ is, the lower $\varepsilon$ must be. The part which has the higher reduced sensitivity

**Table 1.** Results of the TOMS method on the robot arm for the path 1 and 2 with $\varepsilon = 0.1$.

| | Path 1 | | | | | | Path 2 | | | | | |
|---|---|---|---|---|---|---|---|---|---|---|---|---|
| Part n° | Part 2 | | Part 3 | | Part 1 | | Part 3 | | Part 2 | | Part 1 | |
| Loop n° | 1 | 2 | 1 | 2 | 1 | 2 | 1 | 2 | 1 | 2 | 1 | 2 |
| $R_{Mb}$ | 2.12 | 1.05 | 2.43 | 1.04 | 1.23 | 1.02 | 1.36 | 1.01 | 3.84 | 1.08 | 1.23 | 1.04 |
| Product mass (kg) | 9.26 | 8.97 | 7.98 | 7.95 | 2.45 | 2.11 | 15.21 | 15.1 | 8.11 | 7.93 | 2.39 | 1.87 |
| Part mass (g) | 1055 | 768 | 274 | 246 | 1466 | 1130 | 274 | 246 | 926 | 749 | 1434 | 912 |
| Inter torque (N.mm) | 299 | 285 | 117 | 112 | 91 | 89 | 466 | 461 | 120 | 112 | 91 | 88 |
| CPU Time (h:min) | 8:37 | 0:26 | 19:28 | 0:45 | 8:58 | 1:02 | 19:28 | 0:45 | 10:20 | 0:32 | 7:48 | 1:09 |
| Total Time (h:min) | 39:19 | | | | | | 40:04 | | | | | |

coefficient is set as the reference part; its sensitivity coefficient is denoted $S_0^*$ and is determined as shown the Eq. (1). The coefficient of convergence is set to 0.1 for the reference part and is denoted $\varepsilon_0$. For the robot arm, the reference part is the part 2.

$$S_0^* = \max(S_i^*) \tag{1}$$

Then, the $\varepsilon_i^{path1}$ of the part $i$ is then inversely proportional to the percentage of the impact on the mechanical behavior Mb as shown in the Eq. (2).

$$\varepsilon_i^{path1} = \varepsilon_0 . \frac{S_0^*}{S_i^*} \tag{2}$$

Analogous reasoning can be established for the second path which begins the optimization of the component which impacts the highest number of equations. The part 3 has feature characteristics in nine equations. The Eq. (3) shows how to determinate $\varepsilon_i^{path2}$ for each part $i$ using the number of equation $n_i$. In the Table 2 the value of $\varepsilon_i^{path2}$ are determined for each part.

$$\varepsilon_i^{path2} = \varepsilon_0 . \frac{\max(n_i)}{n_i} \tag{3}$$

For paths 1 and 2, the last optimized part, which is the support 1, will only undergo one optimization if setting $\varepsilon$ higher than 0.3. The shaded columns of the Table 1 show which loops won't be done for a adaptive $\varepsilon$. In the Table 3, the gains and losses due to this new value of the convergence coefficient in terms of time, mass and torque are summarized. It can be noticed that the total mass obtained with a variable $\varepsilon$ is higher thans that obtained for an $\varepsilon = 0.1$ whatever the path. This is due to one less topological optimization on one or more parts.

For an on-board system used in aerospace, the addition of mass savings will represent a financial gain during operation. In this case, the value of the coefficient of convergence should be close to 0.1.

**Table 2.** Adaptive values of the coefficient of convergence for each part, for the path 1 and 2.

| | Path 1 | | | | Path 2 | | |
|---|---|---|---|---|---|---|---|
| Part n° | Part 2 | Part 3 | Part 1 | Part n° | Part 3 | Part 2 | Part 1 |
| Characteristic $C_i$ | $m_2$ | $m_3$ | $I_{1(O,z0)}$ | Number of equations $n_i$ | 9 | 6 | 1 |
| $S^*(M_b/C_i) = C_i \cdot \frac{\partial M_b}{\partial C_i}$ | 0.34 | 0.21 | 0.03 | | | | |
| $\varepsilon_i^{path1}$ | $\varepsilon_0 = 0.1$ | 0.16 | 0.7 | $\varepsilon_i^{path2}$ | $\varepsilon_0 = 0.1$ | 0.15 | 0.9 |

In the article [8], a third path where all part are simultaneously optimized was also proposed. It can be noted that, for this path, the variation of the convergence coefficient is not applicable.

**Table 3.** Gain and loss compared to $\varepsilon = 0.1$.

| | Path 1 | Path 2 |
|---|---|---|
| Mass | 13.6% loss | 21.5% loss |
| Inter torque (N.mm) | 2.25% loss | 3.4% loss |
| Total CPU time (h:min) | 2.6% gain | 2.5% gain |

## 4  Conclusion and Perspectives

In this article the TOMS method has been briefly presented. This method is useful to topologically optimize a multi-component mechanical system. A loop denoted TOL is used with regard to the modification of the boundary conditions due to the TO. The TOL is applied on a rigid body until the optimizations converge. The coefficient of convergence has an impact on the result as it is explained in this article. The consequences of the choice of the coefficient of convergence – depending on the use of the system - are shown. For the case study of this article, the adaptive convergence coefficient for each part allows to obtain a little saving of time. However, for a product with a large number of parts, the saving in study time could be much greater. Indeed, study time can be saved on parts with a low impact on the mechanical behavior without losing the gain of mass.

In perspective, a generalization of the choice of the coefficient of convergence will allow the designer to overcome this step.

## References

1. Thompson, M.K., et al.: Design for additive manufacturing: trends, opportunities, considerations, and constraints. CIRP Ann. Manuf. Technol. **65**, 737–760 (2016)

2. Tomlin, M., Meyer, J.: Topology optimization of an additive layer manufactured (ALM) aerospace part. In: Proceeding of the 7th Altair CAE Technology Conference, pp. 1–9 (2011)
3. Kumke, M., Watschke, H., Vietor, T.: A new methodological framework for design for additive manufacturing. Virtual Phys. Prototyp. **11**(1), 3–19 (2016)
4. Weis, P., Großmann, A., Clemen, C., Mittelstedt, C.: Optimization and re-design of a metallic riveting tool for selective laser melting – a case study. Addit. Manuf. **31**, 100892 (2019)
5. Orquéra, M., Campocasso, S., Millet, D.: Design for additive manufacturing method for a mechanical system downsizing. Procedia CIRP **60**, 223–228 (2017)
6. Jilich, M., Frascio, M., Avalle, M., Zoppi, M.: Development of a gripper for garment handling designed for additive manufacturing. Proc. Inst. Mech. Eng. Part C J. Mech. Eng. Sci. 095440621985776 (2019)
7. Sossou, G., Demoly, F., Montavon, G., Gomes, S.: An additive manufacturing oriented design approach to mechanical assemblies. J. Comput. Des. Eng. **5**(1), 3–18 (2018)
8. Orquéra, M., Campocasso, S., Millet, D.: Some principles to optimise an additively manufactured multi-component product. J. Eng. Des. **31**(4), 219–240 (2020)
9. Orquéra, M.: Conception pour la fabrication additive: Approche méthodologique pour les systèmes mécaniques multi-corps, Université de Toulon (2019)

# Design for Additive Manufacturing of a Topology Optimized Brake Caliper Through CAD-Platform-Based Systematic Approach

Enrico Dalpadulo, Fabio Pini$^{(\boxtimes)}$, and Francesco Leali

"Enzo Ferrari" Department of Engineering, University of Modena and Reggio Emilia, Via Vivarelli 10/1, 41125 Modena, Italy
{enrico.dalpadulo,fabio.pini}@unimore.it

**Abstract.** To implement the CAD platform-based approach of Design for Additive Manufacturing (DfAM) and validate it in a real case, an entire design optimization process of a Formula SAE front brake caliper has been performed, to be printed by Powder Bed Fusion (PBF) process. The DfAM consists in the use of a Ti6Al4V titanium alloy to better resist at high temperatures and a topology optimized shape allowed by the technology to save weight despite the density increase. Structural and thermal behavior has been discussed. DfAM process-specific techniques have been implemented for internal geometrical features and optimized shapes. The design for additive workflow is presented and finally the exploited design approach based on a CAD platform is synthesized.

**Keywords:** DfAM · Topology optimization · CAD platform · Brake caliper · Automotive

## 1 Introduction

Since Additive Manufacturing (AM) implementation is giving great potentials in many industrial settings, especially those focused on high performance components, DfAM methodologies study is becoming fundamental [1]. As shown in a previous work by the authors [2], the workflow for the development of optimized components to be produced by AM is not always effective. A general framework consisting of the tasks to be performed from concept definition to end-use functional parts has been defined (Fig. 1).

The main operations that make up the workflow are summarized below. Product Planning is the initial phase. At this stage, analyses on parts and assemblies together with objectives and constraints definition are carried out to collect all the input Product Data and define the Requirement List.

The Design phase is an iterative refinement of the models in order to optimize the product. The Industrialization phase is the step required to make reliable part production and optimize the process. Finally, Production covers part printing process, with the related post-processing and control operations. The implementation of integrated CAD-based platforms as backbone tool to speed up the workflow and increase its effectiveness

© The Author(s) 2021
L. Roucoules et al. (Eds.): JCM 2020, LNME, pp. 92–97, 2021.
https://doi.org/10.1007/978-3-030-70566-4_16

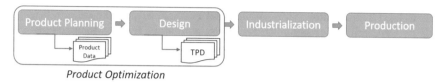

**Fig. 1.** Design for Additive Manufacturing workflow [2].

had been discussed by the authors [2]. The presented work aims to describe in detail the methodologic approach focused on the Design phase (Product Optimization) based on Topology Optimization. A racing automotive brake caliper has been re-designed to be produced by Selective Laser Melting (SLM) AM, with performance improvement objectives, through the application of the CAD platform-based approach. The iterative design refinement required to obtain final models and Technical Product Documentation (TPD) is described on a real-case application and the DfAM approach is analyzed.

## 2    Method

As depicted in Fig. 2, the Design phase relies on a sequence of tasks, that are summarized below. CAD modelling of Design Space (DS) and Non-Design Space (NDS) are mandatory to prepare the geometry considering the design constraints.

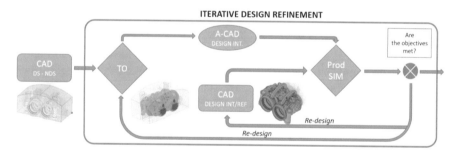

**Fig. 2.** Design phase.

Topology Optimization (TO) is the task to compute part shape according to algorithm settings of target and constraints. CAD design interpretation/refinement is required to obtain the optimized part model and both automatic tools (A-CAD) or manual geometry reconstruction (CAD) can be used. Product Simulation (ProdSIM) is the validation task of the optimized parts with respect to the design objectives and constraints. Re-design loops are required for design optimization.

## 3    Case Study

Components of racing cars are usually subjected to combined stressing factors and in particular braking systems can have issues in use, as they have to work with intense loads at high temperatures [3]. A commercial Formula SAE brake caliper, originally made by CNC milling of Ergal 7075-T6 aluminum alloy, reached temperatures close to 300 °C with problems of strength and deformation, so it has been re-designed to be printed by SLM process. Sergent et al. analyses [4] show critical conditions for a working brake caliper from a structural point of view and how Topology Optimization let its performance improvements. Travi Farias et al. work [5] shows how its thermal management is fundamental, since at high temperatures materials mechanical properties decline. Moreover, the study states that AM enables the construction of complex geometries that can increase model surface/volume ratio and facilitate heat dissipation. Bugatti's full developed case [6] demonstrates with experimental testing how titanium alloys use is feasible and leads to performance improvements. Nevertheless, none of these cases is based on an integrated design approach, whose possible potentials have been discussed by the authors [2]. Referring to the workflow depicted in Fig. 1, the Product Planning and Design tasks are now described. All the tasks are performed on the Dassault Systèmes 3DExperience CAD platform [7] for integrated product-process design.

## 3.1  Product Planning

The analysis of the commercial component is the starting point of DfAM with the aim to define the design features that provide the same working behavior. In particular, that concerns mostly the fluid-dynamic of oil channels. In addition, a make/buy decision step leads to keep standard and commercial elements of the original part, whereas the body and pistons are re-designed. Since original 2D drawing is not available, functional features have been measured by metrology equipment to define the coupling tolerances. Data about mechanical and thermal loads acting on the part have been collected via experimental measurements and analytical models from the vehicle dynamics. Maximum pressure on the oil circuit is 100 bar and tangential load for maximum brake torque at the disc brake is 14 kN. Temperatures reach on average about 200 °C with maximum peaks close to 300 °C. An analysis on the assembly of the front wheel group has been done in order to define physical design constraints related to part fitting, coupling and working. Based on that data-set, FEA of the original caliper returns the structural targets for the project, such as improvement of stiffness at high temperature and weight reduction. A fundamental step is material selection, according to datasheets related to SLM. Since mechanical properties of 7075-T6 alloy suffer of significant drop at high temperature, Ti6Al4V titanium alloy is selected. According to literature data, compared to 7075-T6 alloy, it presents, at working temperature, about 11% higher Young's-modulus/density ratio and even 45% higher yield-stress/density ratio. Moreover, manufacturing analysis, with machining and specific heat and surface treatments is required. Product Data are collected in the Requirement List. The DfAM goals consist in the use of Ti6Al4V to better resist at high temperatures and a topology optimized shape to save weight. An opened and branched geometry can also increase surface/volume ratio with benefits for thermal management in terms of heat dissipation.

## 3.2  Design

Initial step of the Design phase is the DS modeling. It starts from the input data of part and assembly analysis in order to define the maximum volume available for the TO computation. Most functional elements should be removed whereas features for part connections and main couplings must be kept (pistons, brake pads and pins housings, bolt holes, caliper caps) in order to constrain the region. Wheel rim radius, encumbrance of wheel knuckle and brake disk or parts assembly trajectories are also required. NDS volumes must be defined in order to insert regions to keep material, such as bolts holes or internal features.

The second step is the setup of FE model in order to run optimizations and analyses. Hinge restraints are applied for the screws. Maximum 14 kN tangential load (due to braking torque) is applied on a node put on the disk brake midplane, connected to contact surfaces (RBE3) of braking pads. Maximum 100 bar oil pressure is applied on internal surfaces of pistons housings and equivalent calculated reacting forces of 9.8 kN are applied on the thread regions of the caliper caps. Material is created using parameters for build direction (due to AM alloy anisotropy [8]) in favor of security. Discretization of the design space is made by a 1.5 mm tetrahedron (TL4) mesh and refinements. A preliminary static analysis on the design space is run to validate the model setup and check its stiffness (ideal maximum value). TO is setup with target mass reduction and minimization of compliance.

A conceptual iterative design exploration with a symmetry constraint study and a computational refinement is performed. That geometry makes the right and left parts become the same and reduces the modeling to half-body, with potential benefits of design and production time and costs. The results are used to create solid models through improved smoothing conceptual shape generation. These see directly Prod-SIM by FEA, showing that an asymmetrical design plain brings better performance in terms of maximum displacement. Figure 3 shows TO results C1 and C2.

**Fig. 3.**  From left to right, conceptual results (C1, C2) and designs (V1, V2, V3).

Once the TO is finished, an embodiment iterative design starts to improve part shape. The Design Interpretation occurs, modeling the functional geometry with surface design and the branched shape with free-shape design. Moreover, oil channels are introduced and used to contribute to part stiffness working on their shape and position. ProdSIM by FEA is necessary to predict part deformation and stress. Last two refinement cycles involve the detailed design of each part housings (gaskets, o-rings, valves, pins) and the implementation of DfAM guidelines for internal features (self-supporting cross-sections of channels) and the branched shapes (thickness, supporting angles), according to part orientation planned for construction. After validation, results of manufacturing analysis are used to create the raw part model and the final drawings (TPD).

## 4   Results and Discussion

For static load validation, yield stress at 220 °C is considered, with value for build direction (Z) and an additional safety factor 1.2 (racing application), so permissible stress of 560 MPa is calculated. Figure 4 shows results. Moreover, remark on fatigue life and proper processing/treatments on stressed areas is necessary [9].

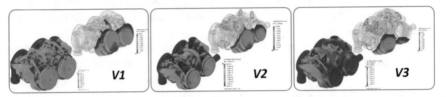

**Fig. 4.** Displacement and stress of V1, V2 and V3.

As reported in Table 1, maximum displacement at 220 °C is 0.675 mm for the original caliper and 0.614 mm for the final design of the optimized one (9% reduction) while the body weight goes from 248 g to 184 g (25,8% saving). Moreover, an analytical thermal study of a braking cycle by Matlab code shows a 20 °C decrease of working temperatures for the final design. One last note is that titanium has worst tribological characteristics (low wear resistance and high size tendency) and thus surface treatment chemical nickel-plating (Niplate) for pistons housings has been defined. Cold and hot tolerances for couplings are re-calculated.

**Table 1.** Comparison between designs.

|                                | Original | V2     | V3     |
|--------------------------------|----------|--------|--------|
| Weight [g]                     | 248      | 176    | 184    |
| Displacement [mm] @ 220 °C     | 0.675    | 0.768  | 0.614  |
| Deformation % @ 220 °C         | –        | +13.78 | −9.03  |
| Displacement [mm] @ 20 °C      | 0.555    | 0.694  | 0.551  |
| Deformation % @ 20 °C          | –        | +25.05 | −0.72  |

The iterative design refinement is now synthesized. Loops for product optimization, each of them performed in the integrated CAD platform, can be outlined at different levels. First loop includes Topology Optimization, A-CAD and Product Simulation and produces conceptual solutions (C1, C2). Automatic tools for design interpretation are used and loops are made fast thanks to the integrated platform. Second loop includes CAD and Product Simulation and lets the development of embodiment solutions (V1) or definitive ones (V2, V3) adding elements of the detail design. Design requires manual (time-consuming) geometry interpretation only for first iteration, whereas re-designs are extremely facilitated by the integrated environment for the subsequent ones.

# References

1. Kumke, M., Watschke, H., Vietor, T.: A new methodological framework for design for additive manufacturing. Virtual Phys. Prototyping **11**(1), 3–19 (2016)
2. Dalpadulo, E., Pini, F., Leali, F.: Assessment of design for additive manufacturing based on CAD platforms. design tools and methods in industrial engineering. In: Proceedings of ADM 2019, pp. 970–981 (2020)
3. Limpert, R.: Brake Design and Safety, S.A.E. International, U.S.A. (1999)
4. Sergent, N., Tirovic, M., Voveris, J.: Design optimization of an opposed piston brake caliper. Eng. Optim. **14**(11), 1520–1537 (2014)
5. Travi Farias, L., Schommer, A., Ziegler Haselein, B., Neumaier, G., Costa de Oliveira, L., Soliman, P., Walter, R.: Design of a Brake Caliper using Topology Optimization Integrated with Direct Metal Laser Sintering. S.A.E. International (2015)
6. Wischeropp, T.M., Hoch, H., Beckmann, F., Emmelmann, C.: Opportunities for braking technology due to additive manufacturing through the example of a Bugatti brake caliper. In: XXXVII. Internationales μ-Symposium 2018 Bremsen-Fachtagung, pp. 181–193 (2019)
7. Dassault Systèmes. https://www.3ds.com/about-3ds/3dexperience-platform/. Accessed 01 2020
8. Simonelli, M., Tse, Y.Y., Tuck, C.: Effect of the build orientation on the mechanical properties and fracture modes of SLM Ti6Al4V. Mater. Sci. Eng. **616**, 1–11 (2014)
9. Denti, L., Bassoli, E., Gatto, A., Santecchia, E., Mengucci, P.: Fatigue life and microstructure of additive manufactured Ti6Al4V after different finishing processes. Mater. Sci. Eng. **755**, 1–9 (2019)

# Product and System Engineering

# Evaluation of the Effects Caused by Mandibular Advancement Devices Using a Numerical Simulation Model

Marco Mandolini[1](✉), Manila Caragiuli[1], Daniele Landi[4], Antonio Gracco[2], Giovanni Bruno[2], Alberto De Stefani[2], and Alida Mazzoli[3]

[1] Department of Industrial Engineering and Mathematical Sciences, Università Politecnica delle Marche, Via Brecce Bianche 12, 60131 Ancona, Italy
m.mandolini@univpm.it
[2] Department of Neurosciences, University of Padua, Via Giustiniani, 5, 35128 Padua, Italy
[3] Department of Materials, Environmental Sciences and Urban Planning, Università Politecnica delle Marche, Via Brecce Bianche 12, 60131 Ancona, Italy
[4] Department of Management, Information and Production Engineering, University of Bergamo, Via Pasubio 7/b, 24044 Dalmine, BG, Italy

**Abstract.** Obstructive sleep apnea syndrome (OSAS) is a sleep disorder that causes pauses in breathing or periods of shallow breathing during sleep. Mandibular advancement devices (MADs) represent a non-invasive treatment for OSAS that has had the highest development in recent years. Nevertheless, literature has not primarily investigated the effects of mandibular advancement. This paper presents a finite element method numerical simulation model for evaluating the stress/strain distribution on the temporomandibular joint (TMJ) and periodontal ligaments caused by advancement devices used for the treatment of OSAS. Results highlight that the mandible lift phase generates significant stress values on TMJ, which cannot be neglected for extended usage of MADs. Furthermore, mandible molar teeth are more loaded than incisor ones.

**Keywords:** Mandibular advancement device · Obstructive Sleep Apnea Syndrome · Finite element method · Virtual prototyping · Computer-Aided Design

## 1 Introduction and Literature Review

Repeated episodes of complete (apnea) or partial (hypopnea) obstruction of the upper airway during sleep characterize the obstructive sleep disorders. The obstructive sleep apnea syndrome (OSAS) may affect the general health of a large percentage of the population, with a defined prevalence of approximately 6% for men and 4% for women, with data increasing in the last few years [1].

Oral Appliances represents a non-invasive alternative treatment for OSAS that had the most considerable development in recent years, and medical literature has demonstrated their effectiveness in reducing the number of respiratory events and in decreasing

© The Author(s) 2021
L. Roucoules et al. (Eds.): JCM 2020, LNME, pp. 101–107, 2021.
https://doi.org/10.1007/978-3-030-70566-4_17

symptoms [2]. Mandibular advancement devices (MAD) are nocturnal intraoral appliances that stabilize the upper airways and increase their diameters, reducing pharynx collapsibility. These devices must be used every night and necessarily have some short-term side effects (e.g., sialorrhea or xerostomia, headache, temporomandibular joints discomfort, dental soreness, and occlusal changes upon awakening). In the long term, on the other hand, there may be overjet and overbite reduction and decrease in the quantity and intensity of posterior occlusal contacts.

In the past, the finite element method (FEM) was used mainly for simulating stress and strain distribution on the temporomandibular joint (TMJ). The masticatory system under normal occlusal loads [3], the mandibular protrusion and opening were the typical configurations analysed. FEM also permits to evaluate periodontal ligaments (PDL) effects on mandibular stiffness [4], on teeth movement [5], or in case of bone loss [6]. Concerning OSAS, Brunzini et al. [7] and Crivellin et al. [8] respectively presented a simulation model toward the study of the effects caused by different MADS and the strength distribution on TMJ using MADs. However, the presented virtual prototype is only a preliminary result that deserves further improvement.

This paper presents a numerical simulation model for studying the stress/strain distribution on TMJ and PDL caused by MADs used for treating OSAS. The simulation model was used to characterize a specific patient and related MAD. This research work that can be intended as a protocol is a step toward understanding and evaluating the side effects of MADs. Results achievable by employing the presented simulation model on a more significant number of patients and MADs will permit practitioners to gain knowledge for improving the current golden rules for prescribing such devices.

## 2  Materials and Methods

The development of the numerical simulation model consisted of a combination of (i) the 3D anatomical model of TMJ and PDL, (ii) the materials, and (iii) the boundary conditions (Fig. 1). The first step of the methodology consisted of gathering computerized tomographic (CT) images of an OSAS affected patient, anatomical atlas (specific for TMJ and PDL), a physical prototype of a MAD, and the plasters casts of the mandibular and maxillary dental arches of the same patient. Such information allowed authors to build the complete 3D geometry of the simulation model. The second step consisted in defining the boundary conditions of the system, strictly related to the mandible roto-translation caused by the MAD. The third step aimed at establishing the constitutive equations of the material to be assigned to the anatomical geometry of the model. After having meshed the model (fourth step), the simulation model was ready to be used.

### 2.1  Anatomical Model

The 3D geometry reconstruction of the anatomical model followed different techniques. Temporal bone, mandible, and teeth were modelled employing an image processing software for 3D design and modelling of geometries from CT images of the patient's skull. Soft tissues (articular disc, joint capsule, anterior ligament, retrodiscal tissue, spheno-mandibular (SPL), and stilomandibular (STL) ligaments), which cannot be univocally

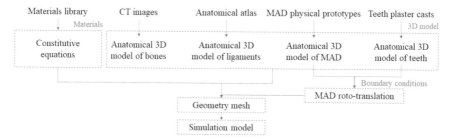

**Fig. 1.** The methodology used for developing the numerical simulation model.

determined from CT images, were modelled considering anatomical atlas and scientific papers (e.g., [9]). PDL ligaments were shaped by offsetting teeth roots of a constant value (i.e., 0.3 mm). The mandibular MAD splint was modelled through a reverse engineering process because such devices are custom-made and handcrafted by practitioners, and the 3D CAD model is commonly unavailable. To be noted that, since the symmetry of a human skull, only half side of the geometry was modelled. Furthermore, the simulation model contains only the mandibular splint; the maxillary one is beyond the scope of this research.

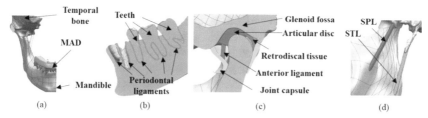

**Fig. 2.** Complete anatomical model. Frontal view of the 3D model (a), sagittal view of PDL (b), sagittal view TMJ (c), and sagittal view of SPL an STL (d).

## 2.2 Materials

Various materials were used for simulating the anatomical geometry. Materials models were retrieved from the literature. In particular, the articular disc, joint capsule, and all the TMJ ligaments were modelled using a Neo-Hookean material model [10]. In contrast, the $2^{nd}$ order Ogden model was adopted to model the PDLs [11]. The teeth and the mandible have been modelled as linear elastic, isotropic, and homogeneous [12]. MAD is a stratified element of three different materials. It consists of a thin inner layer in contact with the teeth and made of soft material (TPU), a middle between the inner and outer layers made of rigid material (PETG), and an outer layer of rigid material (PMMA). Such materials were modelled using a linear elastic model.

## 2.3 Boundary Conditions

The simulation model aims to evaluate the effects caused by the mandible roto-translation when inserting the MAD into the mouth (Fig. 3). First of all, it was necessary to assess the mandible translation, measured in *"lift"* and *"advancement"* according to a reference system based on the mandibular occlusal plane (defined when the mandible is resting). Lift is related to the mandible opening required when dressing the MAD. Advancement, instead, refers to the mandible protrusion caused by the MAD adjustment during the OSAS treatment (measured on the occlusal plane). Mandible *rotation* (Fig. 3c) is the angle between the occlusal plane of the mandible in the resting state and the same plane of the protruded jaw (Fig. 3b).

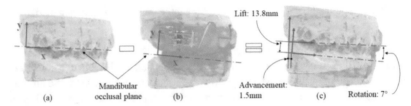

**Fig. 3.** Procedure for evaluating the mandible roto-translation. Teeth plaster casts in the rest position (a), teeth plaster casts combined with mandibular and maxillary MAD splints (b), and calculation of the mandible roto-translation, *lift, advancement,* and *rotation* (c).

Three types of joints were used for connecting anatomical elements: (i) *bonded*: no sliding or separation between surfaces is allowed, (ii) *frictionless*: sliding between surfaces without friction and (iii) *frictional*: surfaces can transmit shearing loads up to a specific value, before sliding. Bonded was used for almost all the ligaments apart between the *articular disc* and *glenoid fossa* (frictionless) and *articular disc* and *joint capsule* (frictional). Moreover, the following contacts were set between these interfaces: (i) *MAD* and *teeth*, (ii) *teeth* and *PDL*, and (iii) *PDL* and *mandible*. Regarding the analysis settings, an iterative solver was selected before running the simulation.

## 2.4 Meshing

The whole model consists of about 120000 10-nodes tetrahedral elements. The element size was adjusted for each geometry for an accurate meshing. The Skewness and the Orthogonal Quality mesh metrics criteria reported an average value $q = 0.29$ (standard deviation 0.19) and $q = 0.71$ (standard deviation 0.19), respectively, resulting in an excellent mesh quality.

## 3   Results and Discussions

The method presented in Fig. 1 was adopted for characterizing stress, strain, deformation and loads of a specific patient's TMJ, PDL, teeth, and MAD (Fig. 4a). The simulation model allowed authors to understand the deformation of the articular disc and joint capsule (Fig. 4b). For a high value of advancement, the articular disc clashes against the joint capsule, and this behaviour determines a higher stress value on the anterior area of the latter (Fig. 4c). SPL and STL ligaments are homogeneously stressed along their axes (Fig. 4d). While the most stressed and displaced teeth are the molars (Fig. 4e and

Fig. 4f). To be noted that practitioners are mainly interested to the stress distribution rather than exact values: this is the reason why Fig. 4 does not contain a scale.

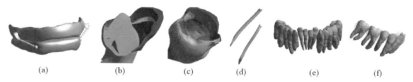

|       |       |       |       |       |       |
|-------|-------|-------|-------|-------|-------|
| (a)   | (b)   | (c)   | (d)   | (e)   | (f)   |

**Fig. 4.** MAD shape (a), TMJ deformation (b), TMJ stress (c), SPL and STL ligaments stress (d), teeth von-Mises stress (e) and teeth displacement (f).

Results also allowed to analyse the trend of TMJ stress (Fig. 5a) and SPL-STL force reaction (Fig. 5b) vs. The mandible roto-translation. It is worth noting that the TMJ stress gets the maximum stress approximately at 50% of roto-translation. Concerning the second set of simulation, mandible molar teeth are the most loaded, while mandible incisors are the least ones (Fig. 5c).

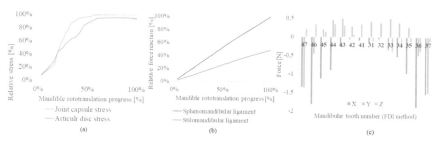

**Fig. 5.** TMJ stress (a) and SPL-STL force reaction (b) vs. Mandible roto-translation, PDL force reaction (c).

## 4   Conclusions

The paper presented a numerical simulation model for evaluating the effects caused by mandibular advancement devices, used for treating OSAS, on the temporomandibular joint and periodontal ligaments. The simulation model was used for characterising a specific patient and MAD (maxillary and mandibular splints connected through two symmetric metallic rods). The results highlight that the mandible lift phase generates significant stress values on TMJ, which cannot be neglected for prolonged usage of MADs. For this specific MAD, mandibular molars are more loaded than incisors.

For the feature, authors are going to analyse also the maxillary splint and arch. Furthermore, the authors will analyse the relationships among MADs (e.g., different materials, advancement system, geometry), patients (e.g., different apnea-hypopnea index - AHI), and the results presented in this paper (i.e., TMJ-PDL stress and teeth load/stress/displacement). By increasing the number of results, it will be possible to improve the current best-practices in prescribing MADs. Practitioners will be capable of advising MADs according to the patient's skeleton and AHI to minimize side-effects of such devices.

# References

1. Chan, C.H., Wong, B.M., Tang, J.L., Ng, D.K.: Gender difference in snoring and how it changes with age: systematic review and meta-regression. Sleep Breath **16**, 977–986 (2012). https://doi.org/10.1007/s11325-011-0596-8
2. Ramar, K., et al.: Clinical practice guideline for the treatment of obstructive sleep Apnea and snoring with oral appliance therapy: an update for 2015. J. Clin. Sleep Med. **11**(7), 773–827 (2015). https://doi.org/10.5664/jcsm.4858
3. Citarella, R., Armetani, E., Caputo, F., Naddeo, A.: FEM and BEM analysis of a human mandible with added temporomandibular joints. The Open Mech. Eng. J. **6**, 100–114 (2012). https://doi.org/10.2174/1874155X01206010100
4. Groning, F., Fagan, M.J., O'Higgins, P.: The effects of the periodontal ligament on mandibular stiffness: a study combining finite element analysis and geometric morphometrics. J. Biomech. **44**(7), 1304–1312 (2011). https://doi.org/10.1016/j.jbiomech.2011.01.008
5. Qian, H., Chen, J., Katona, T.R.: The influence of PDL principal fibers in a 3-dimensional analysis of orthodontic tooth movement. Am. J. Orthod. Dentofac. Orthop. **120**(3), 272–279 (2001). https://doi.org/10.1067/mod.2001.116085
6. Jeon, P.D., Turley, P.K., Ting, K.: Three-dimensional finite element analysis of stress in the periodontal ligament of the maxillary first molar with simulated bone loss. Am. J. Orthod. Dentofac. Orthop. **119**(5), 498–504 (2001). https://doi.org/10.1067/mod.2001.112999
7. Brunzini, A., Gracco, A., Mazzoli, A., Mandolini, M., Manieri, S., Germani, M.: Preliminary simulation model toward the study of the effects caused by different mandibular advancement devices in OSAS treatment. Comput. Methods Biomech. Biomed. Engin. **21**(13), 693–702 (2018). https://doi.org/10.1080/10255842.2018.1511776
8. Crivellin, G., Bruno, G., De Stefani, A., Mazzoli, A., Mandolini, M., Brunzini, A., Gracco, A.: Strength distribution on TMJ using mandibular advancement device for OSAS treatment: a finite element study. Dent. Cadmos **86**(08), 757 (2018). https://doi.org/10.19256/d.cadmos.09.2018.04
9. Sakhavalkar, P.U., Bhoosreddy, A.R., Kotwal, H.J.: Assessment and comparison of the capsular width of temporomandibular joint on ultrasonography and magnetic resonance imaging. J. Ind. Acad. Oral Med. Radiol. **28**, 351–357 (2016). https://doi.org/10.4103/jiaomr.JIAOMR_15_16
10. Savoldelli, C., Bouchard, P.O., Manière-Ezvan, A., Bettega, G., Tillier, Y.: Comparison of stress distribution in the temporomandibular joint during jaw closing before and after symphyseal distraction: a finite element study. Int. J. Oral Maxillofac. Surg. **41**, 1474–1482 (2012). https://doi.org/10.1016/j.ijom.2012.06.005
11. Wu, J., Liu, Y., Peng, W., Dong, H., Zhang, J.: A biomechanical case study on the optimal orthodontic force on the maxillary canine tooth based on finite element analysis. J. Zhejiang Univ. Sci. B (Biomed. Biotechnol.) **19**(7), 535–546 (2018). https://doi.org/10.1631/jzus.B1700195
12. Liu, Z., Qian, Y., Zhang, Y., Fan, Y.: Effects of several temporomandibular disorders on the stress distributions of temporomandibular joint: a finite element analysis. Comput. Methods Biomech. Biomed. Eng. **19**(2), 137–143 (2016). https://doi.org/10.1080/10255842.2014.996876

# Simulation of Dynamic Stresses on High Performance Engine Valve Spring System Considering Coil Clashing Effect

Michele Calabretta, Alessandro Sitta, Salvatore Massimo Oliveri, and Gaetano Sequenzia$^{(\boxtimes)}$

Dipartimento di Ingegneria Elettrica, Elettronica ed Informatica (DIEEI), Università degli Studi di Catania, Viale Andrea Doria 6, Catania, Italy
gsequenzia@dii.unict.it

**Abstract.** The valve train plays a major role in the performance of internal combustion engines by controlling the combustion process and it is therefore one of the key aspects for increasing the efficiency of combustion engines. Considering the dynamics, the spring force must be high enough to reliably close the valve preventing from seating bouncing due to surge modes after the valve closure. On the other side, the spring force should be kept as low as possible in order to reduce the engine friction losses and consequently the fuel consumption. In the high-performance engines, the valve springs have to be designed and optimized for sustaining higher stresses with compact dimensions leading to critical material and manufacturing processes. This requires a reduction of moving masses and a strong focus on design and process optimization of the coil springs for reducing the mechanical load and the friction losses at low engine speed. At the same time, valve train should be reliable at high engine speed. The calculation of stresses and contact forces for moving parts under dynamic load is essential for durability analysis. A method to calculate the contact of moving masses is described and proposed to justify valve motions experimental results. To fully understand the failure mechanism of test bed reliability trials, the dynamic stresses have been calculated modeling the real springs' shape. The contact forces have been reproduced considering the coil clash effects and the dynamic behavior of the flexible spring.

**Keywords:** Valve train · Valve springs · Coil clash · Multibody · FEM

## 1 Introduction

The predictive modeling of the valve train dynamic [1] is essential for durability analysis at the highest engine speed conditions [2]. The method to calculate the complex dynamic of the valve-spring systems allows a correct representation of flexible and complex contacts under full consideration of the system dynamics. The methodology combines the multi-body dynamics modeling [3–5] and the finite element analysis (FEA) [6]. With this approach the valve dynamics, measured with laser interferometric technique

L. Roucoules et al. (Eds.): JCM 2020, LNME, pp. 108–113, 2021.
https://doi.org/10.1007/978-3-030-70566-4_18

[1], is well correlated with multibody analysis. The obtained results have been therefore used for Finite Element Analysis on the self-contacting parts such as the valve springs. In modern high-speed combustion engines progressive springs are typically used for guarantying the maximum load with a compact installation length [7]. The durability of the valve train is typically validated on monitored cylinder head test bed and valve springs are one of the most critical reliability bottle neck. Sometimes the spring breaks directly over the sharp edge of the flattened end of the last coil. In this case the failure mechanism explanation is due to high contact pressure in this area. In many cases the spring breaks between 180°–270° after this point, at a location where a breakage is least expected. Metallurgical analyses of broken springs did not indicate material inconsistencies as reason for the failure. The calculation of the stresses in the spring with traditional FE-analysis does not reveal the location of failure. Dynamic analysis using modal reduction does also not give any useful indication. To fully understand the reason of the failure, true dynamic stresses have to be calculated by considering the correct geometrical shape of the spring, contact forces due to coil clash and the dynamic behavior of the highly flexible spring. The proposed methodology based on experimental and numerical simulations allows the representation of these flexible and complex contacts under full consideration of the spring dynamics.

## 2  Testing Equipment

A motored cylinder head has been used for this investigation (see Fig. 1a). An electric motor connected to the end of the camshaft has been used to drive the intake camshaft controlling speed and torque. The camshaft directly operates the intake valves via hydraulic tappets. Using a laser system, the rig has also been used to take measurements of valve displacement and dynamics. This experimental measurement has been used for the numerical dynamic model validation.

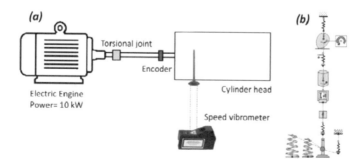

**Fig. 1.** Test rig and measurement system (a) and single valve-train multibody model (b)

## 3  Multibody Dynamic Models

A multi-body dynamic model of a single valvetrain has been developed using Ricardo Software VALDYN as shown in Fig. 1b. This valvetrain has been represented as a series of lumped mass/inertia nodes connected by stiffness and damping values. The cam node has been suspended on a stiffness element representing camshaft bending stiffness and camshaft support stiffness. The hydraulic tappet has been modelled as two mass nodes connected by a special element to account for the action of the high pressure chamber, the expansion spring and the check valve. The tappet has been connected to the valve using a lash stiffness element representing the stiffness of the valve stem between the tip and the center of mass. The valve and spring retainer have been modelled as a single mass node. The valve node has been connected to ground by another lash stiffness element representing the valve head and seat stiffness. The valve train has a double spring pack and each spring has been modelled using 4 mass nodes per coil connected by stiffness [1]. The cam profile has been designed to meet the many conflicting requirements for engine breathing, acceptable durability, high speed dynamics etc. [9, 10]. The spring pack has been designed to maintain contact between cam and the hydraulic tappet at high engine speed. Fig. 2a and Fig. 2b show the calculated dynamic valve lift and dynamic valve acceleration and against crank angle at camshaft speed of 4000 rpm. The calculation model has the capability to make calculations of valvetrain dynamic. This input, correlated with laser results from test bed experimental measurements, has been used to perform and validate a detailed Finite Element Analysis of the valve spring.

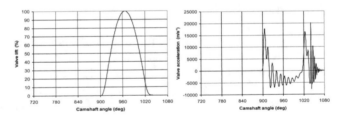

**Fig. 2.** Calculated dynamic valve lift (a) and acceleration (b) at 4000 rpm camshaft speed

## 4   Finite Element Dynamic Model

The simulation output obtained by multibody modelling have been used as dynamic input for a Finite Element model. In particular the valve springs (inner and outer spring) model is generated considering a discretization on section (QUAD 4 element). From the 2D section mesh it has been created an expansion on the valve spring helix with esaedric element. The considered FE model consists in the valve spring system composed by inner and outer spring both in contact with the upper retainer that impose the motion and with contact definition between each coil [11] (Fig. 3).

**Fig. 3.** Finite element model of the valve spring system (a) and details of the high density mesh region to focus on coil clash issues (b)

## 5 Results

The aim of this work is to show the phenomenon of coil clashing and the dynamic behavior of the highly flexible spring. The valve train system has been tested in overspeed condition, with the camshaft speed of 4750 rpm, in order to emphasize the phenomenon.

Considering the valve acceleration calculated by multi-body model and correlated by measurement, the aim of this work is to show in overspeed conditions the phenomenon of coil clashing and the dynamic behavior of the highly flexible through the FE analysis. The phenomenon of auto contact between wires is evident on internal and external wire but the absolute entity is stronger for the inner linear spring. The higher non-linearity of the outer spring (progressive stiffness) with variable natural frequency, is acting as mitigation effects.

The analysis shows the major contribution of coil clashing contact on spring stress value, as plotted in Fig. 5.

From the experimental testing it has been shown a breakage on the inner spring (1.75 wire from the bottom side) and it was also justified from analytical point of view as consequence of coil clashing effect (Fig. 4).

**Fig. 4.** Valve spring after testing in overspeed condition (breakage on inner spring from bottom side).

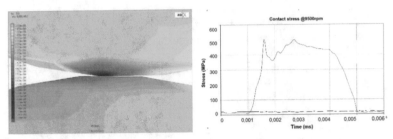

**Fig. 5.** Detail of area of coils clashing area. (a) and Max stress (at 9500 rpm) on outer spring (1.75 wire from the bottom side) for normal contact (blue line) and coil clash one (red line) (b).

## 6  Conclusion

The aim of this work has been the creation of a closed loop between modeling, analysis and measurement for a clear understanding of a High Performance Engine valve train dynamics. In particular the spring breakage phenomena have been studied with a deep analysis on the inner springs self contact "coil clashing" effect. An experimental bench has been used for testing the valve train up to overspeed conditions, to monitor the valve motion and the overall dynamics. The correlation of the experimental results has been done combining the multi body dynamics with the finite elements. Especially the capability to get stresses directly even for components which experience large deformations and high impact contacts under highly dynamic conditions such as the valve springs has been a big advantage for failure analysis and durability studies. The proposed methodological approach allows to study the dynamics of each component of the valve train, to validate the reliability and, at the end, to optimize the system by an engineering and technological point of view already in a preliminary stage of work.

## References

1. Calabretta, M., Cacciatore, D., Carden, P.: Valve train friction-modeling, analysis and measurement of a high performance engine valve train system. SAE Int. J. Engines 3(2), 72–84 (2010)
2. Corvino, C., Calabretta, M., Zanasi, R.: Modeling of variable valve timing on high performance engine using power-oriented graphs method. Technical Report, SAE Technical Paper (2011)
3. Oliveri, S.M., Sequenzia, G., Calì, M.: Flexible multibody model of desmodromic timing system. Mech. Based Des. Struct. Mach. 37(1), 15–30 (2009)
4. Calì, M., Oliveri, S.M., Sequenzia, G.: Geometric modeling and modal stress formulation for flexible multi-body dynamic analysis of crankshaft. In: Proceedings of the 25th Conference and Exposition on Structural Dynamics, pp. 1–9 (2007)
5. Li, F., Zhuang, J., Cheng, X., Wang, Q., Chen, C., Zhou, S.: Multi-body dynamics simulation study on a valve train system. In: 2017 International Conference on Computer Technology, Electronics and Communication (ICCTEC), pp. 116–119. IEEE, December 2017
6. Calì, M., Sequenzia, G., Oliveri, S.M., Fatuzzo, G.: Meshing angles evaluation of silent chain drive by numerical analysis and experimental test. Meccanica 51(3), 475–489 (2016)

7. Sequenzia, G., Oliveri, S., Calabretta, M., Fatuzzo, G., Calì, M.: A new methodology for calculating and modelling non-linear springs in the valve train of internal combustion engines. Technical Report, SAE Technical Paper (2011)
8. Calabretta, M., Cacciatore, D., Carden, P., Plail, J.: Development of a timing chain drive model for a high speed gasoline engine. SAE Int. J. Engines **4**(1), 432–440 (2011)
9. Beloiu, D.M.: Modeling and analysis of valve train, part i-conventional systems. SAE Int. J. Engines **3**(1), 850–877 (2010)
10. Gzal, M., Gendelman, O.V.: Edge states in nonlinear model of valve spring. In ASME International Mechanical Engineering Congress and Exposition, vol. 59414, p. V004T05A087. American Society of Mechanical Engineers, November 2019
11. Gu, Z., Hou, X., Keating, E., Ye, J.: Non-linear finite element model for dynamic analysis of high-speed valve train and coil collisions. Int. J. Mech. Sci. **174**, 105476 (2020)

# Thread Couplings Stress Analysis by Radial Basis Functions Mesh Morphing

Michele Calì[1]([✉]) [ID], Salvatore Massimo Oliveri[1] [ID],
and Marco Evangelos Biancolini[2] [ID]

[1] Department of Electric, Electronics and Computer Engineering, University of Catania,
Catania, Italy
michele.cali@dieei.unict.it
[2] Department of Enterprise Engineering "Mario Lucertini", University of Rome "Tor Vergata",
Rome, Italy

**Abstract.** Traditional analytical methods are approximate and need to be validated when it comes to predict the tensional behavior of thread coupling. Numerical finite element simulations help engineers come up with the optimum design, although the latter depends on the constraints and load conditions of the thread couplings which are often variable during the system functioning. The present work illustrates a new method based on Radial Basis Functions Mesh Morphing formulation to optimize the stress concentration in thread couplings which is subject to variable loads and constraints. In particular, thread root and fillet under-head drawings for metric ISO thread, which are the most commonly used thread connection, are optimized with Radial Basis Functions Mesh Morphing. In metric ISO threaded connection, the root shape and the fillet under the head are circular, and from shape optimization for minimum stress concentration it is well known that the circular shape becomes seldom optimal. The study is carried out to enhance the stress concentration factor with a simple geometric parameterization using two design variables. Radial Basis Functions Mesh Morphing formulation, performed with a simple geometric parameterization, has allowed to obtain a stress reduction of up to 12%; some similarities are found in the optimized designs leading to the proposal of a new standard. The reductions in the stress are achieved by rather simple changes made to the cutting tool.

**Keywords:** Thread root · Fillet under-head · Feature-based modeling · Metric ISO thread · Radial basis functions

## 1 Introduction

All industrial applications commonly use threaded couplings as, for example, in transport, mining, automotive, offshore industry and in manufacturing process and more recently in dental implants. The evaluation of the stress distribution in threaded coupling is a classic topic in mechanical design. The oldest scientific papers [1–3] date from the first half of the 20th century; they focus correctly the main problems: the non-uniform load distribution on the engaged threads, the stress concentration at the root of

L. Roucoules et al. (Eds.): JCM 2020, LNME, pp. 114–120, 2021.
https://doi.org/10.1007/978-3-030-70566-4_19

the threads and under the head of the screw. As reported in [4, 5], failures in screws can be so divided: 15% at the fillet under the head, 20% at the first thread and 65% at the first engaged thread.

Analytical evaluation and numerical finite element (FE) simulations of the stress distribution on the coil are applied in different fields, such as in the design of machinery organs and in their assembly [6, 7], in the biomedical field to assess the influence of thread shape on the mechanical behavior of bone implant systems [8], in the reliability field [9, 10].

In this work Radial Basis Functions (RBF) Mesh Morphing (MM) formulation was applied to construct, directly in the FE environment, a parametric model of a bolted couplings with axial and sheared variable loads with the aim of optimizing the stress concentration in thread root and fillet under-head. RBF were employed in literature to carry shape-based optimization [11], but their use was also explored for advanced studies, such as steady [12, 13] and unsteady [14] fluid structure interaction problems, evaluative shape-based optimizations [15, 16], etc.

The advantages of this approach are: no need to regenerate the grid; the preservation of robustness of the procedure; the capability to support different mesh typologies thanks to its meshless nature and the high parallelizable smoothing process. The morphing action integrated in a solver updating the computational domain "on the fly" during the progress of the computation can be executed in RBF Morph across three steps: 1) setup: that is the manual definition of the morphing targets, i.e. the portions of the FEA mesh that will be updated, and morphing sources, i.e. the portions of the FEA mesh controlled, and the definition of the required movements (design parameters) of the points driving the shape deformation; 2) fitting: the solution of RBF system obtained by collecting the morphing sources; 3) smoothing: by propagating the displacement prescribed on sources to the volume mesh target. Using the commercial tools ANSYS Mechanical® and the companion *RBF Morph™ ACT Extension* plugin, the mentioned three steps were performed as explained in the following sections.

## 2 Discretized Models

Two FE models were built for M12 and M16 bolt-nut assembly. The models employ 128116 second order tetrahedrons (TET 10) elements *SOLID 187* with 254359 nodes for the bolt and 98116 TET 10 elements with 153259 nodes for the nut. In the thread root and fillet under-head contact zone a more dense mesh with patch conformity was employed using element *CONTA 174* and *TARGA 170*. These elements allow to simulate the friction forces in the contacts. Using Remote Point formulation axial F and transversal T external loads (variable between 50 kN and 80 kN and 10 kN and 20 kN, respectively) were applied on a dependent node (center of the spherical zone) and distributed on 36 independent nodes (within the spherical area). In this way the loads can be applied in a realistic way without stress peaks. A preload calculated with the formula:

$$F_{preload} = 0.8\sigma_s A_t \qquad (1)$$

was applied in the assembly systems. The preload was applied such that the stress due to it is equal to 80% of the maximum admissible stress. With this preload the assembled elements do not have relative movement due to the application of external loads acting in the normal direction, and do not have relative movement of slipping due to the shear forces. The discretized models are shown in Fig. 1(a) and 1(b); in Fig. 1(c) loads, preload and constrains section scheme is shown.

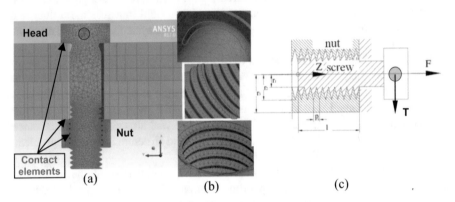

**Fig. 1.** (a) Bold-nut connection model; (b) optimization zones; (c) load section scheme.

## 3  Parameterization

Geometric parameterization based on RBF-MM implements shape modifiers directly on the computational domain. The new geometric configurations resulted from the displacement of a mesh region, through the use of algorithms, based on RBFs, which smoothly propagated the imposed displacement to the surrounding volume. The fitting size in the thread root and fillet under-head, according to current standard (UNI EN ISO 4016), are calculated as follows:

$$R_{min} = \frac{H}{8} = 0.1082\,P \tag{2}$$

being H the height of the non-fillet profile (Fig. 2a): $H = sin60° \, P = 0.866\,P$

Triple multicenter circular fillets were used as guide curves for profile at thread root and for profile under-head. Five parameters ($R_1$, $R_2$, $R'_0$ and $R'_1$, $R'_2$) defining the curves radii were varied at given boundary conditions and mutual relationships to generate a different profile allowed by standards. The RBF interpolation functions drove the mesh morphing of the discretized domain of computational models by applying predefined displacements in the set of 2600 nodes (source points) near the fillet zones without changing their number of the topology of the mesh. During the mesh morphing the nodes on the profile moved according to the defined constrain varying the $R_1$ and $R_2$ fillet radius parameters (Fig. 3).

Then the RBF action was extended in the entire volume, smoothing the grid up to the prescribed surrounding limits. The obtained RBF solution was sent to the ANSYS Mechanical solver and an iterative calculation strategy was chosen.

$H = 0.86603P \quad h_3 = 0.61343P \quad H_1 = 0.54127P \quad R = H/6$

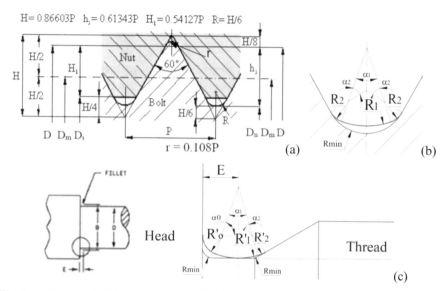

**Fig. 2.** (a) Geometrical features of metric ISO thread coupling; (b) particular of fillet at thread root; (c) particular of fillet under-head.

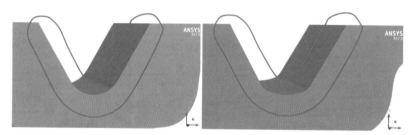

**Fig. 3.** Source points in the thread root.

## 4 Results and Discussion

According to the RBF-MM formulation and parameterization described above four standard metric ISO thread couplings: M12 × 1.25; M12 × 1.75; M16 × 1.5; M16 × 2 were morphed under axial F and transversal T variable loads (50kN ÷ 80kN and 10 kN ÷ 20 kN, respectively). The set of parameters and their variability shown in Table 1 were chosen for thread root analysis. The set of parameters and their variability shown in Table 2 were chosen for fillet under-head root analysis. The theoretical approach used to calculate the maximum von Mises stress is the method of Heywood [4] and Pilkey [5]. The maximum von Mises stress value was calculated evaluating the four components of the stress state: axial, radial, azimuthal, and shear stresses using respectively the following formulas:

$$\sigma_{ax} = \frac{F}{A} K_t(Ax); \quad \sigma_{Rad} = \left( \frac{F\left( \frac{H}{2} + \frac{P}{4} \right)}{W} + \frac{F}{A} \right) K_t(Rad);$$

$$\sigma_{az} = v(\sigma_{ax} + \sigma_{Rad}); \quad \tau = \frac{F^2}{A} \frac{k(Ax)}{1 + q(\frac{\sigma_{ax}}{\sigma_{Rad}})} \tag{3}$$

where $K_t(Ax)$, $K_t(Rad)$ and q values are taken from the Pilkey's diagrams [5]. In this way it was possible to calculate the von Mises stress using the formula:

$$\sigma_{VM} = \sqrt{\sigma_{ax}^2 + \sigma_{Rad}^2 + \sigma_{az}^2 - \sigma_{ax}\sigma_{Rad} - \sigma_{ax}\sigma_{az} - \sigma_{Rad}\sigma_{az} + 3\tau^2} \tag{4}$$

In correspondence with the maximum values of F and T (F = 80 kN and T = 20 kN) were calculated maximum von Mises stress in thread root and in fillet.

**Table 1.** FEM and theoretical values of maximum von Mises stresses in thread root.

| Case studies | $R_1$ [mm] | $R_2$ [mm] | FEM values | Theoretical values |
|---|---|---|---|---|
| M12 × 1.25 | 0.135 ÷ 0.406 | 0.02 ÷ 0.06 | 558.60 | 565.36 |
| M12 × 1.75 | 0.189 ÷ 0.568 | 0.03 ÷ 0.09 | 548.26 | 544.55 |
| M16 × 1.5 | 0.162 ÷ 0.487 | 0.025 ÷ 0.075 | 520.18 | 514.59 |
| M16 × 2 | 0.216 ÷ 0.649 | 0.033 ÷ 0.1 | 482.5 | 489.08 |

**Table 2.** FEM and theoretical values of maximum von Mises stresses in fillet under-head.

| Case studies | $R'_0$ [mm] | $R'_1$ [mm] | $R'_2$ [mm] | FEM values | Theoretical values |
|---|---|---|---|---|---|
| M12 × 1.25 | 0.135 ÷ 0.406 | 1.5 ÷ 4 | 0.4 ÷ 2 | 448.64 | 454.23 |
| M12 × 1.75 | 0.189 ÷ 0.568 | 2 ÷ 6 | 0.6 ÷ 3 | 419.82 | 424.48 |
| M16 × 1.5 | 0.162 ÷ 0.487 | 1.7 ÷ 4.5 | 0.5 ÷ 2.2 | 468.99 | 473.07 |
| M16 × 2 | 0.216 ÷ 0.649 | 2.5 ÷ 7 | 0.8 ÷ 3.5 | 393.67 | 389.16 |

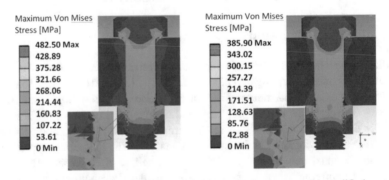

**Fig. 4.** Maximum equivalent von Mises stress: (a) standard geometry; (b) modified geometry.

The increase in the fillet radii led to a better load distribution and a decrease of maximum von Mises stress of about 15%. The greatest improvement (20%) occurred in the M16 × 2 thread couplings using the values of $R_1 = 0.649$; $R_2 = 0.033$; $R'_0 = 0.216$; $R'_1 = 7$ and $R'_2 = 0.8$; the maximum von Mises stress was reduced to 385.90 MPa corresponding to the 60% of limit stress ($\sigma_r = 800$ N/mm$^2$; $\sigma_s = 640$ N/mm$^2$ E = 200 kN/mm$^2$) for 8.8 class of bolt. Figure 4 shows the values of the maximum von Mises stress and their reduction after optimization for this thread.

## 5  Conclusions

A Radial Basis Functions Mesh Morphing formulation was adopted in order to create parametric models of bolted couplings with axial and shear variable loads directly in the FE environment, aimed at optimizing the maximum stress in thread couplings. The optimal drawing of thread root and fillet under-head was found. Four case studies demonstrated its effectiveness optimizing the maximum equivalent von Mises stress. It was possible to reduce the peak load and the maximum stress, enlarging the internal diameter of the thread root with triple multicenter circular fillets so improving the strength and fatigue life by means of a simple and not expensive modification. The comparison with theoretical data showed a really negligible maximum error on load distribution and an improvement of the threaded coupling. Finally, the calculations performed on the two diameters seemed to demonstrate that the improving effect was greater for the largest diameter.

**Acknowledgments.** The research work herein reported was funded by Università degli Studi di Catania within the CRUI-CARE Agreement (research path PIA.CE.RI. 2020-2022 Linea 3 STARTING GRANT 2020) and by "Smart main. of ind. Plants and civ. Struct. by 4.0 monit. Tech. And pron. Appr.-MAC4PRO".

## References

1. Stromayer, C.E.: Stress distribution in bolts and nuts. Trans. Last. Nav. Arch. **60**, 112–121 (1918)
2. Den Hartog J.P. The mechanics of plate rotors for turbo-generators, Trans of ASME, 51(1), 1–10 (1929)
3. Goodier, J.N.: The distribution of load on threads of screws. J. Appl. Mech. 93–100 (1940)
4. Heywood, R.B.: Designing by Photoelasticity. Chapman and Hall, London (1952)
5. Pilkey, W.D., Pilkey, D.F.: Peterson's Stress Concentration Factors. J. Wiley and S, N.Y. (2008)
6. Pedersen, N.L.: Optimization of bolt thread stress concentrations. Arch. Appl. Mech. **83**(1), 1–14 (2013)
7. Ball, R.S.: A Treatise on the Theory of Screws. Cambridge University Press, Cambridge (1998)
8. Zanetti, E.M., Ciaramella, S., Calì, M., Pascoletti, G., Martorelli, M., Asero, R., Watts, D.C.: Modal analysis for implant stability assessment: sensitivity of this methodology for different implant designs. Dent. Mater. **34**(8), 1235–1245 (2018)
9. Birger, I.A.: Load distribution on thread in threaded joints. Bull. Mach. Eng. I (1944)

10. Biancolini, M.E.: Fast radial basis functions for engineering applications. Springer (2017)
11. Biancolini, M.E., Costa, E., et al.: Glider fuselage-wing junction optimization using CFD and RBF mesh morphing. Aircraft Eng. Aerosp. Technol. **88**, 740–752 (2016)
12. Pascoletti, G., Calì, M., Bignardi, C., Conti, P., Zanetti, E.M.: Mandible morphing through principal components analysis. Lecture Notes in Mech. Engineering, pp. 15–23. Springer (2020)
13. Calì, M., Oliveri, S.M.: Application of an effective SIMP method with filtering for topology optimization of motorcycle tubular frame. Int. Rev. Mech. Eng. **11**(11), 836–844 (2017)
14. Di Domenico, N., Groth, C., Wade, A., Berg, T., Biancolini, M.E.: Fluid structure interaction analysis: Vortex shedding induced vibrations. Procedia Struct. Integrity 422–432 (2018)
15. Calì, M., Oliveri, S.M., Cella, U., Martorelli, M., Gloria, A., Speranza, D.: Mechanical characterization and modeling of downwind sailcloth in fluid-structure interaction analysis. Ocean Eng. **165**, 488–504 (2018)
16. Calì, M., Speranza, D., Cella, U., Biancolini, M.E.: Flying shape sails analysis by radial basis functions mesh morphing. Lecture Notes in Mechanical Engineering, pp. 24–36. Springer (2020)

# Computer-Aided Tolerancing Analysis of a High-Performance Car Engine Assembly

Andrea Petruccioli, Francesco Gherardini$^{(\boxtimes)}$, Davide Panari, and Francesco Leali

Department of Engineering "Enzo Ferrari", University of Modena and Reggio Emilia, Via Pietro Vivarelli 10, 41125 Modena, Italy
francesco.gherardini@unimore.it

**Abstract.** This paper proposes the analysis of the tolerances (values, types, datum) and their effects on a mechanical assembly, as a high-performance car engine, by means of a Computer-Aided Tolerancing software. The 3D tolerance stack-ups are investigated to assess the fulfillment of the functional requirements as well as the performance specifications of the assembly. Moreover, after identifying the tolerances that mainly affect the product variability, we finally propose some corrective actions on the tolerances and assess their functional allocation, tightening or relaxing their values, ensuring assemblability and cost reduction.

**Keywords:** Tolerance analysis method · Computer-Aided Tolerancing · Geometric dimensioning and tolerancing (GD&T) · Automotive assembly · Car engine

## 1 Introduction

In the global marketplace, the search for a perfect balance between performance fulfillment and economic feasibility is mandatory. Therefore, the assembly optimization as well as the reduction of the variation of the key dimensions of parts are primary goals to guarantee the product quality with cost and time savings. The application of geometric dimensioning and tolerancing (GD&T) specifications to a product still represents a typical industrial issue [1]. The difficulties in the definition of the correct GD&T scheme and in the tolerance allocation choice are due to the complexity of the real industrial cases, in which different requirements of product life cycle phases must be considered [2, 3]. Usually, the choice of the Datum Reference Frame (DRF) and of the tolerance specifications derives from previous experience and background trial and error tests, whose effectiveness is not always guaranteed. The literature presents some useful references, even if sometimes they are difficult to link with the real application. Thanks to advanced simulation such as Computer-Aided Tolerancing (CAT) software, the tolerance analysis approach improves the knowledge about the process and enables the optimization of the GD&T scheme and the assembly process. These tools represent a support during the early design phases to validate the product tolerance design. For industrial fields as automotive, the product optimization through CAT simulation can produce a consistent increase of quality levels and a reduction of unnecessary costs [4, 5].

© The Author(s) 2021
L. Roucoules et al. (Eds.): JCM 2020, LNME, pp. 121–127, 2021.
https://doi.org/10.1007/978-3-030-70566-4_20

For these reasons, this work aims at performing the tolerance analysis of an automotive assembly, suggesting corrective actions on the tolerance values and allocation schemes. A high-performance engine assembly is selected as a case study for its complexity and representativeness among industrial applications.

The final goal is to identify a set of general allocation schemes of typical industrial cases to be used as guidelines for the GD&T definition and for the selection of tolerances, considering the specific functional requirements of each application.

## 2   Method

To reach the expected levels of customer's satisfaction, quality has become a design approach based on a group of procedures, first the Tolerance Design process. Tolerance Design process is mainly performed by CAT-based approach, which is integrated in the early design phases [6]. The general workflow has an input phase of preparation and setting-up, a modelling and simulation phase and the final output analysis and synthesis phase [6, 7], with their subphases here reported:

- Preparation: Data collection, CAD models import in the CAT software, identification of parts geometry, elimination of non-contributing parts.
- Functional analysis: Functional and technological requirements definition, identification of the measurements consistent with the targets, identification of the functional features of the parts and Datum Reference Frames (DRFs).
- Assembly sequence definition: Relative positioning, clamping and fixing operations of the parts.
- Tolerance set-up: Identification of the dimensional and geometrical tolerances (GD&T) on the parts, definition of the tolerance values.
- Measurements set-up (i.e. stack-up analysis responses to be controlled).
- Analysis and results: Simulation run and data analysis with the target values.
- Synthesis: If necessary, optimization of tolerance values and types, final verification of the requirements.

## 3   Case Study: V12 Engine Assembly

We selected a high-performance V12 engine assembly as a case study, performing a tolerance analysis to verify the effectiveness of its assembly operations. We focused on the bolting operations between the upper (UC) and lower crankcase (LC), and between the cylinder heads and the UC (Fig. 1).

These operations are crucial during the assembly phases since they affect the performance and reliability of the engine itself. To fulfil the design requirements, we performed the 3D tolerance analysis of the engine assembly, then we assessed the adopted GD&T and tolerance allocation schemes, and the tolerance values.

The CAT software adopted is Cetol 6σ (Sigmetrix®), an add-in workbench integrated in CATIA V5 (Dassault Systemes®) platform, based on a vector loop approach [8]. The components to be simulated are treated as rigid models, without parts distortion.

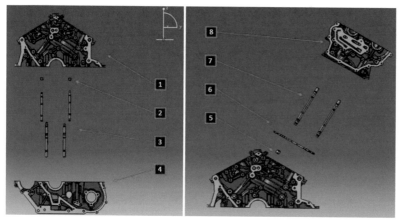

**Fig. 1.** V12 engine assembly: 1. UC, 2. LC bushings, 3. LC studs, 4. LC, 5. Cylinder head bushings, 6. Head gasket, 7. Cylinder head studs, 8. Cylinder head.

After the simulation, the statistical and the sensitivity analysis of the tolerances permit to identify the main contributors to variation.

In accordance with the proposed method, the following steps are performed.

**Preparation and Setting Up:** The CAT model includes the following main components: the UC, the LC, the head gasket and the cylinder heads, as well as the fixture elements, i.e. the crankcase bushings and the studs. There are 14 bushings and 28 studs on the LC side, and 2 bushing and 14 studs on the cylinder heads side. Since they achieve the positioning and bolting operations of the assembly, their interface with the other components is crucial. Non-contributing parts as the head gasket and one of the cylinder heads were not considered in the simulation.

The assembly sequence is identified: the UC is fixed (i.e. it is held in position by an external fixturing system), the bushings are inserted in their housings on the UC with interference fit, then the studs are inserted in their housings on the UC. Once these components are positioned, the LC and the cylinder heads are inserted. The bolting operations on the studs complete the assembly of the engine.

**Modelling and Simulation:** With respect to this assembly sequence, the Datum Reference Frames (DRFs) for the components (Fig. 2) are analysed.

Starting from the UC, the first reference A is the surface between the parts. Then, the holes corresponding to the two opposed external bushings are defined as references B and C. Finally, a position tolerance with respect to A, B, C locates the other housings of both bushings and studs. All the tolerances are transferred from the technical drawings to the 3D models, with respect to the DRF.

As functional measures, we set the gaps of both bushings and studs with respect to their housings in the engine components. We checked the absence of physical interferences between the parts (gap $< 0$), assessing a target value of $\pm 3\sigma$.

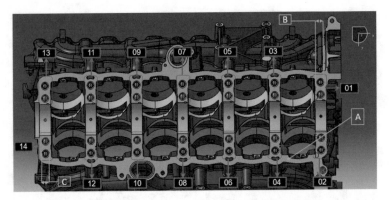

**Fig. 2.** Datum Reference Frame (DRF) identification on the UC.

**Analysis and Synthesis:** The analysis of the simulation outputs provides the following results. We divided them in two groups: (A) the gap measurements between bushings and their housings, and (B) the gap measurements between studs and their housings, for both the UC-LC (1) and the UC-cylinder heads sides (2):

A.1 The target condition is not verified, with an average value of ±1.95σ (Fig. 3). The sensitive analysis identifies the main contributors of variation, the position tolerance of the bushing housings on both the UC and the LC.

A.2 The target condition is not verified, with an average value of ±1σ. The sensitive analysis identifies the main contributors of variation, the position and dimensional tolerance of the bushing housings on the cylinder head.

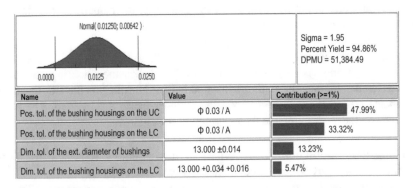

**Fig. 3.** Simulation outputs of A.1 measurements, with main contributors to variation.

B.1 The target condition is completely verified, with a large safety margin, and an average value of ±4.63σ (Fig. 4). The main contributors of variation are the position tolerance of the stud threaded seats and of the stud housings on the LC.

B.2 The target condition is verified, with a lower safety margin compared to B.1, with an average value of ±3.5σ, near the limit of acceptance. The main contributors of variation are the position tolerance of the stud threaded seats and the dimensional tolerance of the stud housings on the cylinder head.

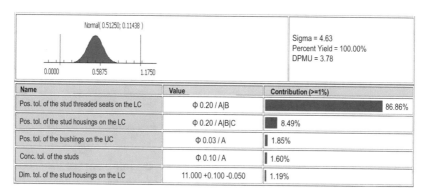

| Name | Value | Contribution (>=1%) | |
|------|-------|---------------------|---|
| Pos. tol. of the stud threaded seats on the LC | Φ 0.20 / A\|B | | 86.86% |
| Pos. tol. of the stud housings on the LC | Φ 0.20 / A\|B\|C | 8.49% | |
| Pos. tol. of the bushings on the UC | Φ 0.03 / A | 1.85% | |
| Conc. tol. of the studs | Φ 0.10 / A | 1.60% | |
| Dim. tol. of the stud housings on the LC | 11.000 +0.100 -0.050 | 1.19% | |

**Fig. 4.** Simulation outputs of B.1 measurements, with main contributors to variation.

The outputs analysis leads to the following corrective actions (Tables 1 and 2).

**Table 1.** Tolerance values of the main contributors need to be tightened (A.1 and A.2)

| | Tolerance | Original value | Corrected value |
|---|-----------|----------------|-----------------|
| A.1 | Position tolerance of bushing housings on the UC and LC | 0.03 | 0 with M.M.C. |
| | Dimensional tolerance of bushing housings on the LC | F7 | E7 |
| A.2 | Dimensional tolerance of bushing housings on the UC | K6 | P6 |
| | Dimensional tolerance of external diameter of the bushings | +0.025 +0.012 | js8 |
| | Dimensional tolerance of bushing housings on cylinder heads | H7 | E6 |
| | Position tolerance of bushing housings on cylinder heads | 0.05 | 0.03 |

**Table 2.** Tolerance values of the main contributors are relaxed (B.1 and B.2)

| | Tolerance | Original value | Corrected value |
|---|---|---|---|
| B.1 and B.2 | Concentricity tolerance of studs | 0.1 | 0.3 |
| | Position tolerance of stud housings on the LC | 0.2 | 0.6 with MMC |
| | Dimensional tolerance of stud housings on the LC | +0.1 −0.05 | +0.2 −0.1 |
| | Dimensional tolerance of stud threaded seats on the LC | 0.2 | 0.2 with MMC |
| | Position tolerance of stud housings on the cylinder head | 0.4 | 0.4 with MMC |

After these corrective actions, the simulation shows the fulfillment of the functional targets for both the bushings and the studs. This result is also due to the introduction of the maximum material condition (MMC) and the relaxation of the tolerance ranges for the stud-related tolerances.

## 4 Discussion and Conclusions

The comparison between the resulting measurement ranges and the functional targets identifies two different scenarios. The outputs of A.1 and A.2 show a reduced capability of the engine assembly process compared to the target of $\pm 3\sigma$, therefore we tight the tolerance ranges of the main contributors to variation to improve the assembly performance. On the other hand, B.1 and B.2 show that the values of the stud-related tolerances are too tight, so it is possible to relax the tolerance values of the main contributors to variation, reducing the manufacturing cost of parts, and reaching the expected production target. Thanks to CAT simulation, the tolerance design of the assembly is checked and validated, providing corrective actions. Thanks to the sensitivity analysis, the most influencing tolerances and reference elements are identified, with respect to the functional requirements of the case study. These outputs will be useful for future works, to extract guidelines for the GD&T definition and for the allocation of tolerances in similar applications, developing an archetype for industrial cases of bolt assembly.

## References

1. Wang, H., Thamma, R.: A scheme for functional tolerancing: a product family in 3D CAD system. Int. J. Ind. Eng. Comput. 3(1), 81–92 (2012)
2. Zhu, Z., Qiao, L., Anwer, N.: An improved tolerance analysis method based on skin model shapes of planar parts. Procedia CIRP 56, 237–242 (2016)
3. Wang, Y., Calhoun, S., Bosman, L., Sutherland, J.W.: Tolerance allocations on products: a life cycle engineering perspective. Procedia CIRP 80, 174–179 (2019)
4. Zhang, X.D., Zhang, C., Wang, B., Feng, S.C.: Unified functional tolerancing approach for precision cylindrical components. Int. J. Prod. Res. 43(1), 25–47 (2005)

5. Zhang, Y., Hu, W., Rong, Y., Yen, D.W.: Graph-based set-up planning and tolerance decomposition for computer-aided fixture design. Int. J. Prod. Res. **39**(14), 3109–3126 (2001)
6. Gherardini, F., Panari, D., Leali, F.: Identification of the main contributors in the 3D tolerances assessment in mechanical transmissions. In: Advances on Mechanics, Design Engineering and Manufacturing II. Lecture Notes in Mechanical Engineering, pp. 152–161. Springer (2019)
7. Vergnano, A., Gherardini, F., Petruccioli, A., Bonazzi, E., Leali, F.: Robust parameter analysis of compliant part models for computer aided tolerancing. In: Design Tools and Methods in Industrial Engineering. ADM 2019. Lecture Notes in Mechanical Engineering, pp. 241–254. Springer (2019)
8. Glancy, C.G., Chase, K.W.: A second order method for assembly tolerance analysis. In: ASME Design Engineering Technical Conference, Las Vegas, Nevada, DETC99/DAC-8707, September 1999

# Review of Product Design and Manufacturing Methods for Prosumers

L. Asión-Suñer[(⊠)] and I. López-Forniés

Universidad de Zaragoza, 50018 Saragossa, Spain
lauraasion@gmail.com

**Abstract.** Product design is a process developed by specialized profiles such as designers or engineers. For this reason, most of product design and/or manufacturing methods developed are aimed at experienced professionals. However, in the current social and technological context, we observe that the number of *prosumers*, users who partially produce the products they consume, is increasing. These users are involved in design, manufacturing or assembly phases of the product to obtain final results that respond to specific needs and desires. Given this emerging trend, we wonder if there are methodologies focused on these users in particular, both in the academic field and in practice. This work aims to clarify this issue through a review of research papers and real cases. The results obtained differentiate between the methodologies that have been defined in the academic field and those that have not been compiled but whose repeated use has agreed its application and existence in practice. The methodologies identified are analyzed in two tables that summarize how they are applied and what their main objective is. The results and conclusions offer both the scientific community and the prosumers a series of product design and manufacturing methods focused on non-specialized profiles.

**Keywords:** Prosumer · Product design · Methodology · Manufacturing

## 1 Introduction

The multiple advances that have taken place in fields such as Information and Communication Technologies (ICT) or Agile Manufacturing [1], as well as the democratization of the creative process and its approach to non-specialized users [2], have resulted in an increase in *prosumer* users. These types of users, according to the original definition of Alvin Toffler, are those who are involved in the design, assembly or manufacturing process of the products they finally consume [3]. That is, they are consumers of what they produce.

This emerging trend, closely related to the maker culture, without being the same, has made increasingly users decide to design or manufacture their own products. However, its intervention is not restricted only to the previous phases of the product's life cycle. An example of this is *hacker* users, who also make modifications to make the product grow and obtain new functionalities that they did not have before [4]. For this reason,

© The Author(s) 2021
L. Roucoules et al. (Eds.): JCM 2020, LNME, pp. 128–134, 2021.
https://doi.org/10.1007/978-3-030-70566-4_21

the number of both physical and digital tools set aside for this purpose has increased exponentially in recent years [5, 6], causing many users to use them frequently unconsciously. These tools are not only intended for individual use, as we can also find tools such as consumer-based coopetition, crowdsourcing and value co-creation [7].

However, despite the number of tools available, there is a lack in the definition of design and manufacturing methodologies focused on prosumer or maker users. Although there are co-design methods that include the participation of experts and end users, their application is only carried out in the phases of product design and development and they are not adapted to each particular user, but they take into account the opinion global to implement improvements. In addition, it has been detected that, in actual practice, there are a number of consolidated methods that are frequently used by prosumers to intervene at different levels in the design or manufacture of their products [8]. For these reasons, this article aims to clarify which methods specifically focused on this type of users currently exist, both in the academic and practical fields.

## 2  Methodological Description

The main objective of this study is to collect and analyze the design and manufacturing product methods focused on prosumer users that have been developed until now. To achieve this, the study is divided into two parts: a bibliographic analysis and a field study. The bibliographic analysis shows works published in the academic field that deal directly with this type of methods, while the field study is based on real cases that prove the existence of consolidated methods frequently used by prosumers.

To perform the bibliographic analysis, a search was carried out in various databases (Web of Science, Scopus, Wiley Online Library, Science Direct) by combining two keywords: first, about prosumer user (*end user, co-design, maker, lead user, design manufacturer*); and, second, in reference to the design or manufacturing method (*method, methodology, tool, guideline + product design, manufacturing*). Due to different definitions of prosumer, the search was skewed, discarding the results related to *professional consumers* or *self-consumption*, especially in relation to the production of electricity. Methods that are focused only on specialized profiles (as designers) or co-creation processes where the users involved are not those who finally consume the product or service created, were also ruled out.

On the other hand, the field study is based on a previous work carried out within this line of research that compiles real cases of prosumer users [9]. This study shows various real cases in industrial fields such as furniture, textile and electronic, among others, where it is detected that we can define methods based on the use of common tools and processes between them.

Final results have been synthesized and compared in tables, that includes a description of each method, as well as its objective and main data. A total of 15 methods have been identified: 9 corresponding to the bibliographic search and 6 to the case analysis. Finally, this work offers a series of conclusions about the current design and manufacturing methods for the prosumer, including the opportunity to any new prosumer or maker to have enough information to join this trend.

## 3  Results

The results have been divided into two parts: *methodologies* and *real cases*. The basic information necessary to understand each methodology separately has been included in Table 1 (bibliographic analysis) and Table 2 (real cases). In the field study we find several examples for each methodology. However, only one representative case has been cited in each one in order to serve as an example of application to facilitate its understanding to the reader.

To facilitate the visualization and comparison of academic methodologies, these are grouped and summarized in the same table that shows the name of each methodology, the author and year of creation, a brief description and what is its main objective: Product Design (PD), Manufacturing (M) or Assembly (A).

**Table 1.**  Summary of academic methodologies.

| Name | Author and year | Description | Objective |
|------|-----------------|-------------|-----------|
| Method to work with end users to create personalized products by CAD modelling | Campbell and Bernabei, 2017 [10] | To increase product emotional attachment through custom design of additively manufactured products | PD + M |
| Lead-User Method | Franke, Von Hippel, and Schreier, 2006 [11] | Theory upgrade to exploit the value of user innovations | PD + M |
| User-centered translation method | Gardan, 2017 [12] | It includes the user in the design process by basing on their perception of the product | PD |
| Guidelines for Finding Lead User for Latent Need Discovery | Hölttä-Otto and Raviselvam, 2016 [13] | To find Lead-Users to express latent needs that are not found in regular users | PD |
| Mixed reality tool for end-users participation in early design tasks | Maurya, Arai, Moriya, Arrighi, and Mougenot, 2019 [14] | It allows end-users to be immersed in a virtual environment, to interact with a virtual prototype and to modify it | PD |
| New Service Development Method for Prosumer Environments | Alcarria, R., Robles, T., Dominques, A.M., Conzales-Miranda, 2012 [15] | To develop prosumer services by providing creation tools, used by prosumers to create final services | PD (Service Design) |

*(continued)*

**Table 1.** (*continued*)

| Name | Author and year | Description | Objective |
|---|---|---|---|
| Method to optimize assemblability of industrial product in early design phase | Favi and Germani, 2012 [16] | To create adaptable and customizable products by improving manual assembly through modular design | PD + A |
| Using elderly as lead users for universal engineering design | Raviselvam, Noonan, and Hölttä-Otto, 2014 [17] | The main idea is that, the elderly may be able to articulate more needs compared to general population | PD + M |
| Design Evaluation and Assessment System | Morita, 2007 [18] | It allows makers (designers, technicians), providers and end-users to evaluate products using the same criteria | PD |

In the following table (Table 2), a series of methodologies based on the study and analysis of real cases are cited. Each methodology is based primarily on the use of a digital or physical tool that allows users to intervene in the design or manufacturing phase of the products that they subsequently consume.

**Table 2.** Summary of methodologies based on real cases.

| Name | Case | Description | Objective |
|---|---|---|---|
| Online Platform to interchange ideas | Ikea Hackers [19] | Users contribute new ideas and modifications on designs that they then share and manufacture | PD + M |
| Design Software, 3D and CAD software | Vectary [20] | It allows to visualize and materialize a design by using tools such as 3D modeling | PD + M |
| Tutorials and instructions | Instructables [21] | These are steps to copy, manufacture and assemble the product, although the user can also influence its design including modifications | M + A |

(*continued*)

**Table 2.** (*continued*)

| Name | Case | Description | Objective |
|------|------|-------------|-----------|
| Open Source, Cloud and Co-design | Ordermade WholeGarment [22] | Free access tools where users share files on which other users can intervene (co-design) | PD |
| Vote System and crowdfunding | LEGO Ideas [23] | Collective participations to take design or market decisions about a product before its manufacture and commercialization | PD |
| Observation, analysis, trial and error | Maker Faire [24] | Extended method in the maker culture based on observing and analyzing a product, to modify it by trial and error | PD + M + A |

## 4 Conclusions

The bibliographic analysis shows nine methods focused on the prosumer with different objectives: to increase the attachment of the product, to exploit the value of users' innovations, to take into account the user's perception, to discover latent needs, to improve the prototype, to evaluate the product and to create adaptable and customizable results. Most of them look for product improvement to obtain a greater final acceptance in the market, only two take into account the real interests of the prosumer: the search for latent needs and the creation of adaptable and customizable products. Furthermore, the nine methods focus on product design and only four of them also focus on manufacturing (3) and assembly (1).

The analysis of cases has allowed to identify a series of methods based on tools because, although the prosumers and makers do not recognize the use of any specific method, the use of tools among them is highly agreed. Therefore, six methodologies can be clearly defined: exchange of ideas on online platforms; 3D and CAD design software; tutorials and instructions; co-design and open source resources; voting and crowdfunding systems; and, finally, the process based on observation, analysis, trial and error. On the other hand, the academic methodologies have not yet been sufficiently extended in practice. However, its evolution and development highlights the growing academic interest in prosumer users and their involvement in product design and manufacturing.

Regarding the main objective of the methods analyzed, we can see that most are focused on product design (14/15) and half of them on manufacturing (7/15), while only 3 are also focused on assembly. This is due to the widespread presence of digital tools before physical ones, a fact that can be verified in the methods analyzed in Table 2. In addition, despite the fact that in most cases prosumers are able to intervene in the phase of product design, only half of them also take care of its manufacture, since in the rest of the cases it is a third person who takes care of it.

About the differences between academic and real cases methods, we find that the objectives of the academic methods don't take into account the real interests of prosumers, who look for unique products made by themselves that respond to their wishes and needs, as has checked in the field study. Furthermore, academic methods only focus on product design, while the cases study shows that users are looking to get involved in assembly and manufacturing as well. This shows that what the user wants (field study) is not the same as what is offered (bibliographic analysis), giving rise to a niche in this field of research.

Given the need to include prosumers in the process of creating their own products by focusing on their true interests and taking into account not only design, but also assembly and manufacturing, finally two potential lines of future research are identified. Firstly, the development of a method specifically aimed at new non-specialized prosumer or maker users who are not yet familiar with the practice. And, secondly, a methodological evolution that includes new design tools that can enrich and facilitate user intervention, such as modular design.

# References

1. Halassi, S., Semeijn, J., Kiratli, N.: From consumer to prosumer: a supply chain revolution in 3D printing. Int. J. Phys. Distrib. Logistics Manag. **49**(2), 200–216 (2019)
2. Val Fiel, M.: Prosumer e impresión 3D: La democratización del proceso creativo / Prosumer and 3D printing: democratization of the creative process. Revista 180. Arquitectura, Arte, Diseño **37**, 17–22 (2016)
3. Toffler, A.: The Third Wave, pp. 282–306. Bantam books, New York (1980)
4. Knott, S.: Design in the age of prosumption: the craft of design after the object. Des. Cult. **5**(1), 45–67 (2013)
5. Ertekin, Y., Husanu, I.N.C., Chiou, R., Konstantinos, J.: Interdisciplinary senior design project to develop a teaching tool: Dragon conductive 3D printer. In: ASEE Annual Conference and Exposition, Conference Proceedings (2014)
6. Jeong, Y., Kim, H.J., Cho, H., Nam, T.J.M.: Integrator: a maker's tool for integrating kinetic mechanisms and sensors. Int. J. Interactive Des. Manuf. (2019)
7. Rusko, R.: The redefined role of consumer as a prosumer: value co-creation, coopetition, and crowdsourcing of information goods. Small Medium Enterprises: Concepts, Methodol. Tools Appl. **4**, 1980–1992 (2013)
8. Anzalone, G.C., Wijnen, B., Pearce, J.M.: Multi-material additive and subtractive prosumer digital fabrication with a free and open-source convertible delta RepRap 3-D printer. Rapid Prototyping J. **21**(5), 506–519 (2015)
9. Asión-Suñer, L., López-Forniés, I.: Prosumer and product design through digital tools. In: International Conference on the Digital Transformation in the Graphic Engineering, pp. 506–519. Springer (2019)
10. Campbell, R.I., Bernabei, R.: Increasing product attachment through personalised design of additively manufactured products. In: Proceedings of the International Conference on Engineering Design, ICED, vol. 5, no. DS87-5, pp. 71–79 (2017)
11. Franke, N., Von Hippel, E., Schreier, M.: Finding commercially attractive user innovations: a test of lead-user theory. J. Prod. Innov. Manag. **23**(4), 301–315 (2006)
12. Gardan, J.: Definition of users' requirements in the customized product design through a user-centered translation method. Int. J. Interact. Des. Manuf. **11**(4), 813–821 (2017)

13. Hölttä-Otto, K., Raviselvam, S.: Guidelines for finding Lead user like behavior for latent need discovery. In: Proceedings of Nord Design, vol. 2 (2016)
14. Maurya, S., Arai, K., Moriya, K., Arrighi, P.A., Mougenot, C.: A mixed reality tool for end-users participation in early creative design tasks. Int. J. Interact. Des. Manuf. **13**(1), 163–182 (2019)
15. Alcarria, S., Robles, R., Dominques, T., Conzales-Miranda, A.M.: New service development method for prosumer environments. In: ICDS 2012: The Sixth International Conference on Digital Society, vol. 7, January 2012
16. Favi, C., Germani, M.: A method to optimize assemblability of industrial product in early design phase: from product architecture to assembly sequence. Int. J. Interact. Des. Manuf. **6**(3), 155–169 (2012)
17. Raviselvam, S., Noonan, M., Hölttä-Otto, K.: Using elderly as lead users for universal engineering design. Assistive Technol. Res. Ser. **35**, 366–375 (2014)
18. Morita, Y.: Research on a design evaluation and assessment system from the perspective of the relationship between universal design and good design. In: International Conference on Mechanical Engineering and Mechanics, pp. 40–47 (2007)
19. Ikea Hackers. https://www.ikeahackers.net/. Accessed 14 Jan 2020
20. Vectary. https://www.vectary.com/features/. Accessed 14 Jan 2020
21. Instructables. https://www.instructables.com/howto/. Accessed 14 Jan 2020
22. Peterson, J., Larsson, J., Mujanovic, M., Mattila, H.: Mass customisation of flat knitted fashion products: simulation of the co-design process. Autex Res. J. **11**(1), 6–13 (2011)
23. Lego Ideas. https://ideas.lego.com. Accessed 14 Jan 2020
24. Maker Faire. https://makerfaire.com. Accessed 14 Jan 2020

# State of the Art for Evaluation of Inventive Design Solution Concepts

Muhammad Irshad Yehya[1]([⊠]), Remy Houssin[2], Amadou Coulibaly[1], and Hicham Chibane[1]

[1] ICUBE Laboratory-CSIP Team, INSA of Strasbourg, Strasbourg, France
muhammad.irshad_yehya@insa-strasbourg.fr
[2] ICUBE Laboratory-CSIP Team, University of Strasbourg, Strasbourg, France

**Abstract.** The concepts of solutions resulting from the inventive design process measures are generally described in a declarative manner, which does not allow having a shareable formal or visual representation between partners of the project. In addition, the absence of a model does not allow evaluation and compare competing concepts. In this paper, most relevant work regarding inventive design solution concept presented along with potential merits and demerits and highlighted the need of a systematic method to evaluate behavioral performance of solution concepts.

**Keywords:** Solution concept · Evaluation model of solution concepts · Inventive design · TRIZ

## 1 Introduction

In today's fast growing technological world companies, inventors and R&D facing with new challenges regarding its capacity to quickly and accurately respond to society's development demands or problems. To cope with the acceleration of new artefacts' demand, innovative activities need to be more assured in terms of optimal solution with respect to many aspects (functionality, maintainability, reliability, security, cost, quality, etc.) [1].

Ideas are defined in two ways i.e. useful or not useful. New ideas which are useful, constructive and profitable or solve a problem are referred as creative ideas while only fancy, not realistic and not useful ideas called mistakes [2]. With reference to the inventive design techniques like TRIZ (theory of inventive problem solving) where ideas are defined more properly than conventional terms of creative activities. After going through properly defined process, creative idea is called "Solution concept" [3]. A solution concept should be taken as an idea which has its all necessary information or characteristics most probably documented in sufficient details for an artefact (concrete way with a description template) so that an objective estimation becomes possible. In inventive design when ideas are characterized in a concrete way with description template, the concepts called Solution Concepts.

L. Roucoules et al. (Eds.): JCM 2020, LNME, pp. 135–142, 2021.
https://doi.org/10.1007/978-3-030-70566-4_22

Inventive design is the field which develops methodologies for problem solution regarding the development of innovative artefacts [4]. There are many methods created which are able to successfully carry out inventive design steps to produce solution concepts or ideas and solution concept ranking processes [5]. However has the drawback "inability to evaluate optimum solution concept". Herstatt highlighted the inability of routine design methods for creating inventive ideas [6].

## 2  Aim and Scope

The aim of this research is to identify, interpret and summaries the literature currently available on evaluation of inventive design solution concepts. Within the scope of this research, the focus has been on relevant research articles, and other authentic literature specifically for evaluation of solution concepts.

## 3  Inventive Design Solution Concepts

### 3.1  Inventive Design Definition

Several attempts have been made to define inventive design by both scientific and artistic terms. Inventive design is a thought-provoking engineering activity associated to all areas of human life. It follows the laws of science, fulfill the requirements for realization of a solution concept and needs professional integrity and responsibility. It is an attempt to possible realization of concepts by meeting particular constraints [2]. Inventive design is acquisition of knowledge and capitalization [7]. Inventiveness here is in terms of invention which means "the action of creating or designing (something that has not existed before), typically a process or device" [8]. T. Chikatham one of the authors [2] compared inventive design and routine design as given in Table 1: In Inventive Design, the contradicted features of a situation or problem by TRIZ tools is highlighted and these contradictions are then considered as a key source to solve problem [9].

**Table 1.**  Inventive design vs routine design [2]

| Routine design | Inventive design |
| --- | --- |
| Manage what is known | Discover what is unknown |
| Optimization of existing data for best result | Moving further ahead from the optimized result of existing data |
| Accept compromise as a potential solution | Refuse compromise as a possible solution |

## 3.2 Our Position in Inventive Design Process

The first article on TRIZ was by Genrich Altshuller in 1956 [10]. Its application has been observed in various fields like technic, management, sociology, education and marketing [11]. During 80s Altshuller work developed into a set of tools to carryout systematic creativity called TRIZ "Theory of inventive problem solving" and in 90s to OTSM "General theory of strong thinking" and ZhSTL "Lifetime strategy for creative person" [10]. The research progress in field of TRIZ as a result gave rise to organized data in order to understand TRIZ and its extension as IDM (Inventive Design Method).

The Inventive Design Method is the output of many years of research by INSA Strasbourg France [12]. With the intention to build an abstract model with initial problem formulation using the theoretical approach of TRIZ C. Zanni Merk [7] proposed an ontology for computer based innovation. In further progress of TRIZ, Francois Rousselot [13] proposed an important formal definition of contradiction and its possible applications in inventive design with respect to the TRIZ fundamentals. Following the previous work the extension of TRIZ called inventive design method IDM proposed and developed and framed into a computer software named STEPS by INSA Strasbourg [14]. Currently there are different TRIZ based computer-aided innovation (CAI) systems, which support design phase inventive design [15]. These CAI systems provide suitable tools for designer to generate solution concepts and their rankings on the basis of multi criteria decision methods but still lack in evaluation of generated solution concepts. Keeping all these in consideration our focus will be regarding need of a confident evaluation model along with specific criteria (quality, maintainability etc.) of solution concepts, which come from the previous steps of inventive design using different TRIZ tools and methods.

Inventive design can be subdivided into four main steps shown in Fig. 1: initial situation analysis, contradiction formulation, solution concept generation, and solution concept selection or evaluation of solution concepts [5].

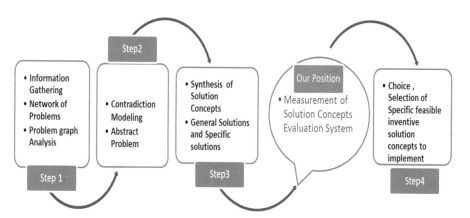

**Fig. 1.** Inventive design solution concepts evaluation status

The primary steps of inventive design development known by the term 'concept development'. Concept development contains a sequences of divergent steps and convergent steps, completed at different levels of solution concept [16]. In concept development, generation of solution concepts belongs to divergent steps while evaluation and selection of the best solution concepts are related to convergent steps.

### 3.3 Inventive Design Solution Concept Evaluation

A concept of any product is a predicted representation of the shape of product, its working standards and technology. It is a well explained definition of product characteristics that how the product will satisfy the voice of customer. Generally a concept is considered as a sketch or 3D model following a detail description. The success of any product depends on the selection of best concept. Success here means the commercialization of the quality product.

With reference to inventive design solution concepts, the arrival of TRIZ and its extension to IDM has been considered as a systematic methodology to produce set of solution concepts as compared to earlier unstructured methods [17]. As mentioned in the introduction that a solution concept should be describe in a concrete way with a description template. The description template for solution concept characterizes into five categories [13]. These five categories are: 1) an abstract of idea describing the general properties, all the relevant performing functions, extra note, merits/demerits. 2) A complete explanation of problem model (in which contradiction contain action parameter and evaluation parameter or SU-field model) and solution model (using tools like inventive principle, inventive standard, physical effect database). 3) Possible outcomes, trends and developments from the outcomes of related projects in term of hypothesizes and laws of technical systems evolution, 4) keeping the objective of the project point out all the inacceptable conditions. 5) A sketch of the solution concept, which is synthesized by model of solution and hypothesis of solution concept. After the set of solution concepts are identified and ranked using Pugh matrix by the inventive design methodology for the next step to select the solution concept to develop or solution concept evaluation.

The evaluation phase of solution concept is the key challenge for the designers, as well as the customers. Particularly in solution concept selection phase qualitative methods are used to evaluate the generated solution concepts. Although, to facilitate the inventive design solution concepts evaluation steps, there are many methodologies used by designers and customers [14, 19]. However, these methodologies are generally described in a qualitative, declarative manner, which does not allow to choose the best solution concept neither to have a shareable formal or visual representation between partners of the project. In inventive design the process stops after ranking of solution concepts and final selection of solution concept to develop depends on R&D or the top management of company. That is why, the absence of a confident model does not allow evaluation and compare competing concepts thereby making a challenge for researchers and designers to develop a confident model for evaluation of inventive design solution concepts.

### 3.4   Research Gaps in Inventive Design Solution Concept Evaluation

In this section, an overview of the most prominent research work regarding inventive design solution concept is given in the Table 2. The purpose of this research is to highlight the need for inventive design solution concepts evaluation model.

**Table 2.**  Literature review on inventive design solution concepts

| Title of paper | Focus area | Research gap | Ref. | Year |
|---|---|---|---|---|
| Improving Inventive Design Methodology's Agility | Focus on elimination of non-value added activities to reduce long time taking in inventive design solution concepts process | After solution concepts making the final selection depends on the R&D or decision makers of company, no semantic model available for evaluation of solution concepts | [20] | 2019 |
| Contribution to Formalizing Links Between Invention and Optimization in the Inventive Design Method | Prevention of rejection of good solution concepts at early phases of inventive design and ranking of solution concepts for decision makers to select | Need for quantitative evaluation method to give adequate confident to decision makers regarding selection of solution concept to develop | [2] | 2017 |
| An Approach to Identify the Readiness Level of a Solution Concept in the Inventive Design Method | Solution Concept Feasibility | Selection of Solution concept depends of expertise of decision maker | [3] | 2016 |
| On Solution Concept Evaluation/Selection in Inventive Design | Improve early decision processes for feasibility of solution concepts | Lack of evaluation or selection method for solution concepts | [14] | 2015 |
| A Software Framework to Support Engineering Analysis for Inventive Solution Concepts | Pre dimensioning and feasibility criteria of solution concept | No proper facility for solution concept selection | [21] | 2015 |
| Towards a formal definition of contradiction in inventive design | Formalization of contradictions to reach relevant solution concept synthesis stage | After Ranking of Solution concepts using Pugh Matrix, final selection depends on R&D or top management of company and no confident model for evaluation of solution concepts | [13] | 2012 |
| A research agenda for computing developments associated with innovation pipelines | Computer added innovation and future agenda proposal | Need innovative methods for concept selection and solution concept evaluation | [22] | 2011 |

*(continued)*

**Table 2.** (*continued*)

| Title of paper | Focus area | Research gap | Ref. | Year |
|---|---|---|---|---|
| Monitoring the Impact of Solution Concepts within a Given Problematic | Using Pugh's approach for qualitative evaluation of concepts of for producing solution concepts | After given list of solution concepts the method stops further input in selection of solution concepts | [23] | 2009 |

Not every method is used in every inventive design process. Only those that seem appropriate for the problem situation and that contribute to a successful outcome are used. Based on the most relevant and latest research articles considered in this state of the art, we clearly identify the research gap regarding evaluation of inventive design solution concepts. More clearly stating that there is not any certain tool, model etc., for evaluation of solution concepts in inventive design to suggest a specific solution in the solution concept building phase or solution evaluation phase as showing our position in Fig. 1.

## 4   Conclusion and Future Research

Instead of all the features included in the IDM there is still a gap in providing a model to the best selection of solution concept. The evaluation of solution concepts is no doubt the most important step of inventive design process because of its effect on all following steps in terms of performance, quality, maintainability, cost, safety, etc., of the selected solution concept for development. As it is obvious that failure of a selected solution concept for development can barely be compensated at next phases of advance design and development by resulting long time of redesign and rework expense. According to S.R Daly [24] almost 80% of the cost is committed at solution concept selection phase and figured out that 80% of a manufactured product's cost occur during the process of product design.

To date most of research in inventive design has been made in from the very beginning steps to solution concept ranking. One of the toughest, critical and complex problems in inventive design process is the evaluation of best solution concept to go for manufacture or implement [24]. Following the research gaps through this research, we are considering to propose a semantic modeling approach and method to evaluate behavioral performance of solution concepts in inventive design along with the proper definition of solution concept.

## References

1. Okudan, G.E., Tauhid, S.: Concept selection methods & ndash; a literature review from 1980 to 2008. Int. J. Des. Eng. **1**(3), 243 (2008)
2. Chinkatham, T.: Contribution of the formulation of the links between Invention - Optimization in Inventive conceptual design To cite this version (2017)

3. Chinkatham, T., Souili, A., Taheri, A., Cavallucci, D.: An approach to identify the readiness level of a solution concept in the inventive design method. Procedia CIRP **39**, 179–184 (2016)
4. Yan, W., Zanni-Merk, C., Rousselot, F.: Matching of different abstraction level knowledge sources: the case of inventive design. In: Lecture Notes Computer Science (including Subseries Lecture Notes Artificial Intelligence Lecture Notes Bioinformatics), vol. 6884, LNAI, no. PART 4, pp. 445–454 (2011)
5. Shah, J.J., Kulkarni, S.V., Vargas-Hernandez, N.: Evaluation of idea generation methods for conceptual design: effectiveness metrics and design of experiments. J. Mech. Des. Trans. ASME **122**(4), 377–384 (2000)
6. Herstatt, P.C.: The ' Fuzzy Front End' of innovation, no. 4, pp. 1–24 (2001)
7. Zanni-Merk, C., Cavallucci, D., Rousselot, F.: An ontological basis for computer aided innovation. Comput. Ind. **60**(8), 563–574 (2009)
8. Oxford Lexico Dictionary. https://www.lexico.com/en/definition/invention
9. Altshuller, G., Al′tov, G., Altov, H.: And Suddenly the Inventor Appeared: Triz, The Theory of Inventive Problem Solving. Technical Innovation Center, Inc, Worcester (1996)
10. Altshuller, G.S., Shapiro, P.B.: On the psychology of inventive creativity. Moscow (1956)
11. Ilevbare, I.M., Probert, D., Phaal, R.: A review of TRIZ, and its benefits and challenges in practice. Technovation **33**(2–3), 30–37 (2013)
12. Cavallucci, D., Strasbourg, I.: From TRIZ to inventive design method (IDM): towards a formalization of inventive practices in R & D departments, 2–3 (2012)
13. Rousselot, F., Zanni-Merk, C., Cavallucci, D.: Towards a formal definition of contradiction in inventive design. Comput. Ind. **63**(3), 231–242 (2012)
14. Chinkatham, T., Cavallucci, D.: On solution concept evaluation/selection in inventive design. Procedia Eng. **131**, 1073–1083 (2015)
15. Innvention-machine. http://invention-machine.com/factsheet_GoldfireInnovator.html
16. Backhouse, C.J.: A review of: 'Total Design.' By STUART PUGH. (Addison Wesley, 1991) [Pp. 278] Paperback, £16.95. Level: Designer/design lecturer. Int. J. Comput. Integr. Manuf. **4**(5), 321 (1991)
17. Altshuller, G.S.: To find an idea, Introduction to theory Invention Problem solving.- Novosibirsk. Nauk. Publ. (1986)
18. Chinkatham, T., Cavallucci, D.: Early feasibility evaluation of solution concepts in an inventive design method framework: approach and support tool. Comput. Ind. **67**, 1–16 (2015)
19. Stalnaker, R.: On the evaluation of solution concepts. Theory Decis. **37**(1), 49–73 (1994)
20. Hanifi, M., Chibane, H., Houssin, R., Cavallucci, D.: Improving inventive design methodology's agility. In: International TRIZ Future Conference, pp. 216–227 (2019)
21. Chinkatham, T., Cavallucci, D., Knittel, D.: A software framework to support engineering analysis for inventive solution concepts. Procedia Eng. **131**, 626–634 (2015)
22. Cavallucci, D.: A research agenda for computing developments associated with innovation pipelines. Comput. Ind. **62**(4), 377–383 (2011)
23. Cavallucci, D., Rousselot, F., Zanni, C.: Monitoring the impact of solution concepts within a given problematic. IFIP Adv. Inf. Commun. Technol. **304**, 1–8 (2009)
24. Daly, S.R., Yilmaz, S., Christian, J.L., Seifert, C.M., Gonzalez, R.: Design heuristics in engineering. J. Eng. Educ. **101**(4), 601–629 (2012)

# Systemic Approach for Local Energy Mix Assessment

Nathanael Dougier, Pierre Garambois[✉], Julien Gomand, and Lionel Roucoules

Arts et Metiers Institute of Technology, LISPEN, HESAM Université, 13617 Aix-en-Provence, France
pierre.garambois@ensam.eu

**Abstract.** Whereas energy mainly comes from main national power plants, distributed energy resources and storage technologies would allow local territories to choose their energy sources and increase their autonomy. This paper presents a decision-support tool that propose to find new system architecture based compromises between economic, technical and environmental objectives. Based on a systemic approach, it takes into account a broad range of technologies and assesses multi-scale territories thanks to a physical modelling. Numerical simulations show the influence of different parameters on the ability of a system to balance power demand.

**Keywords:** Multi-energy system · Microgrids · Systems modelling · Optimization · Decision support

## 1 Introduction

Initially independent from a location to another, energy production has been aggregated and huge networks were developed in the previous century. This has increased the ability to balance energy production between regions/countries and thus improved the supply reliability. Nevertheless this centralisation has also caused a dependence for local territories to the national/continental production. Also, a majority of the primary energy still comes from fossil resources (oil, coal and gas) that needs to be imported by many countries [1]. Therefore, local territories do not control the origin or the management of their energy. On the opposite, the recent increase in distributed energy resources (DER), like wind turbines or photovoltaic (PV) panels, and storage facilities could allow a local management of energy. Thus, it is possible to develop a local network that can be managed independently or connected to the main network, also called a microgrid.

The subject of the work presented here is the development of local energy systems, focusing on the electricity vector. A local energy system is defined in this paper as a set of energy production, storage and transport technologies installed in the local territory (from a building to a region), with known control strategies. The chosen method consists in modelling and simulating systems in order to compute their performance indicators, defined here-after, and thus compare them.

© The Author(s) 2021
L. Roucoules et al. (Eds.): JCM 2020, LNME, pp. 143–148, 2021.
https://doi.org/10.1007/978-3-030-70566-4_23

Currently, the development of DER technologies serves the national interest and lacks a local strategy. The point of view of this work consists in finding a satisfying combination of technologies that balance the consumption and the production at almost any time – according to the level of autonomy from the main grid granted – while responding to today's economic, environmental and social challenges. Nevertheless, studies on this topic mainly focus on economic performance to choose the best system [2, 3]. When several indicators are taken into account, they are usually pre-weighted so only a single solution is proposed [4, 5]. However, due to the coexistence of antagonist indicators, there is no best combination but only compromises between the different indicators.

Comparing a huge number of energy systems and get an accurate description of the physics of the energy generation and transport requires much computation time. A trade-off is usually made between the range of energy systems considered and the accuracy of the modelling. Two kinds of studies on energy systems can be defined: studies on control strategies considering a fixed system and analysing its dynamic behaviour [6, 7], and studies on the system design comparing multiple energy systems for a given consumption but with less accuracy in technologies' modelling [8]. At a large scale (region, state), it seems easier to focus on the main power exchanges and to aggregate the detailed parameters to get a correct approximation. However, at a small scale (group of buildings), where every technology/strategy/transport choice is significant, technological variables take a greater importance. The multi-scale challenge this work intends to answer is to optimise quickly a large number of energy systems and keep at the same time a detailed physical modelling. In order to link local parameters to global performances and underline the systemic consequences of each technological choices, various models are developed and related to fit the simulation and optimization goal.

This paper presents a tool using a systemic approach to support decision-making and raise awareness about the complementarity of energy sources and the good use of storage at local scale. The purpose is to propose several optimums among energy mix, trade-offs between different economic, environmental and technical objective-functions, without prior weighting. To be adapted to multi-scale territories (from building to region size), the analysis offers the most exhaustive range of potential energy systems (production/storage technologies and control strategies) while keeping an accurate technological and physical modelling.

The paper is organised as follows. Section 2 describes the chosen methodology. Section 3 presents the case study. Eventually, the results are discussed in Sect. 4 and perspectives are underlined.

## 2   Methodology

Considering the multi-scale challenge, different levels of modelling are needed from an energy system point of view, allowing a quick resolution, to a detailed modelling of each element. This study considers an energy system as a combination of four main components, as shown on Fig. 1: technologies, control strategies, demand and losses. Each of these is developed in several modelling levels going from a global pattern to detailed models close to technological parameters. Models are represented by intercon-nected blocs. The simulation and the optimisation of energy systems can use different levels in order to answer specific questions. The advantages of this systemic approach are its modularity (one bloc can be easily replaced by another), its ability to show the influence of a technological parameter on the whole energy system and thus to analyse the physical and technological frontiers of energy systems. The tool designed can adapt to all situations.

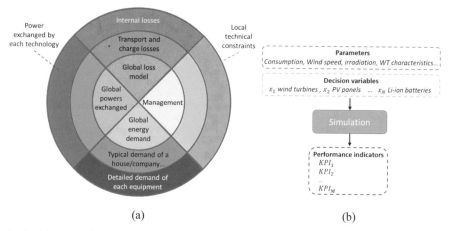

(a)

(b)

**Fig. 1. (a)** Basic elements of an energy system (core) and the different modelling layers; **(b)** Simulation inputs and outputs

## 2.1 Simulation

As explained above, Fig. 1. (b), decision support is based on economic, environmental and technical indicators. Decision variables are the number of element of each technology or more technological parameters according to the scale of the study. A simulation is used to determine indicators' values. The latter computes the powers produced, stored and consumed by each technology and the state of energy of all storage systems at each time step based on parameters, such as consumption power curve, meteorological data (wind speed, solar irradiation…) and technologies' characteristics. Time steps can range from minutes to days: without considering the dynamic behaviour of the system, it allows an accurate description of phenomena.

## 2.2 Control Strategies

An example of technological parameter that can be modified with the systemic approach is the control strategy. A control strategy is a set of rules describing the operation of an energy system, i.e. which power should be generated by each producer/storage system at every time. Even though this works isn't about the analysis of control strategies, they have to be considered due to their influence on the efficiency of an energy system.

In this work, all control strategies are based on the principle of a priority order. The power balance over a time step is sequential: each production system is called following a defined order until the consumption and sometimes the storage capacity are balanced or the production capacity is reached. Most studies consider a fixed priority order, often sorting technologies by increasing marginal cost [8]. For each power plant, a strategy defines the power that has to be produced/stored considering the technical parameters and the power balance. Whereas many articles analyse a case study with a fixed strategy, it is here possible to compare a same combination of technologies with different strategies.

The presented tool thus allows to consider the priority order and the control strategies as new decision variables of the optimization.

## 3 Case Study

The purpose of this case study is to show the ability of the systemic approach to underline the influence of technological parameters – here the presence of storage devices, the priority order and the control strategy – on global performance indicators, here the resources consumption.

Technologies considered are one 2 MW-wind turbine, 20000 m$^2$ of PV panels, a run-of-river dam (r-dam), a 1 MW-biomass power plant, a 1 MW-gas power plant, a 2 MWh-battery and 10 MWh-pumped hydroelectric energy storage (PHES). The chosen control strategies presented here are: technology can generate power to balance consumption and charge storage devices (Strategy 1) and technology can only generate power to balance consumption (Strategy 2).

Figure 2 represents the cumulated power produced/stored along with the power demand of 6000 typical homes – the size of a small town – isolated from the main grid. It shows that the production sources alone (a) does not balance the consumption at all time steps – (21:00; 00:00) period. Adding storage devices (b) and changing the priority order allows to balance the consumption all the time with time steps when production is above consumption to charge the storage system and others when the storage systems discharge. The operation of the same set technologies with different control strategies (c) is then presented. Changing the control strategies modifies the sollicitation of the storage devices, ends the renewable energies' restrictions and therefore reduces the gas and biomass consumption. Indeed, approximately 5.3 tons of resources (gas and biomass) are consumed in situation (a), 6.4 tons in (b) and 3.6 tons in (c) and greenhouse gases emissions are respectively 99, 121 and 69 geqCO2/kWh.

Therefore, the designed tool's ability to assess the global impact of technical parameters is proven. The choice of an energy system then relies on the worth given to each indicator.

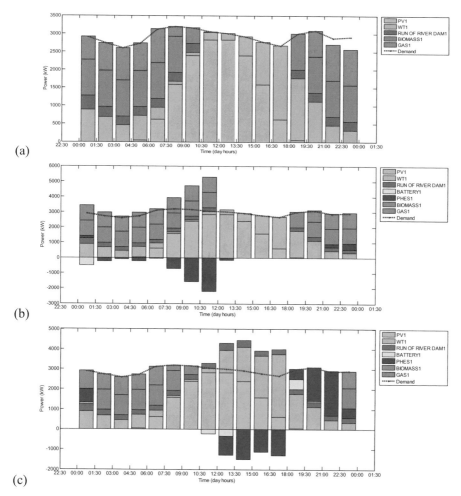

**Fig. 2.** Cumulated power (kW) produced/stored along with the power demand of 6000 typical homes isolated from the main grid: (**a**) System with production technologies only; (**b**) System with producers and then storage technologies; (**c**) System with storage technologies before controllable producers and a change in control strategy;

## 4 Conclusion

In order to support the choice of local energy systems, this article presents a systemic approach based on physical modelling. The developed tool proposes compromises between antagonist objectives representing environmental, economic and technical concerns. Systems ability to balance energy demand over one day have been tested through several simulations. Moreover, results underline the flexibility of the tool and its capacity to assess technical parameters usually considered as fixed – e.g. technologies and management strategies. However, comparing manually energy systems takes time and does not ensure to find all the systems able to balance the demand. To support decision-making, the goal of the future work will be to automatize the procedure in order to first sort the energy systems corresponding to the local needs and then to compare their performances, with appropriate indicators.

**Acknowledgments.** The research work reported here was made possible by the Region Provence-Alpes-Côte d'Azur (France).

# References

1. Key World Energy Statistics 2016 (2016)
2. Nikmehr, N., Ravadanegh, S.N.: Optimal power dispatch of multi-microgrids at future smart distribution grids. IEEE Trans. Smart Grid **6**(4), 1648–1657 (2015). https://doi.org/10.1109/tsg.2015.2396992
3. Hernandez-Aramburo, C.A., Green, T.C.: Fuel consumption minimization of a microgrid. In: 39th IAS Annual Meeting in Conference Record of the 2004 IEEE Industry Applications Conference, 2004, vol. 3, pp. 2063–2068, October 2004. https://doi.org/10.1109/ias.2004.1348751
4. Carpinelli, G., Mottola, F., Proto, D., Russo, A.: A multi-objective approach for microgrid scheduling. IEEE Trans. Smart Grid **8**(5), 2109–2118 (2017). https://doi.org/10.1109/tsg.2016.2516256
5. Ross, M., Abbey, C., Bouffard, F., Jos, G.: Multiobjective optimization dispatch for microgrids with a high penetration of renewable generation. IEEE Trans. Sustain. Energy **6**(4), 1306–1314 (2015). https://doi.org/10.1109/tste.2015.2428676
6. Gergaud, O.: Modélisation énergétique et optimisation économique d'un système de production éolien et photovoltaïque couplé au réseau et associé à un accumulateur, École normale supérieure de Cachan - ENS Cachan (2002)
7. Gabler, H., Luther, J.: Wind-solar hybrid electrical supply systems. Results from a simulation model and optimization with respect to energy pay back time. Solar Wind Technol. **5**(3), 239–247 (1988). https://doi.org/10.1016/0741-983x(88)90021-5
8. Assoumou, E.: Modélisation MARKAL pour la planification énergétique long terme dans le contexte français, École Nationale Supérieure des Mines de Paris (2006)

# Towards a Better Understanding of Open Source Hardware Design Reuse in Company-Community Collaboration

Amer Ezoji[1(✉)], Romain Pinquie[2], and Jean François Boujut[2]

[1] Univ. Grenoble Alpes, CNRS, Grenoble INP (Institute of Engineering Univ. Grenoble Alpes),
G-SCOP, 38000 Grenoble, France
`amer.ezoji@grenoble-inp.fr`
[2] Univ. Grenoble Alpes, CNRS, Grenoble INP, G-SCOP, 38000 Grenoble, France

**Abstract.** The open-source-software movement that emerged in the late 90s has recently extended to hardware. In this paper, we try to better understand how the reuse of design solutions facilitates company-community collaboration. On the one hand, based on existing research studies, we analyze three fundamental questions - 1) who wants to reuse a design? 2) Why do they want to reuse a design?, and 3) How do they reuse a design? - from a company perspective and a community perspective. On the other hand, we identified that companies and communities must create a common understanding of the design problems and solution and they could benefit in reusing design artifact to speed up the development time and improve the quality and transferability of the results. However, this research shows that companies and communities don't use the same type of tools and methods to reuse design knowledge which may cause some problems for collaboration [2].

**Keywords:** Open design · Open source hardware · Design reuse · Design collaboration

## 1 Introduction

Open source refers to a product development movement where the source of the product (software, hardware, or hybrid) is publicly accessible so as to create derived works. The open source movement has its origin in the software community. Open source software is a term for software published under licenses that do not give any private intellectual property rights to the developers. The source code is publicly and freely accessible so that anyone can reuse it with or without prior modification [3]. The transfer of open source principles from software development to the world of physical objects gave birth to the phenomenon of open source hardware (OSH). OSH is piece of hardware whose design is made publicly available so that anyone can study, modify, distribute, make, and sell the design or hardware [5]. OSH falls in the category of open source innovation. The development of OSH relies on open design. Open design is "the openness of all accompanying documents in a product development process, with the aim of collaborative development of tangible objects" [4]. One major benefit of open design is

L. Roucoules et al. (Eds.): JCM 2020, LNME, pp. 149–154, 2021.
https://doi.org/10.1007/978-3-030-70566-4_24

to provide designers with the ability to reuse solutions in an open collaboration[1]. Design reuse which is an activity that consists in using an existing design, in whole or in part, to fulfill a new set of requirements is a critical issue for the development of OSH. So, the goal of this research is to identify the main issues of design reuse practices in the company and community and provide significant conditions to facilitate design reuse in company-community collaboration.

## 2   Design Reuse in Company and Community

In this section, we review the current methods and technologies supporting design reuse in two contexts. First, we consider the work of design reuse in business. Second, we focus on the design reuse in OSH communities.

### 2.1   Design Reuse in Companies

To better understand the existing design reuse practices in a company, we analyze "who" reuse a design artifact, "why" and "how" do they reuse it.

- **Who in a company want to reuse the design?**

In a company the main stakeholder for design reuse is a design engineer. Design engineers use up to 70% of paste design knowledge to solve new design problems [16] while minimizing resources and efforts.

- **Why a company reuses design?**

Given the competitive pressures in business, design reuse enables companies to reduce costs and time-to-market while maximizing customer satisfaction [16] by finishing the design task in a shorter time, reducing of delivery times, shortening the testing phase and improving the quality [13, 15].

- **How does a company facilitate the reuse of design?**

Companies use a mix of techniques to facilitate design reuse. The first technique aims at using a design rationale to capture the reasons and arguments leading to a design decision [7, 11]. The second technique consists in organizing the design data and using data structures such as ontologies [14]. The third technique corresponds to careful management of data using information systems such as PDM/PLM software [9]. When such database-based information systems fail to retrieve existing designs, information retrieval techniques such as search engines can facilitate design reuse. In computer-aided design environment, the use of data exchange formats, the storage of models in well-structured libraries, and the capitalization with templates facilitate design reuse. Design methods, especially modular product architecture approach may, also foster design reuse [12].

---

[1] Open collaboration is "any system of innovation or production that relies on goal-oriented yet loosely coordinated participants who interact to create a product (or service) of economic value, which they make available to contributors and noncontributory alike" (Levine, S. 2014).

## 2.2 Design Reuse in Communities

Design reuse in OSH communities is hardly documented. Thus, to better understand the characteristics of reuse in open design, we rely on:

- **Who in a community want to reuse the design?**

In the previous project *"OPEN - Methods and tools for community-based product development"*, a qualitative research approach aimed at deriving personas. Personas are ideal types that represent a category of user. A persona is described in all its dimensions (i.e. social, family, occupation, etc.) and can be integrated in user stories or scenarios. The qualitative analysis of 23 interviews led to 5 personas: 1) The participant in open source development [5] (OSD) projects (A) 2) The inventive creator (B) 3) The entrepreneur and open source process [6] (OSP) founder (C) 4) Community manager & Visionary (D) 5) The maker & Geek (E)

|  |  | A | B | C | D | E |
|---|---|---|---|---|---|---|
| Motivation to create | Proble solving |  | X |  |  |  |
|  | Empowerment through technology |  |  | X | X |  |
|  | Collective challenge | X | X | X |  | X |
|  | Ideological |  |  | X | X |  |
|  | Fun | X |  |  |  | X |
|  | Teaching & Learning | X |  |  | X | X |
| Motivation for openness | Sharing | X |  | X |  | X |
|  | Innovation |  | X | X | X | X |
|  | Access to expertise | X | X |  | X | X |
|  | Social impact | X | X | X | X |  |
| Vision | Fun | X | X |  |  | X |
|  | Educate | X |  |  | X | X |
|  | Make money |  |  |  |  |  |
|  | Have an impact |  | X | X | X |  |
|  | Create own product | X | X | X |  | X |
|  | Create an alternative system |  |  | X | X |  |

**Fig. 1.** Qualitative analysis of 23 interviews in community for design reuse

Among the 5 personas, the Maker & Geek, the participant in OSD projects and potentially the inventive creator are the main stakeholders for design reuse. However, these personas may have potentially different needs in terms of reuse and we need to characterize them.

- **Why a community should reuse designs?**

A member who is involved in a community-based and open source product development process shares some fundamental motivations with a company since, when reusing a design, he is willing to reduce the design time and costs, to cover a lack of expertise, or to avoid design errors, that is, choices that make certain designs "not allowed" or inappropriate for their intended use. However, open design breathes life into product design since individual objectives - e.g. personal development and process learning [2], fame, or political and ethical actions - drive community members [7]. In the previous project OPEN, the analysis of the 23 interviews concluded that the motivations for being active in a community-based OSH are development project were:

- CAT 1: problem solving, empowerment through technology, collective challenge, ideological, fun, teaching and learning.

- CAT 2: sharing, innovation, access to expertise, social impact.
- CAT 3: have fun, educate, make money, have in impact, create our own product, and create an alternative system.

- **How does a community facilitate the reuse of design?**

The strong influence of ICT on OSH development has brought new mechanisms that facilitate design reuse. For instance, web-based repositories such as Wikifactory, Hac kster or Hackaday that support the development of OSH provide anyone with a free and open access to the design artifacts protected by open source designs. Such platforms also integrate some basic but essential capabilities supporting the three functional spaces of collaboration [6]:

- *Communication* with social features and online documentation (stories, collections, activity and news feeds, forums, chats, wiki, etc.),
- *Cooperation* with configuration management features (concurrent text editing, issue tracking, version control, etc.)
- *Coordination* with project management capabilities (markdown task lists, notifications, milestones, task allocation through workflows, etc.)

The recent adoption of linked data (semantic web) also facilitates the reuse of designs because the retrieval and navigation across existing related OSH projects on the Web becomes easier than a few years ago. For instance, the Open Source Hardware Observatory is a semantic search engine that crawls the web and makes its content searchable. However, the reuse of the design artifacts required to recreate or modify OSH is often challenging because most design knowledge (assumptions, rules, choices…) remains tacit knowledge (e.g. commonsense) or unrecognized knowledge (e.g. expertise and skill). When one accesses the repository of an open source project, the quantity of recognized and codified knowledge in models is relatively limited. Indeed, in most cases, the repository only includes a snapshot of the solution at a given time, that is, a set of models, codes, […], and documents without the underlying design rationales that conducted to the actual solution. The lack of organization and traceability in the design process of OSH makes difficult the reuse of models and the collaboration between companies and communities.

# 3   Design Reuse in Company-Community Collaboration

The necessary conditions for a collaborative design are: (1) have a common vocabulary/understanding about the design content and design process (communication), (2) have a common goal/aim, (3) use compatible technologies/tools, and (4) be engage in knowledge creation by sharing of design knowledge and integration [1, 8]. In the following sub-sections, we identify the enabling design reuse in company-community collaboration according to the key conditions for collaborative design and comparison between Sects. 2.1 and 2.2.

- **Who in company-community collaboration could reuse design?**

Company-community collaboration requires stakeholders to share a common understanding of the design and shall consequently an engineering background. Therefore, the actors who are prone to reuse a design are potentially design engineers from company side and makers/geek from community side.

- **Why a company-community member could reuses design?**

Companies and communities could reuse design artifacts because they share common goals. Common goals include, but are not limited to, the learning of new skills - either to improve the current situation or to compensate for a lack of expertise -, the motivation to avoid design errors and rework iterations, the need for boosting creativity, and - for some makers - the willingness to develop new business opportunities by limiting costs, time-to-market, and risks.

- **How a company-community collaboration member could facilitate the reuse of design?**

Design reuse for the development of OSH in company-community collaboration is neither limited by the technical skills of collaborators nor by the lack of shared goals. Unfortunately, from a technological point of view, companies and communities do not use the same type of practices and software to capture and reuse design knowledge. Although both actors use ICT to manage their design artifacts (models, code, videos, photos, etc.), communities use open social platform that do not support a structured industrial processes, such as configuration management, traceability, change management, or validation and verification. Consequently, to collaboratively develop OSH, companies and communities need new knowledge capture and reuse technologies blending the openness and flexibility of open source community-based product development with the quality and traceability of technologies used in industry.

## 4 Conclusions

Design reuse, which is a hot research topic in industrial product design, is also a core activity in the development of community-based OSH. In this paper, we try to better understand how design reuse can facilitate the development of OSH in company-community collaboration. First, our analysis shows that companies and communities include actors - makers and design engineers - who have the engineering background require to collaborate. Second, companies and communities share common goal for design reuse including the minimization of costs and time, but makers have different aim, such as personal development, improvement and personalization of a working tool, collective challenge, or simply have fun. Third, both entities use ICT to support their collaborative design activities; communities prefer to use web-based technologies for their openness and flexibility whereas companies adopt heavy information systems because they can meet their needs. Future work will focus on the specification and development of new methods and tools that could guide and support designers of OSH in the reuse of existing designs among and across the communities.

# References

1. Achten, H.H.: Requirements for collaborative design in architecture (2002)
2. Aksulu, A., Wade, M.R.: A comprehensive review and synthesis of open source research. J. Assoc. Inf. Syst. **11**(11), 6 (2010)
3. Balka, K.: Open Source Product Development: The Meaning and Relevance of Openness. Springer Science & Business Media, Berlin (2011)
4. Bonvoisin, J., Boujut, J.F.: Open design platforms for open source product development: current state and requirements. In: DS 80–8 Proceedings of the 20th International Conference on Engineering Design: Innovation and Creativity, Milan, Italy, vol. 8, pp. 011–020 (2015)
5. Bonvoisin, J., Buchert, T., Preidel, M., Stark, R.G.: How participative is open source hardware? insights from online repository mining. Des. Sci. **4** (2018)
6. Brissaud, D., Garro, O.: Conception distribuée, émergence. In: Tollenaere, M. (ed.), Conception de Produits Mécaniques: Méthodes, Modèles et Outils, pp. 105–114 (1998)
7. Hisarciklilar, O., Boujut, J.F.: A speech act theory-based information model to support design communication through annotations. Comput. Ind. **60**, 510–519 (2009)
8. Kleinsmann, M.S.: Understanding collaborative design (2006)
9. Levandowski, C., Bokinge, M., Malmqvist, J., Johannesson, H.: PLM as support for global design reuse–long term benefits and immediate drawbacks. In: IFIP International Conference on Product Lifecycle Management, pp. 488–500. Springer, Berlin (2012)
10. Levine, S.S., Prietula, M.J.: Open collaboration for innovation: principles and performance. Organ. Sci. **25**(5), 1414–1433 (2014)
11. Matta, N., Ducellier, G.: How to learn from design project knowledge. Int. J. Knowl. Learn. **9**(1–2), 164–177 (2014)
12. Meehan, J.S., Duffy, A.H., Whitfield, R.I.: Supporting 'Design for Re-use' with modular design. Concurr. Eng. **15**(2), 141–155 (2007)
13. Pakkanen, J., Huhtala, P., Juuti, T., Lehtonen, T.: Achieving benefits with design reuse in manufacturing industry. Procedia CIRP **50**, 8–13 (2016)
14. Rezgui, Y., Boddy, S., Wetherill, M., Cooper, G.: Past, present and future of information and knowledge sharing in the construction industry: towards semantic service-based e-construction? Comput. Aid. Des. **43**(5), 502–515 (2011)
15. Sanghee, K., Bracewell Rob, H.: Improving design reuse using context. In: Guidelines for a NPD Processes, the 16th international conference of design engineering, DS42_P-54 (2007)
16. Sivaloganathan, S., Shahin, T.M.M.: Design reuse: an overview. Proc. Inst. Mech. Eng. J. Eng. Manuf. **213**, 641–654 (1999)

# Comfort Driven Redesign: The Case of Library Chairs

Rosaria Califano, Iolanda Fiorillo[✉], Giovanni Baglivo, Claudia Chirico,
Antonietta Dello Russo, Jose Garro, Michele Leo, Conrado Pacheco, Gianluca Vitolo,
and Alessandro Naddeo

University of Salerno, Fisciano 84084, SA, Italy
ifiorillo@unisa.it

**Abstract.** University students spend most of their time in a sitting position. Prolonged sitting on ill-fitted furniture and the resulting lousy posture is making students having different musculoskeletal disorders and is strictly related to students learning outcomes. This study aims to improve postural comfort of chairs placed inside the Science & Technology Library at the University of Salerno. A previous study about these library chairs showed that the lumbar area was the most suffering part while perceived (dis)comfort was dependent on time. Based on this, an ergonomic redesign and, consequently, manufacturing of the chair has been done. A perceived-comfort comparison between the library chair and the redesigned one has been performed. A statistical sample of 28 healthy students performed a 20-min experiment two times, alternatively on the library chair and the redesigned one. The 20-min experiment was divided into two 10-min tasks ("Reading & Writing" and "Laptop use") to simulate a study day. The participants' postures were acquired non-invasively using cameras and processed by Kinovea; questionnaires were used to rate the perceived subjective (dis)comfort. A procedure for improving an existing product through a comfort-driven redesign is proposed. Results showed the redesigned library chair lead on increasing postural comfort (particularly in the lumbar area) thanks to the new design and modifications.

**Keywords:** Design method · Postural comfort · (dis)comfort · Redesign · Student seat

## 1 Introduction

The chair is considered a critical element for postural comfort [1–3], especially for who conducts a sedentary life, such as students [4]. Indeed, uncomfortable and awkward postures can decrease students' interest and outcomes in learning [5, 6].

A previous study [7] conducted a postural analysis among the chairs inside the Science and Technology Library (S&T Library) [8] at University of Salerno (UNISA), designed by architect Nicola Pagliara [9]. A brief investigation showed the un-suitability of chairs for students that were frequently forced to change postures. Small movements around the static posture witnessed the increase of postural discomfort [10]. In this previous experiment [7], a postural comfort analysis had been done in order to identify

© The Author(s) 2021
L. Roucoules et al. (Eds.): JCM 2020, LNME, pp. 155–161, 2021.
https://doi.org/10.1007/978-3-030-70566-4_25

the critical factors that influenced the postural comfort sitting on the library-chair. The adopted method was merely statistical and experimental, with the aim to get the basis for a future improved chair design. According to the results, the thoracic and lumbar areas were confirmed as critical. Furthermore, the lacking of lumbar support resulted in comfort decay over time and lowest comfort values for thoracic and lumbar areas: this confirmed the essential role of lumbar support on perceived discomfort [11, 12]. Thus, improvements in these parts were needed, particularly reducing the gap between the seat pan and backrest, a continuous contact in the lower back could lead to a considerable reduction of lumbar pain [13]. This paper focuses on the redesign of the library chair developed in collaboration with the MGR Group S.r.l., an Italian company specialized in the upholstery sector. A comparison between two chairs (the original and redesigned ones) was made through experiments using subjective and objective data [14, 15].

The research question is: Which design procedure is needed to improve the perceived comfort while seating on library chairs, and how can we rate the improvement?

## 2    Materials and Method

### 2.1    Design from Experiment

In the previous experiment [7], the analysis of the usability of the library chair has been performed; this analysis aimed to investigate the chair characteristics that affect either the perceived overall comfort and postural comfort. As far as feedbacks registered during the experiments, all participants accused pain in the lumbar region and the main factor was identified as they were unable to lean their back on the backrest to sit properly and to unload the weight of the head and the back [13, 16–19]. The software CaMAN [20] was used to process postural angles and to compare subjective and objective postural data and to correlate perceived discomfort to chair's characteristics. The thoracic and the lumbar areas were confirmed as critical even through the statistical analysis. Thus, the proposed improvements were focused on these parts. The best choice (out of three) of adding a lumbar/thoracic support was based on the simulations made by DELMIA® and the virtual postural analysis made by CaMAN [7, 15, 20].

### 2.2    Redesign of Library Chair

The redesign of library chair has been developed in order to suit the P50 of population [21]. After a brainstorming session, the following steps have been deployed:

A cardboard model of the back-seat was designed to realize the physical prototype; a sheet steel was manufactured by plasma cut and curved to fit the structure of the chair;

The back-seat was coated with a 3 cm high-density foam, shaped for best fitting with the backrest; the seat pan was also coated with a 1 cm low-density foam to prevent the buttocks slipping;

The back-seat and the seat pan were coated with black eco-leather and then riveted to the chair.

The final prototype is shown in Fig. 1.

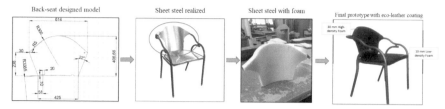

**Fig. 1.** Redesigned chair with details

## 2.3  Experiment Setup

The comparison between the Library Chair and the Redesigned Chair have been made through experiments set up at S&T Library of the Salerno University. In order to simulate a study-day, an online survey was spread before experiments to find out the main study tasks in S&T Library: 98 students declared to perform mainly "Reading & Writing" and "Laptop use" activities. For postural angles detecting, the video acquisition system was equipped with two phone-cameras (FHD $1920 \times 1080$) fixed on tripods to obtain the lateral views. Videos from the two perspectives were registered simultaneously. The angles were detected for counting the macro-movements of human joints [22] that are related to perceived discomfort.

## 2.4  Participants

Twenty-eight students, 11 females (height between 152 cm and 177 cm; weight between 50 kg and 73 kg) and 17 males (height between 168 cm and 186 cm; weight between 55 kg and 100 kg), aged between 23 and 30, were recruited among the Salerno University population. All students enjoyed good health and 50% of them slightly frequented the S&T Library.

## 2.5  Questionnaires

For the acquisition of the subjective data, a two-sections questionnaire was used.

The first questionnaire section is about anthropometric data, frequency of use of S&T Library study-areas (never, once a week, 2–3 times a week, more than 3 times a week) and expected perceived comfort on the chair to test [16] (that is, how the chair to test seemed comfortable at first sight). The second questionnaire section is designed to rate the Localized Postural (Dis)comfort perception [23] related to the different body parts: neck, right and left shoulders, thoracic area, lumbar area, buttock, right and left thighs, right and left ankles. The last question regarded the global comfort evaluation. The expected perceived comfort, the postural (dis)comfort perception and the global comfort were evaluated on a 7-point Likert Scale, from $-3 = $ "Extremely Uncomfortable" to 3 $= $ "Extremely Comfortable" [24].

## 2.6  Experiment Protocol

Students were asked not to wear wet clothes or short trousers in order to avoid clothing influence on comfort perception [24] and were invited to sign the "Informed Consent".

Experiments were split into two different days in order to evaluate each chair independently from the other one. In the first day, participants tested the "Library Chair", while on the other day the "Redesigned Chair". For each day, the following procedure has been deployed:

1. Participants were asked to fill the first section of the questionnaire and invited to have a look at the study area (the ensemble of chair and desk) to rate the expected comfort;
2. Participants sat on the chair and started to perform the two 10-min activities with a pause in between. At the end of each task, the second questionnaire section has been filled by subjects;
3. After the experiments, the videos were processed by the software Kinovea® in order to acquire the angles of body joints.

## 3   Results

Figure 2 shows results from the Localized Postural (Dis)comfort questionnaire: the redesigned chair always scored higher comfort values than the previous library chair. In particular, there is a postural comfort improvement in the thoracic (50%) and lumbar (60%) areas. Thus, the main goal of this work has been reached.

**Fig. 2.** Results from the localized postural (Dis)comfort questionnaire

Furthermore, participants expected low postural comfort for the "Library Chair" than the "Redesigned Chair" (Table 1). Besides, after performing the two tasks, the "Library Chairs" lead on the decreasing of Global Comfort (Table 1), while, the "Redesigned Chair" showed an increase of Postural Comfort (Table 1). As far as postural angles, Table 2 shows the percentages of movement reduction of participants: the more movement reductions, the higher postural comfort.

**Table 1.** Results from questionnaires - expected and global comfort values. Rated on a 7-point Likert scale ($-3 =$ Extremely Discomfortable; $3 =$ Extremely Comfortable)

| | Expected comfort | Global comfort for "Reading&Writing" task | Global comfort for "Laptop use" task |
|---|---|---|---|
| Library chair | 0.500 | $-1.036$ | $-1.464$ |
| Redesigned chair | 1.889 | 2.036 | 2.179 |

**Table 2.** Percentages of movement reductions

| | Reading and writing | Laptop using |
|---|---|---|
| Thoracic zone | 57% | 68% |
| Lumbar zone | 71% | 71% |
| Hips | 64% | 68% |

## 4 Conclusions

The proposed procedure for responding to the research question is the following:

- redesign of the chair and realization of prototype;
- comparison analysis between the original library chair and the redesigned one throughout experiments to collect subjective data (questionnaires) and objective data (video recording to detect postural movements);
- statistical analysis and discussion of outcomes for improvement assessment.

Questionnaires analysis reveals that the Redesigned Chair gave benefits in terms of postural comfort thanks to the thoracic-lumbar support. This result is proved by the highest comfort ratings scored by the redesigned chair. Besides, a reduction of movements number (and angles ranges) thus an addition of comfort perception has been detected. A future improvement of this kind of analysis can be implemented through the evaluation of the suitable backseat shape for this chair.

**Acknowledgments.** Authors want to thank the management and the operator MGR GROUP S.r.l., for the support in this study development.

## References

1. Asundi, K., Odell, D., Luce, A., Dennerlein, J.T.: Changes in posture through the use of simple inclines with notebook computers placed on a standard desk. Appl. Ergon. **43**(2), 400–407 (2012)

2. Netten, M.P., Van Der Doelen, L.H.M., Goossens, R.H.M.: Chair based measurements of sitting behavior a field of sitting postures and sitting time in office work. In: International Conference on Digital Human Modelling and Applications in Health, Safety, Ergonimics and Risk Management (2013)
3. Helander, M., Zhang, L.: Field studies of comfort and discomfort in sitting. Ergonomics **40**(9), 895–915 (1997)
4. Da Silva, L.B., Coutinho, A.S., da Costa Eulalio, E.J., Soares, E.V.G.: School furniture and work surface lighting impacts on the body posture of Paraiba's public school students. Work: J. Prev. Assess. Rehabil. **42**(4), 579 (2012)
5. Al-Hinai, A., Al-Kindi, M., Shamsuzzoha, A.: An ergonomic student chair design and engineering for classroom environment. Int. J. Mech. Eng. Robot. Res. **7**(5), 534–543 (2018)
6. Hira, D.S.: An ergonomic appraisal of educational desks. Ergonomics **23**(3), 213–221 (1980)
7. Fiorillo, I., Anzisi, F.J., Carbone, A., Califano, R., Naddeo, A.: A comfort evaluation tool for sitting postures: the case of Library chairs. In: WORK (accepted for publication) (2019)
8. The Science and Technology Library of UNISA (website): https://www.biblioteche.unisa.it/biblioteca_scientifica/biblioteca_scientifica
9. Pagliara, N.: Serie Fisciano (website): https://nicolapagliara.wordpress.com/design/serie-fisciano/
10. Grondin, D.E., Triano, J.J., Tran, S., Soave, D.: The effect of a lumbar support pillow on lumbar posture and comfort during a prolonged seated task. Chiropractic Manual Ther. **21**(1), 1–9 (2013)
11. Grujicic, M., Pandurangan, B., Xie, X., Gramopadhye, A.K., Wagner, D., Ozen, M.: Musculoskeletal computational analysis of the influence of car-seat design/adjustments on long-distance driving fatigue. Int. J. Ind. Ergon. **40**(3), 345–355 (2010)
12. Van Geffen, P., Reenalda, J., Veltink, P.H., Koopman, H.F.J.M.: Decoupled pelvis adjustment to induce lumbar motion: a technique that controls low back load in sitting. Int. J. Ind. Ergon. **40**(1), 47–54 (2010)
13. Vergara, M., Page, Á.: Relationship between comfort and back posture and mobility in sitting-posture. Appl. Ergon. **33**(1), 1–8 (2002)
14. Piro, S., Fiorillo, I., Anjani, S., Smulders, M., Naddeo, A., Vink, P.: Towards comfortable communication in future vehicles. Appl. Ergon. **78**, 210–216 (2019)
15. Comentale, M., Naddeo, F., Contrada, A., Forlone, G., Saturno, G.: Comfort and ergonomics evaluation of a checkout workstation. ARPN J. Eng. Appl. Sci. **13**, 4117 (2018)
16. Vink, P.: Vehicle Seat Comfort and Design. 1st ed. (2016)
17. Naddeo, A., Califano, R., Vink, P.: The effect of posture, pressure and load distribution on (dis)comfort perceived by students seated on school chairs. Int. J. Interact. Des. Manuf. **12**(4), 1179–1188 (2018)
18. Naddeo, A., Califano, R., Vallone, M., Cicalese, A., Coccaro, C., Marcone, F., et al.: The effect of spine discomfort on the overall postural (dis)comfort. Appl. Ergon. **174**, 194–205 (2019)
19. Naddeo, A., Di Brigida, L., Fontana, C., Montese, J., Quartuccia, M., Nasti, M., et al.: A body-shaped lumbar-sacral support for improving car-seat comfort. In: Work (accepted for publication) (2019)
20. Cappetti, N.: Proposal of a new quantitative method for postural comfort evaluation. Int. J. Ind. Ergon. **48**(25), 25–35 (2015)
21. DINED (website): https://dined.io.tudelft.nl/en/database/tool
22. Bouwens, J.M.A., Fasulo, L., Hiemstra-van Mastrigt, S., Vink, P.: Effect of in-seat exercising on comfort perception of airplane passengers. Appl. Ergon. **73**, 7 (2018)

23. Grinten, M.P.: Development of a pratical method for measuring body part discomfort. In: Advances in Industrial Egonomics and Safety IV (1992)
24. Joshi, A., Kale, S., Chandel, S., Pal, D.: Likert scale: explored and explained. Br. J. Appl. Sci. Technol. (2015)

# Constraint Analysis and Optimization of NES System Based on Variable Pitch Spring

Zhenhang Wu, Manuel Paredes[✉], and Sébastien Seguy

Institut Clément Ader UMR CNRS 5312, Université de Toulouse, UPS, INSA,
ISAE-SUPAERO, MINES-ALBI, CNRS, 3 rue Caroline Aigle, Toulouse, France
`paredes@insa-toulouse.fr`

**Abstract.** This study proposes the realization of a device with a pure cubic stiffness mechanism to suppress a wide range of vibrations, which is known as the Nonlinear Energy Sink. Deciding how to construct a light, reliable NES device is always a challenge. According to our design, the device can counterbalance the undesirable linear stiffness that emerges from the intrinsic property of a variable pitch spring. Our goal is to reduce the mass of the spring while keeping the same cubic stiffness. Through the multifaceted analysis of the nonlinear constraint, we try to explore the full potential of NES device to reduce its mass. Meanwhile, a global search method, *Multi Start*, is applied by repeatedly running a local solver. Finally, a new design with different variable pitch distribution is proposed.

**Keywords:** Nonlinear Energy Sink (NES) · Variable pitch spring · Constraint analysis · *Multi Start* method

## 1 Introduction

Vibration is an inevitable phenomenon and can be harmful to structures in the domains of civil engineering or the machine industry. In recent decades, novel systems composed of a primary linear structure and a strongly nonlinear attachment have attracted attention because of their ability to absorb a wider range of vibration frequencies and their relatively small attached mass [1]. Because of the existence of a nonlinear component, which is always referred to as cubic stiffness, the system demonstrates localization and irreversible transient transfer of energy from the primary system to the attached mass [2]. This new concept is called a Nonlinear Energy Sink (NES). Several methods have been proposed for achieving a cubic stiffness. G. Kerschen used a straight air tract to support the primary system and NES. An essential cubic stiffness nonlinearity was obtained by means of a thin hat without tension [3]. Our idea is to use a variable pitch spring to produce the desired cubic stiffness by designing the distribution of the pitch.

The structure of the current paper is as follows. In Sect. 2, a NES system design process will be presented based on the variable pitch spring. In Sect. 3, the nonlinear constraint will be defined according to the various conditions. In Sect. 4, *Multi Start*, a global search method, will be applied and the optimization result will be compared. Section 5 contains some concluding remarks and discussion.

© The Author(s) 2021
L. Roucoules et al. (Eds.): JCM 2020, LNME, pp. 162–168, 2021.
https://doi.org/10.1007/978-3-030-70566-4_26

## 2 NES Design Process

The most significant component in a NES system is the cubic stiffness, which is obtained through the variable pitch spring in our study. A core characteristic of the variable pitch spring is its progressively increasing pitch, which results in a variable stiffness. Sequenzia G provides an iterative calculation method to define precisely the geometry of helix line [6]. Another design of a variable pitch spring coupled with a negative stiffness mechanism has been proposed by D. Qiu [5] and used in our study. The objective displacement-force relation of a variable pitch spring can be defined by the following equation, where $k_0$ is the linear stiffness before the spring reaches its transition point $s_t$, $a_1$ is the spring rate in the nonlinear part, and $a_3$ is the cubic stiffness value.

$$F = \begin{cases} k_0 u & (u \leq s_t) \\ a_3(u - s_t)^3 + a_1(u - s_t) + k_0 s_t & (u > s_t) \end{cases} \tag{1}$$

However, when we apply the nonlinear spring, the force displacement characteristic inevitably includes a linear part [5]. To skip the linear part, a method of combining two pre-compressed springs at the transition point is adopted, as in part (b) of Fig. 1. So the composed stiffness curve $f_K = (a_1 + k_0) u + a_3 u^3$ is obtained. To counterbalance the linear phase $a_1 + k_0$, a negative stiffness mechanism is introduced, as in part (a) of Fig. 1. When the two linear springs are subjected to horizontal pre-compression, they have a negative stiffness in the vertical direction. The force-displacement relation can be expressed as a Taylor expansion. It is important to note that the Taylor expansion is valid only when the displacement remains close to the initial balancing point. By adding the force of the pre-compressed variable pitch spring combined with a negative stiffness mechanism, a composed force displacement relation is expressed:

$$F = \left( a_1 + k_0 - 2k_l \frac{l_p}{l_{0l} + 2l_c - l_p} \right) u + \left( a_3 + k_l \frac{l_{0l} + 2l_c}{(l_{0l} + 2l_c - l_p)^3} \right) u^3 \tag{2}$$

Here, $l_{0l}$ and $l_p$ are the free length of the linear spring and its length of pre-compression. $l_c$ is the length of the connectors, which are connected to the ground and the central mass by hinges. The connector is marked in Fig. 2. A schematic diagram of this NES system is proposed in Fig. 1. The linear component of Eq. (2) can be counterbalanced by defining the stiffness of cylindrical spring $k_l$. To design a cylindrical spring with the target stiffness, the diameter of a linear cylindrical spring, $D_l$, and the diameter of its wire, $d_l$, are necessary to fix its active number of coils through the following equation:

$$n_l = \frac{G d_l^4}{8 D_l^3 k_l} \tag{3}$$

A method using a special coordinate point is adopted to manufacture a variable pitch spring with the required stiffness curve. Groups of coils with different pitches are symmetrically distributed on both sides of the spring. The detailed process to determine the distribution of variable pitch can be found in Donghai's work [4]. Through the above analysis, the whole NES system is determined by 11 independent variables ($d$, $D$, $l_f$, $d_l$, $D_l$, $l_{0l}$, $k_0$, $a_1$, $st$, $a_3$, $p_c$). $p_c$ is a dimensionless coefficient of pre-compression $l_p = p_c * l_{0l}$. The mass of the total of 4 springs is easily obtained:

$$M = 2\left(\left(\sum_{i=1}^{n_a} n_i + 1.5\right)D\frac{d^2\pi^2}{4} + (n_l + 1.5)D_l\frac{d_l^2\pi^2}{4}\right)\rho \qquad (4)$$

The sum of $n_i$ and $n_l$ are the total numbers of active coils in the variable pitch spring and the cylindrical spring, the extra 1.5 coils being added as end coils at both spring sides to support the active coils. The lower boundary ($Lb$) and upper boundary ($Ub$) of the parameters are presented in Table 1.

**Table 1.** Lower and upper boundaries for the parameters of the NES system.

|  | $d[mm]$ | $D[mm]$ | $l_f[mm]$ | $d_l[mm]$ | $D_l[mm]$ | $l_{0l}[mm]$ | $k_0[N/m]$ | $a_1[k/m]$ | $s_t[mm]$ | $a_3[N/m^3]$ | $p_c$ |
|---|---|---|---|---|---|---|---|---|---|---|---|
| Lb | 0.1 | 20 | Course + 2toler | 0.1 | 20 | 20 | 100 | 100 | 0.5Course + toler | 0.2e5 | 0.3 |
| Ub | 5 | 60 | 95 | 5 | 60 | 70 | 800 | 800 | 60 | 8e5 | 0.6 |

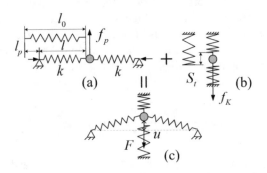

**Fig. 1.** Schematic of NES mechanism

## 3  Nonlinear Constraint Analysis

In order to maintain reliable operation of the device in complex situations, the design of the spring and the parameter values should be strictly chosen to meet the different constraints, e.g. design constraints, buckling conditions, space between coils of the nonlinear spring, and static maximum constraint. These nonlinear constraints involve all the design parameters, and they can hardly be separated individually. For the sake of convenience, we sort the constraints in Table 2.

**Table 2.** Definition of the optimization constraint

| No | Description | Constraint |
|---|---|---|
| 1 | Avoid the occurrence of negative numbers of coils during the calculation | $k_0 < a_1$ |
| 2 | Ensure the accuracy of the fitting | $n_a > 8$ |
| 3 | Maintain spring characteristics | $n_l > 4$ |
| 4 | Strengthen the spring and induce low internal stress. Facilitate the manufacturing process | $5 < \frac{D}{d} < 20$ |
| 5 |  | $5 < \frac{D_l}{d_l} < 20$ |
| 6 | Ensure spring free length is smaller than its critical length $L_k = v_l l_{0l}/D_l$ to avoid buckling. Sitting factor for linear spring where both ends are connected with hinge joints $v_l = 1$ | $v_l \frac{l_{0l}}{D_l} < 2.6$ |
| 7 | Sitting factor $v_{nl} = 0.5$ for nonlinear spring that is embedded | $v_{nl} \frac{L_l}{D} < 2.6$ |
| 8 | Constant helix angle of linear spring | $\arctan\left(\frac{p_l}{\pi D_l}\right) < 7.5°$ |
| 9 | The helix angle is variable for the nonlinear spring, its maximal helix angle is considered | $\arctan\left(\frac{\max(p_{nl})}{\pi D}\right) < 7.5°$ |
| 10 | Ensure the maximum corrected main stress does not exceed the acceptable maximum stress. Maximum stress: $\tau_{max} = 8FDcos(\alpha)(1 + 2d/(3D))/(\pi d^3)$. Stress correction factor $k = (w + 0.5)/(w - 0.75)$. Maximum allowable stress $\tau_{zul}$ is equal to 0.48Rm. For stainless steel $Rm = 1919 - 255.86ln(d)$ | $\tau_{zul} > k\tau_{max}$ |
| 11 | Reduce the risk of a crash under alternating load. Compressed solid length of spring $ln = d_l\left(n_{lin-active} + 1.5\right) + Sa$. Sum of minimum spaces between active coils. $Sa = n_{lin-active} (0.0015D_l^2/d_l + 0.1d_l)$ | $l_p \leq l_{0l} - l_n$ |
| 12 | Avoid nonlinear spring 1 or 3 being released central mass (in Fig. 2 right) | $s_t > \frac{Course}{2} + toler$ |
| 13 |  | $l_f - s_t > \frac{Course}{2} + toler$ |
| 14 | Avoid the linear spring 2, 4 separation in Fig. 2 left. Distance between hinges in extreme position $D_{max}^2 = \left(l_{0l} - l_p + 2l_c\right)^2 + \left(l_f - s_t\right)^2$, free distance $D_f = l_{0l} + 2l_c$ | $D_f - D_{max} > toler$ |
| 15 | Error control. Taylor expansion with higher order term $\Delta F(u) = 0.75k_l(l_{0l} + 2l_c)u^5/\left(l_{0l} - l_p + 2l_c\right)^5$. Maximum force $F_{max} = A_3(Course/2)^3$ | $\Delta F(u) < 0.05F_{max}$ |

When the central mass moves to the upper maximal displacement, two potentially dangerous conditions appear. The separated position is indicated in the dotted red circle in Fig. 2, which corresponds to the 12th/13th/14th constraints.

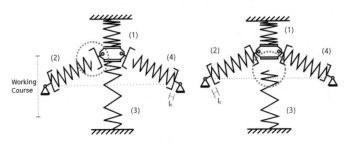

**Fig. 2.** Detachment conditions (1) left: spring 2, 4 separation; (2) right: spring 1 or 3 released

## 4 Optimization Result

The function *fmincon* of *Matlab* is used to obtain a local minimum value with respect to the nonlinear and linear constraint. This kind of nonlinear optimization problem is extremely sensitive to the initial starting point. To generate a global solution, one idea is to repeatedly run a local solver with the various starting points. The best solution is then picked among the multiple local minima. The *Multi Start* method was applied in our study.

Firstly, we should define the working course, *Course = 80 mm*, and the tolerance, *toler = 5 mm*, to prevent the center of mass falling outside the working course. The target cubic stiffness, $A_3$ is *4e5 N/m³*. The initial value is chosen with the parameters given in [5].

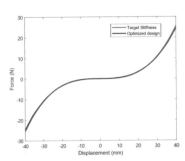

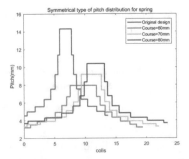

**Fig. 3.** Comparison of target stiffness and optimized design stiffness.

**Fig. 4.** Symmetrical type of pitch distribution for different optimized designs (80/70/60 mm) and original design.

When the number of local solvers is 1000, the lightest quality design parameter is obtained in Table 3 and a new distribution of pitch is found in Fig. 4. Our mass is reduced to 88.2 g. So, compared with the original mass of 182.98 g, there is a mass saving of 51.8%. At the same time, the error of the force-displacement relation of optimization (4.28%) is kept below the tolerance level (5%) presented in Fig. 3. In other side, the different design targets for example $Course = 70\,mm$ and $Course = 60\,mm$ are performed. The mass result reduce to 69.9 g and 60.3 g respectively. The detailed optimal parameters are presented in Table 3.

**Table 3.** Parameter of optimization result for $Course = 80\,mm/70\,mm/60\,mm$

| $d[mm]$ | $D[mm]$ | $l_f[mm]$ | $d_l[mm]$ | $D_l[mm]$ | $l_{0l}[mm]$ | $k_0[N/m]$ | $a_1[N/m]$ | $s_t[mm]$ | $a_3[N/m^3]$ | $p_c$ |
|---|---|---|---|---|---|---|---|---|---|---|
| 1.6 | 25.1 | 90 | 1.71 | 26.9 | 70 | 165.9 | 165.9 | 48.7 | 2.73e5 | 0.3 |
| 1.5 | 22.2 | 80 | 1.75 | 26.9 | 70 | 185.7 | 185.7 | 40 | 2.58e5 | 0.3 |
| 1.4 | 20 | 72.7 | 1.87 | 26.9 | 70 | 231.4 | 253.4 | 35 | 2.15e5 | 0.3 |

## 5  Conclusions

This article has studied the realization and optimization of a NES system with a variable pitch spring. The main challenge is to construct a pure cubic stiffness system. The variable pitch springs are combined axially with pre-compression and negative stiffness mechanism are used to counterbalance the linear phase and leave a pure cubic stiffness. By using *Multi Start* method, it is possible to obtain global resolution with respect to the nonlinear constraint and minimize the mass of the system. The stiffness curve of optimized design basically coincides with the target stiffness curve, while the mass is reduced effectively. For the next stage of the study, we will analyze the optimization result and other constraints still further to reduce the number of parameters.

## References

1. Gourdon, E., Alexander, N.A., et al.: Nonlinear energy pumping under transient forcing with strongly nonlinear coupling: theoretical and experimental results. J. Sound Vibr. **300**(3–5), 522–551 (2007)
2. Gendelman, O.V.: Transition of energy to a nonlinear localized mode in a highly asymmetric system of two oscillators. Nonlinear Dyn. **25**(1–3), 237–253 (2001)
3. Kerschen, G., et al.: Experimental demonstration of transient resonance capture in a system of two coupled oscillators with essential stiffness nonlinearity. J. Sound Vibr. **299**(4–5), 822–838 (2007)
4. Qiu, D., Paredes, M., Seguy, S.: Variable pitch spring for nonlinear energy sink: application to passive vibration control. Proc. Inst. Mech. Engi. Part C J. Mech. Eng. Sci. **233**(2), 611–622 (2019)

5. Rodriguez, E., Paredes, M., Sartor, M.: Analytical behavior law for a constant pitch conical compression spring. 5 (2005)
6. Sequenzia, G., Oliveri, S., et al.: A new methodology for calculating and modelling non-linear springs in the valve train of internal combustion engines. SAE Technical Paper (2011)

# Generative Design in the Transmission System of a Skateboard

Mikel Iturrate[1]([✉]), Xabier Amezua[2], Xabier Garikano[2], Jose Antonio Oriozabala[2], Iñaki Martin[3], and Eneko Solaberrieta[3]

[1] Department of Business Management, Gipuzkoa Faculty of Engineering, University of the Basque Country UPV/EHU, Donostia, Spain
mikel.iturrate@ehu.eus
[2] Department of Graphic Design and Engineering Projects, Gipuzkoa Faculty of Engineering, University of the Basque Country UPV/EHU, Donostia, Spain
[3] Department of Graphic Design and Engineering Projects, Gipuzkoa Faculty of Engineering, University of the Basque Country UPV/EHU, Eibara, Spain

**Abstract.** Generative design (GD) is a new way of designing products that respond to the current characteristics of the 4.0 industry era. It allows the customization of designs and helps to minimize process times and material quantities. In the case study, two parts of the transmission system of a skateboard have been designed using GD of Solid Edge ST10. The design has been customized for 67 kg users who will use the skateboard for riding and not for jumping or doing tricks. On the other hand, the aim was to minimize the weight and the use of material for its manufacture. Once the initial solid is defined, the software generates specific designs for different levels of mass reduction or product quality. An axle with a mass reduction of 70% from the initial solid and a base with reduction of 60% have been proposed. The GD more than an alternative is a requirement to respond to the demands of the 4.0 industry in which customization and cost reduction are one of the fundamental pillars. This type of design, at the same time, forces to adapt the manufacturing processes so that it is necessary to manufacture in additive manufacturing (AM) machines.

**Keywords:** Generative design · Additive manufacturing · 4.0 Industry · Customisation

## 1 Introduction

The concept of industry 4.0 refers to a new way of organizing productive resources. The aim is to set up smart factories capable of offering greater adaptability to production needs and processes and making the allocation of resources more efficient, paving the way for this new industrial revolution [1]. It will join technological achievements made in recent years with the modern use of information technologies [2, 3]. Industry 4.0 seeks to guarantee the flexibility and efficiency of production but at the same time facilitating the strategy known as mass customization (MC) that focuses on satisfying the individual requirements of each client [1, 4].

L. Roucoules et al. (Eds.): JCM 2020, LNME, pp. 169–174, 2021.
https://doi.org/10.1007/978-3-030-70566-4_27

Studies show that approximately 80% of design time is spent on routine tasks with little or no added value [5]. However, current design tools such as CAx tools (Computer Aided Technologies) make possible rapid design with automated tools and define technical specifications of a product without the participation of a design engineer [2]. These technologies are penetrating the manufacturing industry, making it intelligent and adaptable to current market requirements of customization, better product quality and shorter time to market [6, 7]. Topology optimization has also emerged as the mayor light weighting strategy in order to greater efficiency or lower energy consumption, best exploiting the design freedoms offered by generative design (GD) [8]. GDs main goal is to optimize the distribution of material within boundaries and respecting the requirements of loads or functional surfaces [9]. This new way of designing has coexisted for a while but it has been better known in non-industrial fields [10].

Using GD engineers can analyse the different options presented in a more dynamic way and select the one that best responds to the requirements. In addition, it offers solutions that would not be possible to design with conventional design tools. However, due to the shapes of these designs, it is not possible to manufacture the parts entirely using conventional technologies and additive manufacturing (AM) is required. However, with AM the surfaces of the parts are currently of high roughness and have imprecise finishes making the use of conventional technologies still necessary.

AM allows the manufacture of customized objects with sophisticated designs and in new materials. This technology has undergone an important evolution, improving aspects such as the precision or the production speed in addition to a considerable decrease in costs [11]. Thus, in some cases AM has already replaced conventional manufacturing techniques, although there are still reasonable doubts regarding its use in mass production [12].

The present work aims to show the possibilities offered by GD. For this purpose, two pieces of a skating board have been designed using this technology and the results have been compared with a conventional design of these same pieces.

## 2  Materials and Methods

GD has been used (Solid Edge ST10) to design two pieces that participate in the rolling system of a skateboard. The rolling system is a set divided into different parts: wheels, axle, intermediate joints, the base and different elements of union (Fig. 1A). Obviously, it is not suitable to apply GD in all these parts. Therefore, GD has only been applied to design the axle (Fig. 1B) and the base (Fig. 1C).

First of all, the boundaries are defined by designing the solid of each part.

**Fig. 1.** The rolling system (A) and the solids of the axle (B) and the base (C) Functional surfaces in red in B and C.

It must be defined the main dimensions of the parts and the functional surfaces, which in the case of the axle, are the wheel seats and the hole (In red in Fig. 1B). In the case of the base, the functional surfaces are the joint seats and the surfaces in contact with the shaft (In red in Fig. 1C).

After defining the solids and their functional surfaces, the next step will be the stress calculation. The skateboard is designed for 67 kg users that use it for displacements and not for jumping or do tricks. Even so in the design will be considered that the skateboard can be used by a person of 80 kg and it was applied a safety coefficient of 1.2. The material chosen for fabrication in both, the axle and the base is 5805 10NiCr5-4 steel alloy (Brinell Hardnesss 140–170, Elastic (Young's, Tensile) Modulus 190 GPa, Poisson's Ratio 0.29, Shear Modulus 73 GPa, Tensile Strength: Ultimate (UTS) 460 to 1180 Mpa).

Several tests have been carried out with each part without altering the restrictions in terms of stress and material on functional surfaces but varying the quality and the percentage of weight reduction from 20% to 70%.

## 3   Results

Table 1 show different design solutions for both parts of the skateboard (axle and base) when applying different mass reduction. In addition, the figures in the tables show the critical surfaces represented in colour maps. Blue areas represent non-critical surfaces while green, yellow or red areas represent critical ones.

**Table 1.** Different axle and base design results by applying varying degrees of mass reduction.

| REDUCTION | PROPOSED AXLE | PROPOSED BASE |
|---|---|---|
| 40% | | |
| 60% | | |
| 70% | | |

In the case of the axle, it is possible the design that reaches the 70% of mass reduction. The mass has been reduced from 0.974 kg of the initial solid to 0.292 kg of the proposed solution. Its' admissible maximum stress is 246.328 MPa. However, in the case of the base, the proposed design for a 70% mass reduction is clearly unsatisfying. The base has been splitted in two parts becoming useless. Thus, the best solution is the one that reaches the 60% of mass reduction. The base mass has been reduced from 0.391 kg to 0.156 kg. The maximum admissible stress is also 246.328 MPa.

Besides designing the two parts, 2D plans of the axle and the base have been drawn up in order to define the functional surfaces. To machine these functional surfaces, both parts must be placed on specifically designed tooling that also must have been designed and represented in 2D plans.

## 4  Discussion

Results section show shapes of the designed parts far from simple geometries. In addition, GD help reducing the participation of design engineers by automatizing, contributing in some purposes of industry 4.0 such as the reduction of designing processes times [2] or minimising time spent in routine tasks [5]. Besides, GD make it possible to easily design personalized products responding to MC requirements [1, 4, 6, 7].

Topology optimization has also contributed improving efficiency by reducing both pieces mass [8]. Comparing the originally designed solids with the result, a large material removal is observed. It is also noticeable how the program respects the requirement to not remove material in areas where removal has been restricted. These areas are represented in dark blue, showing that they are under minimal stress.

On the other hand, these shapes can be restrictive because they cannot be manufactured using conventional technologies, and it could condition the implementation of GD. AM enable the manufacture of these shapes, but on the other hand, have their limitations, especially in terms of surface finishing or mass production. Consequently, some functional surfaces and threads must be necessarily mechanised in traditional machine tools.

## 5  Conclusions

GD not only offers a different alternative to traditional design methods, also facilitates product customization, reduces design time and optimizes the amount of material required to manufacture. However, GD does not imply the disappearance of 2D dimensional drawings since all functional surfaces must continue to be manufactured according to traditional methods and these require 2D drawings. Instead, it significantly simplifies the creation of these drawings.

Parts designed using GD will need 3D printing machines to be manufactured. These machines have the advantage of versatility in terms of manufacturing complex shapes or better use of materials, but are limited in terms of obtaining good surface finishes or manufacturing large quantities.

In the case study, it was possible to design two parts of a skateboard using GD. These pieces have a certain level of customization due to the fact that they have been designed for a certain user profile. In addition, the material and weight of the pieces have been minimized by 60 to 70%.

**Acknowledgements.** The authors of this paper thank the Faculty of Engineering Gipuzkoa for locating the DEHI laboratory in their facilities, and the Country Council of Gipuzkoa (Grant n° 70/19) and MINECO Ministry of Economy and Competitiveness (grant number PID2019-108975RA-I00) for financing this project.

# References

1. Pasetti Monizza, G., Bendetti, C., Matt, D.T.: Parametric and generative design techniques in mass-production environments as effective enablers of industry 4.0 approaches in the Building Industry. Autom. Constr. **92**, 270–285 (2018)
2. Zawadzki, P., Zywicki, K.: Smart product design and production control for effective mass customization in the industry 4.0 concept. Manag. Prod. Eng. Rev. **7**(3), 105–112 (2016)
3. Lee, J., Bagheri, B., Kao, H.A.: A cyber-physical systems architecture for Industry 4.0-based manufacturing systems. Manuf. Lett. **3**, 18–23 (2015)
4. Shrouf, F., Ordieres, J., Miragliotta, G.: Smart factories in Industry 4.0: a review of the concept and of energy management approached in production based on the Internet of things paradigm. In: IEEE International Conference on Industrial Engineering and Engineering Management, vol. 2015, pp. 697–701, January 2014
5. Stokes, M.: Managing Engineering Knowledge: MOKA: Methodology for Knowledge Based Engineering Applications. Strategy, p. 257 (2001)
6. John, J.F.R., Rittinghouse, W.: Cloud Computing Implementation, Management, and Security (2010)
7. Zheng, P., et al.: Smart manufacturing systems for Industry 4.0: conceptual framework, scenarios, and future perspectives. Front. Mech. Eng. **13**(2), 137–150 (2018)
8. Plocher, J., Panesar, A.: Review on design and structural optimisation in additive manufacturing: towards next-generation lightweight structures. Mater. Design **183**, 108164 (2019)
9. Tyflopoulos, E., Flem, D.T., Steinert, M., Olsen, A.: State of the art of generative design and topology optimization and potential research needs. Proc. Nord. Des. Era Digit. Nord. **2018**, 1–5 (2018)
10. Jana, G., Miroslav, V., Ladislav, G.: Surface interpolation and procedure used in the generative engineering design of surface-based automotive components. Int. J. Veh. Des. **77**(4), 211–226 (2018)
11. Stock, T., Seliger, G.: Opportunities of sustainable manufacturing in industry 4.0. Procedia CIRP **40**, 536–541 (2016)
12. Dilberoglu, U.M., Gharehpapagh, B., Yaman, U., Dolen, M.: The role of additive manufacturing in the era of industry 4.0. Procedia Manuf. **11**, 545–554 (2017)

# Methodology of Product Sustainable Redesign. Case Study: Furniture of a Clothing Retail Store

Natalia Muñoz López[✉], José Luis Santolaya Sáenz, and Anna Biedermann

Department of Design and Manufacturing Engineering, University of Zaragoza, María de Luna 3, 50018 Zaragoza, Spain
nataliam@unizar.es

**Abstract.** Companies awareness of the impact generated by its products increases and motivates them to develop initiatives to improve their sustainability. In this work, a methodology consisting of three main phases: sustainability assessment, redesign process and comparison of designs, is proposed to obtain more sustainable product designs. Methodology is based on the Life Cycle Sustainability Assessment (LCSA) approach, which is applied to simultaneously evaluate environmental, economic and social aspects. In the case study the sustainability improvement of the furniture of a clothing retail store is addressed. A set of indicators are considered to evaluate the sustainability performance of both initial design and redesign. The study concludes that the application of different sustainability strategies allows a significant enhancement of the environmental and economic indicators.

**Keywords:** Sustainable design · Life Cycle Sustainability Assessment · Sustainability indicators

## 1 Introduction

Design for sustainability (DfS) has been mainly focused on the integration of environmental aspects into product design (usually referred to as Eco-design). However, sustainability does not only consist of the environmental impact. Three dimensions: environment (planet), economy (profit) and social well-being (people) should be taken into account according to a worldwide approach [1, 2] and should be simultaneously analyzed to effectively achieve sustainability assessment. Consistent with this perspective, the Life Cycle Sustainability Assessment (LCSA) methodology [3–5] was developed based on the same structure already established in Life Cycle Assessment (LCA) method [6, 7].

On the other hand, eco-design studies [8] are generally aimed at comparing different stages of the entire life cycle of a product, comparing impacts of different alternatives and evaluating impacts in order to propose improved alternatives. Studies which include a new sustainable product redesign, its sustainability assessment and a comparison with the initial design, are usually not carried out.

© The Author(s) 2021
L. Roucoules et al. (Eds.): JCM 2020, LNME, pp. 175–181, 2021.
https://doi.org/10.1007/978-3-030-70566-4_28

In this work, a practical scheme to project more sustainable products considering environmental, economic and social aspects is proposed and applied in the case of the furniture used in a clothing retail store. A redesign phase, in which sustainability strategies consistent with the initial product specifications like operation or appearance, is addressed. In addition, indicators of each sustainability dimension are compared to detect improvements. The methodology applied and results obtained are shown in the following sections.

## 2  Methodology

A methodology composed of three main phases is proposed to achieve more sustainable design [9] (Fig. 1):

- Phase 1- Sustainability of the initial design is assessed according to LCSA approach. Environmental impacts, economic aspects and social issues are obtained compiling inputs and outputs of the system object of study and evaluating a set of suitable indicators.
- Phase 2- Redesign process is carried out incorporating sustainability aspects. Several sustainability strategies can be proposed according to the Life Cycle Design Strategy (LiDS) wheel [10]. A new inventory of the production stage relative to the new product design is performed and the sustainability indicators associated with the redesign are quantified.
- Phase 3- Comparing designs. The sustainability of both initial design and redesign is compared through environmental, economic and social indicators. To interpret the results obtained associated with the redesign and to know if a sustainability improvement has been reached, a parallel representation of indicators [5] is chosen instead of a sustainability aggregated index.

In this work, furniture used in a clothing retail store is the system analyzed and the product creation stage is the object of study. To value social issues, UNEP's guidelines [11] were used. The environmental dimension is evaluated by the Global Warming Potential ($GWP_{100}$) indicator, which represents total emissions of the greenhouse gases calculating the radiative forcing over a time horizon of 100 years. For the economic dimension, the budget of material execution (BME) that expresses the total cost to develop each piece of furniture is the indicator proposed. Finally, workers are selected to analyze the social dimension of sustainability and the indicator used is the time required by the workers in the production process ($T_w$).

The environmental module of Cype software, Probas database [12] and emission factors of electric commercial companies were used to calculate sustainability indicators.

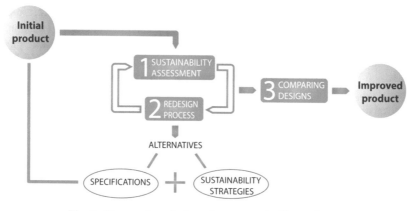

**Fig. 1.** Phases of a redesign of more sustainable products.

## 3 Case Study

The methodology was applied to design more sustainable furniture, which is used in a clothing retail store located in the centre of a big city. As is shown in Fig. 2, a set of furniture pieces is needed to obtain a good arrangement and exhibition of the selling items: three glass and two chipboard shelves on the store sidelines and other in the storefront to hang and expose clothes, a set of bars to show accessories, a central table to put down the tried clothes by the customers and where the shop assistants blend it and finally, two exhibitor furniture to expose costume jewelry, charge and packaging the sold items.

Furniture pieces:
① Accesories bar          ② Warehouse ledge     ③ Chipboard shelf     ④ Glass shelf
⑤ Exhibitor furnit.       ⑥ Central table       ⑦ Storefront furnit.  ⑧ Ground level shelf

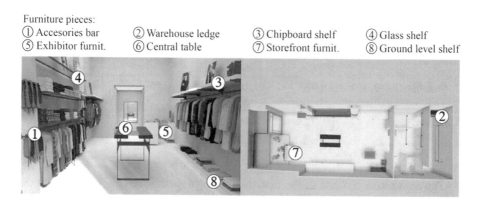

**Fig. 2.** Furniture pieces.

The production process of the furniture pieces is analyzed in detail to obtain energy consumptions, economic costs and working times. Materials used are laminated chipboard, walnut wood, templated glass and corten steel ASTM A588, which is acquired in boards, sheets and profiles of different sizes.

The following inventory data, in reference to one unit of the furniture piece, are shown in Table 1: quantity, mass and material type used by each piece, energy consumption, working time needed in processing and production total costs. If the furniture piece of 3. Chipboard shelf is reviewed, we can observe that 2 units are used at the clothing store, which has a mass of 68,7 kg each one, the materials used to fabricate them are laminated chipboard and corten steel ASTM A388, an energy consumption of 0,37 MJ is required in processing, the working time is 125,6 s and the production costs are 182,7 €. Sustainability indicators of the initial design were calculated. Greenhouse emissions of 494,4 KgCO$_2$eq, a total execution cost of 2153,4 € and an average working time of 2,1 h were obtained. These indicators will be compared with those obtained after the redesign process in Fig. 3.

**Table 1.** Furniture production stage inventory.

| Furniture piece type | Units | Mass (kg) | Material type | Processing | | BME (€) |
|---|---|---|---|---|---|---|
| | | | | Energy (MJ) | Tw (s) | |
| 1. Accessories bar | 5 | 1,2 | ASTM A588 | 0,05 | 56,9 | 4,8 |
| 2. Warehouse ledge | 4 | 3,8 | Walnut wood | 0,07 | 14,7 | 69,5 |
| 3. Chipboard shelf | 2 | 68,7 | Lam. chipboard ASTM A588 | 0,37 | 125,6 | 182,7 |
| 4. Glass shelf | 3 | 29,7 | ASTM A588 Templat. glass | 0,11 | 53,3 | 151,5 |
| 5. Exhibitor furnit. | 2 | 66,2 | Lam.chipboard A1 6061A Templat. glass | 4,94 | 2741,3 | 199,1 |
| 6. Central table | 1 | 58,2 | Walnut wood ASTM A588 Templat. glass | 0,38 | 198,3 | 422,2 |
| 7. Storefront furnit. | 1 | 92,9 | Lam. chipboard ASTM A588 | 0,86 | 985,6 | 121,1 |
| 8. Ground level shelf | 1 | 33,4 | Lam. chipboard | 0,46 | 223,2 | 89,7 |

The next phase consists of the furniture redesign process having into account the specifications of the initial design, keeping the store esthetics unchanged and to guaranty the storing and display of the clothes. Concerning the production stage, the strategies of materials quantity reduction, replacement or elimination of elements and selection of low-impact materials are considered suitable to achieve a sustainability improvement.

Table 2 shows the initial design and the redesign of the furniture pieces 3, 5, 7 and 8. Different sustainability strategies are applied in each case preserving initial design specifications. The production process of the redesigned furniture pieces is analyzed taking into account these strategies and energy consumptions, economic costs and working times are obtained for the new design projected. As is shown in Table 2, significant improvements of the mass of the raw material and thus, in $GWP_{100}$ indicator, are obtained. At the same time, emissions, working times and costs can be reduced in the production process. In all cases, the percentage reduction of indicators according to the sustainability strategies applied, are indicated.

**Table 2.** Redesign of furniture pieces.

| 3.Chipboard shelf | | 5. Exhibitor furniture | |
|---|---|---|---|
| | | | |
| Board thickness reduction | Hook elimination | With drawers | With doors |
| 7. Storefront furniture | | 8.Ground Level shelf | |
| | | | |
| | | Walnut wood   Pine wood | |
| With bottom drawer   Without bottom drawer | | | |

| Type of furniture | Raw material | | Processing | | % BME (€) | Sustainability strategies | | | |
|---|---|---|---|---|---|---|---|---|---|
| | % Mass (kg) | % GWP (kgCO₂) | % GWP (kgCO₂) | % Tw (s) | | A | B | C | D |
| 3.Chipb.shelf | -39,4 | -57,7 | -44,4 | -23,3 | -64,0 | X | X | X | |
| 5. Exhibitor | -19,5 | -53,3 | -75,5 | -66,5 | -20,7 | X | | | X |
| 7. Storefront | -10,4 | -43,4 | -22,2 | -43,4 | -49,2 | X | | X | |
| 8.Ground shelf | -10,4 | -61,4 | -0 | -0 | -46,5 | X | | | |

A: replacement of materials; B: elimination of pcs.; C: reduction of materials; D: pcs. replacement

Finally, the third phase of redesign methodology is carried out. A comparison between the environmental, economic and social indicators of the furniture initial pieces and its redesign are shown in Fig. 3. An improvement of the sustainability indicators is achieved. A reduction of 40% and 20% is reached in the environmental and economic indicators respectively. The working time is decreased in 1,15 h.

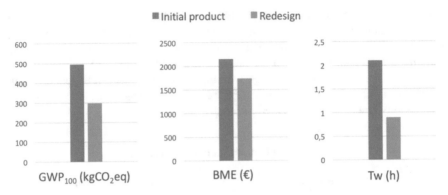

**Fig. 3.** Sustainability indicators comparison

## 4 Conclusions

This work presents a redesign methodology to achieve more sustainable products. Three phases are carried out: sustainability assessment, redesign process and comparing designs. A set of suitable indicators was selected to simultaneously assess the three sustainability dimensions and to compare initial and final designs.

The redesign of furniture used in a clothing retail store was performed. A production inventory was obtained by means of a detailed study of each furniture piece. Data of materials, energy consumptions, production costs and working times were calculated in both initial design and projected redesign of furniture. The early specifications of each furniture piece like operation and appearance were taken into account in the redesign process.

A substantial improvement of the furniture sustainability indicators at the production stage was managed by the application of different sustainability strategies. Particularly, the use of pine wood and the pieces' elimination were the most effective strategies.

A significant enhancement of greenhouse emissions and processing costs and a slight improvement in the working times was achieved.

## References

1. WCED (World Commission on Environment and Development): Our common future. Oxford University Press (1987)
2. UNCED Agenda 21: The Rio Declaration on Environment and Development. United Nations Conference on Environment and Development, Rio de Janeiro (1992)
3. Kloepffer, W.: Life-cycle based sustainability assessments as part of LCM. In: Proceedings of the 3rd International Conference on Life Cycle Management, Zurich (2007)
4. Finkbeiner, M., Schau, E.M., Lehmann, A., Traverso, M.: Towards life cycle sustainability assessment. Sustainability 2(10), 3309–3322 (2010)
5. Valdivia, S., Ugaya, C.M.L., Hildenbrand, J., Traverso, M., Mazijn, B., Sonneman, G.: A UNEP/SETAC approach towards a life cycle sustainability assessment-our contribution to Rio+20. Int. J. Cycle Assess. 18, 1673–1685 (2013)

6. ISO: ISO 14040 International Standard. Environmental management - Life cycle assessment - Principles and framework. International Organisation, Geneva, Switzerland, (2006a).
7. ISO: ISO 14040 International Standard. In: Environmental management - Life cycle assessment - Requirements and Guidelines. International Organisation, Geneva, Switzerland (2006b)
8. Lacasa, E., Santolaya, J.L., Biedermann, A.: Obtaining sustainable production from the product design analysis. J. Clean. Prod. **139**, 706–716 (2016)
9. Santolaya, J.L., Lacasa, E., Biedermann, A., Muñoz, N.: A practical methodology to project the design of more sustainable products in the production stage. Res. Eng. Des. **30**, 539–558 (2019)
10. Brezet, J.C., Van Hemel, C.G.: Eco-design: A Promising Approach to Sustainable Production and Consumption. UNEP. United Nations Publications, Paris (1997)
11. UNEP/SETAC: Guidelines for Social Life Cycle Assessment of Products. United Nations Environment Programme, Paris (2009)
12. UBA: Umweltbundesamt (UBA). German Environmental Protection Agency (2007)

# Enhanced Material-Driven Design Methodology: Luffa Cylindrica's Case

Alejandro Plumed[1], David Ranz[1(✉)], Ramón Miralbes[1], and Gustavo Vargas[2]

[1] Departamento de Ingeniería de Diseño y Fabricación, EINA, Universidad de Zaragoza,
C/ María de Luna s/n, 50018 Zaragoza, Spain
dranz@unizar.es

[2] David L. Hirschfeld, Department of Engineering, Angelo State University,
San Angelo, TX, USA

**Abstract.** As society and regulations demand more ecological materials, we must focus on finding new properties in potential candidates that make them really feasible and open up new possibilities. A new methodology is established to reach that goal, based on already existing ones, but with a broader vision; more focused on the use, and not solely on the product. To put it into practice, it's been applied on Luffa Cylindrica, a plant with interesting properties. A volumetric, mechanical and perceptual characterization of it has been carried out, so that we can create new concepts specially linked to that. One of these paths has been fully developed to the point of getting a physical prototype, while more complex ones have been suggested. The stated methodology may be used as a flexible guide to find new materials not only more environmentally-friendly, but better than what's stablished.

**Keywords:** Natural · Material · Eco-friendly · Luffa · Design methodology

## 1 Introduction

Social pressure leads to the use of more ecological solutions in the industry. The problem is that they don't usually match the features of their traditional counterparts, and end up succeeding by mandatory regulation. Materials choice is one of the fields in which this behavior applies.

The most effective approach to this problem is to find applications in which each new material excels, instead of trying to substitute established solutions with worse alternatives that will not naturally spread in the market.

There are some methodologies that already focus on the material itself to discover its maximum potential. *Material Driven Design* (MDD) [1] is one of those, in this case specifically thought for eco-friendly materials. Material Experience Lab created the MDD Methodology [1], in order to give a clear guide to accomplish innovative products using that kind of eco-solutions. This methodology is highly focused on the experiential characteristics of the materials, leading to final outcomes that accurately reflect the core values of them. A new methodology is established to reach that goal,

L. Roucoules et al. (Eds.): JCM 2020, LNME, pp. 182–187, 2021.
https://doi.org/10.1007/978-3-030-70566-4_29

based on already existing ones, but with a broader vision; more focused on the use, and not solely on the product. Here, a modified version of it is stated, with a broader focus, not only suitable for consumer-oriented products, but for large-scale applications. This enhanced new approach methodology has been applied on *Luffa Cylindrica*, the dried fruit of a plant with interesting properties. A volumetric, mechanical and experential characterization of it has been carried out, so that we can create new concepts specially linked to those properties. One of the processes associated to this methodology was fully developed till a physical prototype was manufactured, while more complex ones have been suggested. The stated methodology may be used as a flexible guide to find new eco-friendly materials applications, which outperform their predecessors.

## 2   Methodology

This new methodology based on the MDD is stated (Table 1). While MDD is more product-guided, its modified version covers more application fields, so it can deliver higher-level industrial uses, while not dismissing the outcomes of the MDD.

**Table 1.**  Comparison between methodologies.

| Stage | MDD | Modified MDD |
|---|---|---|
| Phase I | • Mechanical characterization<br>- Manufacturing processes<br>- Mechanical information<br>- Full experiential characterization<br>- UX Tools<br>- Broad research<br>- Properties<br>- Cost | • Volumetric characterization<br>• Mechanical characterization<br>- Manufacturing processes<br>- Mechanical testing<br>- Material benchmarking<br>- Light experiential characterization<br>- Broad research<br>- Properties<br>- Cost |
| Phase II | "Manifesting materials experience patterns" | • Use cases<br>• Applications/Concepts |
| Phase III | "Manifesting materials experience patterns" | • Development of the product or application |
| Phase IV | • Concepts<br>• Development<br>• Agile prototyping | • Prototyping |

This new approach puts more emphasis on mechanical and volumetric analysis of the material, so that a more specific characterization of it can be reached, and delays the use of user-centered techniques of the MDD to later stages, when the application path is defined and the specific tools to be used are clear.

## 3    Phase I: Luffa Cylindrica Understanding

*Luffa Cylindrica* is a *Cucurbitaceae* plant [2] growth in warm weather. However, the Luffa, commonly referred to as vegetable sponge or luffa cloth, is obtained from its matured dried fruit. Although it's already used for personal care as scrub, and eaten in eastern countries like India, China or Vietnam, no industrial application of it has been found yet.

### 3.1    Volumetric Characterization

The first step was to get a simple model of the Luffa in a CAD file, in order to know its internal structure and potentially get a micromechanical model [3]. Due to its fibrous structure, a medical-oriented program called 3DSlicer was used. It can generate volumes with cross sections as inputs, which was a good approach in this case, as substantial variability can be observed within a few centimeters (Fig. 1).

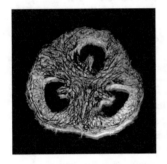 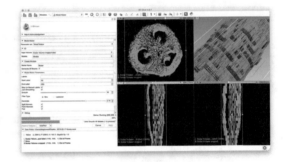

**Fig. 1.** Luffa slice and smoothing process in 3DSlicer.

Due to the superficial smoothness of the Luffa, the minimum thickness that could be achieved for each slice was 10 mm and thus one piece was divided in 25 parts. In that way several full models of Luffa were modelled and analysed.

While conducting a full CT scan would be useful in order to get an exact micromechanical model, stacked up 2D images, along with the smoothing tools provided by 3DSlicer is an economic pathway to create an accurate CAD file.

### 3.2    Mechanical Characterization

Mechanical tests were carried out, both on the material on its own and together with linen as a composite sandwich configuration. The last case is particularly interesting. Bending tests were executed for Luffa-Linen, Cork-Linen and EPS-Linen, according to ASTM C393/C393M [4]. While Cork's showed better values in each measured mechanical property, Luffa's outperformed EPS' in core shear ultimate strength (a mean of 0,261 MPa and 0,198 MPa respectively) and facing ultimate strength (a mean of 12,61 MPa and 10,4 MPa, respectively), with the first one being substantially less dense.

Additionally, testing the Luffa on its own, it was possible to place it in a material benchmarking graph (Fig. 2). Specifically, "Compact Luffa Cores", meaning that the side surface of the material (the least dense part) was removed, while the core material of 4 slices was compacted, so that they made up a less empty block.

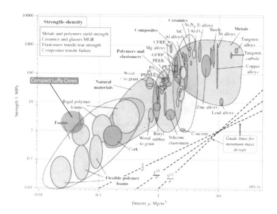

**Fig. 2.** Compact Luffa Cores in material Ashby' chart.

### 3.3  Other Properties

Aside from mechanical properties, Luffa also shows interesting capabilities that could drive the final use of the material. Among them, acoustic characteristics (up to 87% absorption rate as a composite [5]), magnificent permeability and oil dissipation [8], up to 220% higher deflective compression index than traditional alternatives [6] or the possibility to use it as a biocompatible filler together with PLLA, cellulose and HA [7].

As stated before, a light experiential characterization was also carried out with 24 people, concluding that Luffa is seen as a more expensive, heavier and harder material before touching it.

## 4  Phase II: Conceptualization

Once the material was characterized, it was time to associate its properties to random isolated uses, and later evaluate the potential and feasibility of each option to follow one path.

The two most interesting alternatives were: a product based on its oleophilic characteristics to help with oil spills and a saddle that made use of its open-structure and hydrophobic properties.

Although the first concept also seems quite attractive in order to take advantage of material properties. It raised serious doubts about whether, as a single-and-critical purpose product, it could be fully developed and tested in the limited given budget and timespan. Finally, the second option was chosen and fully developed till prototype manufacturing.

# 5   Phase III and IV: Development and Prototyping

The saddle was developed as a sandwich composite using 100% ecological materials, such as a luffa core and linen reinforced faces. This provides an equivalent structural behaviour than traditional one. Besides, the plant sponge features open-core structure, which provide an effective airflow to dissipate the sweat.

Its molds were 3D printed. Inside them, the material was placed and impregnated with eco-epoxy resin, achieving an almost 100% ecological solution (Figs. 3 and 4).

**Fig. 3.** Prototyping process.

**Fig. 4.** Final prototype.

# 6   Conclusions

With the Luffa, it's been checked how the modified version of the MDD can deliver both final products and high-level industrial applications. This methodology may serve as a flexible guide to work with eco-friendly materials, find a suitable application and fully characterize them in many points of view.

**Acknowledgments.** The authors thank Iberluffa (Iberica de esponjas vegetales) for providing material needed for this research work.

# References

1. Karana, E., Barati, B., Rognoli, V., Zeeuw van Der Laan, A.: Material driven design (MDD): a method to design for material experiences. Int. J. Des. **9**(2) (2015). https://www.ijdesign.org/index.php/IJDesign/article/view/1965/693. ISSN 1994-036X. Accessed 21 Jan 2020
2. Luffa.info. https://luffa.info/luffagrowing.htm. Accessed 20 Jan 2020
3. ASTM C393 / C393M – 16. Standard Test Method for Core Shear Properties of Sandwich Constructions by Beam Flexure
4. Islam, M.R., Tudryn, G., Bucinell, R., Schadler, L., Picu, R.C.: Morphology and mechanics of fungal mycelium. Sci. Rep. (2017). https://doi.org/10.1038/s41598-017-13295-2
5. Körük, H., Genc, G.: Investigation of the acoustic properties of bio luffa fiber and composite materials. Mater. Lett. **157**, 166–168 (2015). https://doi.org/10.1016/j.matlet.2015.05.071
6. Chen, Y., Yuan, F., Guo, Y., Hu, D., Zhu, Z., Zhang, K., Zhu, S.: A novel mattress filling material comprising of luffa fibers and EVA resin. Ind. Crops Prod. **124**, 213–221 (2018). https://doi.org/10.1016/j.indcrop.2018.07.074
7. Cecen, B., Kozaci, L.D., Yuksel, M., Ustun, O., Ergur, B.U., Havitciouglu, H.: Biocompatibility and biomechanical characteristics of loofah based scaffolds combined with hydroxyapatite, cellulose, poly-L-lactic acid with chondrocyte-like cells. Mater. Sci. Eng. C **91**, 437–446 (2016). https://doi.org/10.1016/j.msec.2016.07.007
8. Ola, A.: Assessment of raw *luffa* as a natural hollow oleophilic fibrous sorbent for oil spill cleanup. Alexandria Eng. J. **53**(1), 213–218 (2014). https://doi.org/10.1016/j.aej.2013.11.001

# A Well-to-Wheel Comparative Life Cycle Assessment Between Full Electric and Traditional Petrol Engines in the European Context

Filippo Cucinotta, Marcello Raffaele, and Fabio Salmeri[✉]

Department of Engineering, University of Messina, Contrada Di Dio (S. Agata), 98166 Messina, Italy
fsalmeri@unime.it

**Abstract.** Automotive sector is crucial for the economic and social system. Conversely, it also plays an important role in the global emissions balance with strong consequences on the environment. Currently the Research world is engaged in the reduction of the emissions, especially in order to contrast the Climate Change and reduce toxicity on humans and the ecosystem. This study presents a comparative Life Cycle Assessment, Well-to-Wheel, between the most common technology used in the automotive sector, i.e. the traditional petrol Internal Combustion Engine and the full Battery Electric Vehicle. The different configurations have been analysed within 17 different impact categories in terms of climate change, human health, resourced depletion and ecosystems. The Well-to-Wheel approach allows to focus the attention on the use stage of the vehicle, considering the local effects due to the direct emissions in high density urban zones and it mitigates the dependence of usage hypotheses, different scenarios and intrinsic differences between the various models of cars in circulation.

**Keywords:** Life Cycle Assessment · Electric vehicles · Green design · Climate Change

## 1 Introduction

Road transport is the third largest source of European greenhouse gases accounting for around 26% of total emissions [1]. In the UK, e.g., conventional road transport also remains the predominant source of many local emissions including nitrogen oxides (NOx) and particulates (PMs). Within urban areas, the percentage contributions due to road transport are particularly high. For example, although road transport is responsible for around a quarter of particulates on a national level, in London road transport contributes almost 60% of known primary emissions [2]. Worldwide population growth and industrialization have resulted in increases in demand for energy in the transportation sector, among other sectors. As a result, air pollution and anthropogenic greenhouse gas emissions have become key global problems. Various options are available to mitigate

© The Author(s) 2021
L. Roucoules et al. (Eds.): JCM 2020, LNME, pp. 188–193, 2021.
https://doi.org/10.1007/978-3-030-70566-4_30

the effect of the road emissions. The use of alternative fuels, mainly biodiesel, petrol–alcohol blends, natural gas and liquefied petrol gas in vehicular applications has grown in recent years in European Union countries, the United States, Japan, India, Brazil and many other markets. Legislation is also in place to encourage or effectively force further adoption of these fuel types [3]. European Union requirements regarding vehicle emissions for passenger cars and light commercial vehicles, specified as Euro 6, set strong limits for emissions of HC, CO, and NOx. Instead, $CO_2$ emissions are covered by separate legislation. The Life Cycle Assessment (LCA) is a very valid method in order to evaluate the real impact of a process or a product [4, 5]. In particular, the Well-to-Wheel approach (WTW) allows to better emphasize on the usage phase, focusing on aspects relating to emissions in urban areas and more specifically verifying aspects related to engines and the production and refining of fuels [6].

## 2 Materials and Methods

The LCA analysis has been carried out with OpenLCA software version 1.10.1 [7], integrated with the ecoinvent database version 3.6 [8]. The allocation of the environmental burden has been handled by means of the cut-off system. In this way, if a material is recycled, the primary producer does not receive any credit for the provision of any recyclable materials. As a consequence, recyclable materials are available burden-free to recycling processes, and secondary (recycled) materials bear only the impacts of the recycling processes. Recyclable materials have been selected on the European market for products, consequently they include all the upstream burdens included average transports of that product within the geography, as well as inputs of the product itself to cover losses in trade and transport [8].

The consumption is stated, depending on the propulsion concept, in litre petrol per 100 km, for the traditional Internal Combustion Engine (ICE) and in kilowatt-hours per 100 km, for the Battery Electric Vehicle (BEV). For a general comparison, the energy consumption is converted in MJ/100 km/100 installed-kW as well for tank-to-wheel and well-to-tank approach. The consumption per kilometre of a car is strongly influenced by several factors, among which the most important are speed, weight and aerodynamics. In order to make a comparison between the different powertrains, a car model was chosen, looking for those on the market, sedan-type cars (with 3 volumes), with similar, as much as possible, installed power, of the same segment (medium) and of the same year of production (2018). The choice was for a Mercedes C-Klasse with a petrol 155 kW internal combustion engine (1991 $cm^3$) and a Tesla model S with a 193 kW electric engine.

This particularly affects the energy mix referred to in the production of electricity for recharging electric car batteries but also the emission related to the petrol production and distribution. In all cases, reference was made to urban driving in areas with high population density. The electricity market activity ends with the transport of the energy at low voltage (400 V) electricity in the transmission network over aerial lines and cables. The ecoinvent dataset includes: electricity inputs produced in this country and from imports and transformed to low voltage; the transmission network; direct emissions to air; electricity losses during transmission. The dataset is extrapolated from year 2017 to the year of the calculation (2019). The energy mix change greatly in function of the natural resources and the political choices of the country. The Well-to-Tank (WTT) stage includes fuel extraction, refining and distribution for ICE vehicles and, electrical energy production, transformation and distribution for BEVs. The Tank-to-Wheel (TTW) stage includes the fuel combustion for ICE while it has not emissions for BEVs. The Life Cycle Impact Assessment (LCIA) has been conducted according to the ILCD 2018 method [9]. In this work, the midpoint methodology has been adopted. The ILCD midpoint categories are grouped into four areas of protection: Climate Change (CC), Human Health (HH), Resources Depletion (RD) and Ecosystems (EQ). The impact categories of the endpoint method are those of IPCC for CC [10], of Posch et al. [11] for EQ, Henderson et al. [12], Frischknecht et al. [13], for HH and van Oers and Guinée [14], Milà I Canals et al. [15], van Oers et al. [16] for RD.

## 3   Results and Discussion

The emission factors, expressed in g/100 km, have been taken from the specialized site https://ecoscore.be for the two chosen cars. These emissions are related only to the use phase of the vehicle, i.e. they are the pollutants emitted while driving. A share of these emissions depends on engine technology, another part on the fuel type and fuel consumption, and a third part is measured during the official approval tests a vehicle has to undergo before it can enter the European market. The specific TTW emissions for the ICE vehicle are 12800 g/100 km of $CO_2$, 23.3 g/100 km of CO, 0.4 g/100 km of NOx and 2.9 g/100 km of HC. The estimated energy consumption is of 249 MJ/100 km (corresponding to a petrol consumption of 8.4 l/100 km) for the ICE vehicle, and an estimated energy consumption of 78.1 MJ/100 km for the BEV. The results are reported for 100 km and for 100 kW of installed power, according to the ILCD 2018 method. Each emission is reported in terms of equivalent unit and is categorized in the different areas of protection. Furthermore, the data are represented as a percentage comparison, making the ICE vehicle as 100, i.e. $(BEV_{emission} - ICE_{emission})/ICE_{emission}\%$. In this way a positive result indicates that the BEV has a higher impact on the environment and vice versa.

The global results for 100 km and for 100 kW of installed power, according to the ILCD 2018 method, are reported in Fig. 1. Each emission is reported in terms of percentage difference between the BEV and the ICE referring to the European average. The standard deviation shows a great variability of the data, especially in function of the electricity mix of the country. Focusing on some particular impact category, it is possible to see better the difference in terms of country (see Fig. 2) yellow has been attributed to a zero difference between ICE and BEVs, green to a better impact and red to a worst impact.

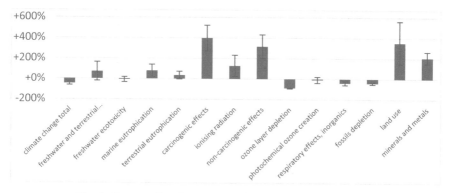

**Fig. 1.** Impact effect for 100 km and 100 kW of installed power

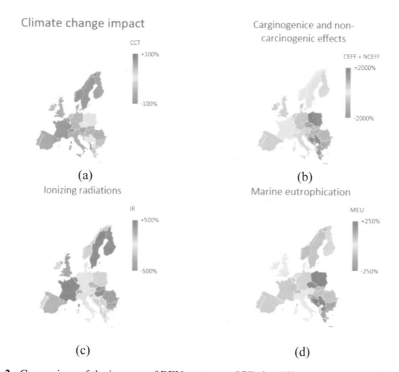

**Fig. 2.** Comparison of the impacts of BEV, respect to ICE, for different categories of impact

## 4   Conclusions

A comparative Well-To-Wheel Life Cycle Assessment among an ICE and a similar BEV, has been carried out in the most relevant European countries. The impact categories analysed have been based on the ILCD 2018 method with the ecoinvent database (year 2019) and the OpenLCA software.

The finding obtained are summarized as follows:

- Electric vehicles play a fundamental role in the battle against Climate Change, since they lead to a WTW reduction until 50% compared to the traditional ICE vehicles, with small exceptions in countries heavily dependent on fossil fuels;
- Conversely, BEVs have a generally higher impact on categories related to human health;
- The eutrophication of the waters (marine and fresh) and of the soil is highly dependent on the energy production methods;
- Countries with electricity produced by means of nuclear power plants have very good performance in terms of Climate Change but have major repercussions on ionizing radiation and dissipated water;
- The WTW approach allowed to focus on the fuel impact but doesn't consider construction, maintenance and dismantling of the vehicles.

## References

1. UNFCCC: Greenhouse gas emissions by country and sector (infographic). https://www.europarl.europa.eu/news/en/headlines/society/20180301STO98928/greenhouse-gas-emissions-by-country-and-sector-infographic
2. DEFRA: Air Pollution in the UK 2015 (2016)
3. Directive 2009/28/EC of The European Parliament and of The Council of 23.04.2009 on the promotion of the use of energy from renewable sources and amending and subsequently repealing Directives 2001/77/EC and 2003/30/EC. Official J. Eur. UN (2009)
4. Barone, S., Cucinotta, F., Sfravara, F.: A comparative life cycle assessment of utility poles manufactured with different materials and dimensions. In: Eynard, B., Nigrelli, V., Oliveri, S., Peris-Fajarnes, G., Rizzuti, S. (eds.) Advances on Mechanics, Design Engineering and Manufacturing, pp. 91–99. Springer, Cham (2017). https://doi.org/10.1007/978-3-319-45781-9_10.
5. Cucinotta, F., Guglielmino, E., Sfravara, F.: Life cycle assessment in yacht industry: a case study of comparison between hand lay-up and vacuum infusion. J. Clean. Prod. (2016). https://doi.org/10.1016/j.jclepro.2016.10.080
6. Campanari, S., Manzolini, G., Garcia de la Iglesia, F.: Energy analysis of electric vehicles using batteries or fuel cells through well-to-wheel driving cycle simulations. J. Power Sources 186, 464–477 (2009). https://doi.org/10.1016/j.jpowsour.2008.09.115.
7. Ciroth, A.: ICT for environment in life cycle applications openLCA - a new open source software for life cycle assessment. Int. J. Life Cycle Assess. 12, 209–210 (2007). https://doi.org/10.1065/lca2007.06.337
8. Wernet, G., Bauer, C., Steubing, B., Reinhard, J., Moreno-Ruiz, E., Weidema, B.: The ecoinvent database version 3 (part I): overview and methodology. Int. J. Life Cycle Assess. 21, 1218–1230 (2016). https://doi.org/10.1007/s11367-016-1087-8
9. European Commission - Joint Research Centre - Institute for Environment and, Sustainability: International Reference Life Cycle Data System (ILCD) Handbook - General guide for Life Cycle Assessment - Detailed guidance, Luxembourg (2010). https://doi.org/10.2788/38479
10. Intergovernmental Panel on Climate Change (IPCC): Climate change 2007: Synthesis report (2008). https://doi.org/10.1256/004316502320517344.
11. Posch, M., Seppälä, J., Hettelingh, J.P., Johansson, M., Margni, M., Jolliet, O.: The role of atmospheric dispersion models and ecosystem sensitivity in the determination of characterisation factors for acidifying and eutrophying emissions in LCIA. Int. J. Life Cycle Assess. 13, 477–486 (2008). https://doi.org/10.1007/s11367-008-0025-9

12. Henderson, A.D., Hauschild, M.Z., Van De Meent, D., Huijbregts, M.A.J., Larsen, H.F., Margni, M., McKone, T.E., Payet, J., Rosenbaum, R.K., Jolliet, O.: USEtox fate and ecotoxicity factors for comparative assessment of toxic emissions in life cycle analysis: sensitivity to key chemical properties. Int. J. Life Cycle Assess. **16**, 701–709 (2011). https://doi.org/10.1007/s11367-011-0294-6

13. Frischknecht, R., Braunschweig, A., Hofstetter, P., Suter, P.: Human health damages due to ionising radiation in life cycle impact assessment. Environ. Impact Assess. Rev. **20**, 159–189 (2000). https://doi.org/10.1016/S0195-9255(99)00042-6

14. van Oers, L., Guinée, J.: The abiotic depletion potential: background, updates, and future. Resources **5** (2016). https://doi.org/10.3390/resources5010016.

15. Milà I Canals, L., Bauer, C., Depestele, J., Dubreuil, A., Knuchel, R.F., Gaillard, G., Michelsen, O., Müller-Wenk, R., Rydgren, B.: Key elements in a framework for land use impact assessment within LCA. Int. J. Life Cycle Assess. **12**, 5–15 (2007). https://doi.org/10.1065/lca2006.05.250.

16. van Oers, L., De Koning, A., Guinée, J.B., Huppes, G.: Abiotic resource depletion in LCA. Public Work. Water Manag. 1–75 (2002)

# Product Eco-Design in the Era of Circular Economy: Experiences in the Design of Espresso Coffee Machines

Claudio Favi[1], Marco Marconi[2(✉)], Marta Rossi[3], and Federica Cappelletti[3]

[1] Department of Engineering and Architecture, Università degli Studi di Parma, Parco Area delle Scienze 181/A, 43124 Parma, Italy
[2] Università degli Studi della Tuscia, Largo dell'Università, 01100 Viterbo, Italy
marco.marconi@unitus.it
[3] Department of Industrial Engineering and Mathematical Sciences, Università Politecnica delle Marche, Via Brecce Bianche, 60131 Ancona, Italy

**Abstract.** Product eco-design includes several methodologies aimed at supporting companies in the development of sustainable products. Currently, this theme is assuming an important role in both the academia and industry worlds due to the increasing attention to environmental problems and the need for a transition toward circular economy business/organizational models. In this context, the present paper focuses on the industrial sector of espresso coffee machines manufacturing which has several unexploited potentialities. The analysis of the sector specificity (internal and external contexts), as well as of the product lifecycle allowed to define an eco-design framework to guide companies involved in the design and production of espresso coffee machines. Effective eco-design strategies should include the combined use of specific methods, tools and metrics to manage all the most important lifecycle phases (beginning of life, middle of life, end of life) during the design activities in order to set preventive actions that avoid future potential environmental impacts. Only in this way, the environmental and economic benefits of the circular economy paradigm (e.g. remanufacturing/reuse of selected components) can be practically exploited in real industrial contexts. The presented case studies confirmed that the application of design for disassembly rules positively contributes to increase the product performances during maintenance and end of life, while a re-design oriented to component modularity could be a key strategy to pursue remanufacturing for boilers, a key and expensive component included in espresso coffee machines.

**Keywords:** Product eco-design · Circular economy · Remanufacturing · Recycling · Coffee machine

## 1 Introduction

Eco-design includes a set of design methodologies that support technical departments in the development of sustainable products. Several key factors are forcing companies

L. Roucoules et al. (Eds.): JCM 2020, LNME, pp. 194–199, 2021.
https://doi.org/10.1007/978-3-030-70566-4_31

to launch in the market products with a minimized environmental load: new legislations, the opportunity to enlarge the market share, economic advantages, brand preservation, increasing environmental awareness of customers [1]. This context stimulated the development of hundreds of general-purpose or sectorial eco-design methods and tools from both the scientific and industrial world [2]. In recent years, the concept of circular economy (CE) acquired importance as a business/organizational model focused on the increasing of product lifetime, minimization of scraps/wastes, implementation of closed-loop end of life (EoL) scenarios [3]. Currently, the implementation of CE models is mainly based on "remedial" actions that foresee the recovery of products and materials discarded at the end of their useful life. Too many decisions are still made without a strategy, resulting in higher costs and lower success rate, while eco-design must be a proactive management approach [4]. However, such approaches have a limited efficiency and only a transition to "preventive" actions can lead to full benefits of the CE paradigm [5]. This is the main reason why in the last years the scientific community is increasingly focused on the proposal of design methodologies aimed at supporting the development of products that are easy to manage and recover at EoL. One of the most widespread is design for disassembly (DfD), a target design methodology focused on the product disassembly, a preliminary but essential phase for both maintenance/repair of products during the use phase, and at the EoL, in order to favor the component reuse/remanufacturing and the material separation/recycling [6]. Even if disassembly occurs during the lifetime or at the EoL, the choices made at the design phases strongly influence the disassembly performance. For this reason, several literature studies are focused on identifying the optimized disassembly sequence for components with high residual value [7], defining calculation methods and new disassemblability metrics (e.g. time, cost), evaluating EoL performance, developing tools for supporting design activities [8].

The present paper aims to investigate how existing eco-design methods, tools and metrics can be used in the context of espresso coffee machines. The characterization of such industrial sector and products provided several eco-design actions that can be re-injected in other field and in the development of other products. The novelty consists in defining a framework to guide companies belonging to this sector in the transition toward eco-sustainability and the exploitation of CE opportunities, through the implementation of eco-design methods and tools.

After the Introduction that clarifies the general context, Sect. 2 explains how eco-design can be applied in case of coffee machine production, detailing the most useful and suitable methods, tools and metrics. Section 3 presents some experiences of eco-design implementation carried out in collaboration with an Italian company. Finally, Sect. 4 discusses outcomes, limitations and future work.

## 2 Product Eco-Design for Coffee Machines

Starting from the analysis of eco-design methods and tools, a customization for professional coffee–machine sector is proposed; a selection of tools is identified, and benefits related to their implementation are discussed considering the life cycle phases and the issues faced.

A professional espresso coffee machine is a complex product whose main functionality is to support the barman in the preparation of coffee-based drinks. Analyzing the product and the different lifecycle phases, it is clear that to effectively pursue eco-design, several internal and external factors need be taken into account. Specific legislations (e.g. EU directives on eco-design or waste of electric equipment), customer requirements (e.g. high-end vs low-end markets, customer awareness on environmental themes), company business objectives, the complexity of certain key components/functional groups (e.g. boiler, brewing groups), the high consumption of energy during the long lifetime (up to 10–15 years), the difficulty to manage the product at EoL and exploit CE potentialities, are some of the aspects to be considered, forcing to take into account the whole product lifecycle. Different eco-design strategies and metrics need to be simultaneously used to have sensible benefits and avoid burden shifting, as depicted in Fig. 1.

Beginning of Life (BoL) includes all the phases needed for the coffee machine manufacturing until its distribution and use. Due to the complexity of such products a high number of heterogeneous materials are commonly used (e.g. steel for frame, copper for boilers, aluminum for other internal parts), brass or aluminum for brewing groups, different typologies of plastics for internal parts and the external case, stainless steel in case of high-end market products). Moreover, it must be considered that material choice strongly influences the performances of the successive phases (e.g. recyclability or possibility to separate material at EoL). Therefore, eco-design in the context of coffee machine BoL requires the use of material selection tools or at least a material database including environmental properties of materials (e.g. *Ansys Granta Selector* tool equipped with the *Ansys Granta MI* database). The use of such tools allows anticipating sustainable matters yet in the first design phase and reduce the environmental impact of products.

Middle of Life (MoL) is a key phase, since it is well known that most of the environmental impacts for energy using products is caused during their use (up to 95–98% in case of coffee machines). The energy consumption is the most important metric concerning the use phase. Its accurate forecast and/or measurement through standard protocols and use scenarios (not existing at the moment in the field of coffee machines) are essential aspects to implement effective eco-design actions (e.g. thermal insulation of boilers and pipes for heat dissipation reduction). Concerning Life Cycle Assessment (LCA) indicators to be considered for the use phase analysis, the Global Warming Potential (GWP measured in [kgCO2eq]) is certainly the most suitable and useful indicator for coffee machines, as also confirmed by recent literature studies [9].

End of Life includes all the phases successive to the product discard at the end of the useful life. The monitoring of the product/component disassemblability during the design phase is a key activity to guarantee the implementation of closed-loop EoL strategies. DfD metrics, as *disassembly time and cost*, must be quantified for target components, on the basis of available design data (e.g. 3D models, product structure, bill of materials), by using dedicated methods and tools [8]. In addition, *EoL indices* should be exploited during the coffee machine design to preliminarily verify the feasibility and convenience of a certain EoL scenario (e.g. remanufacturing), and focus the design for EoL strategy toward the right direction [10].

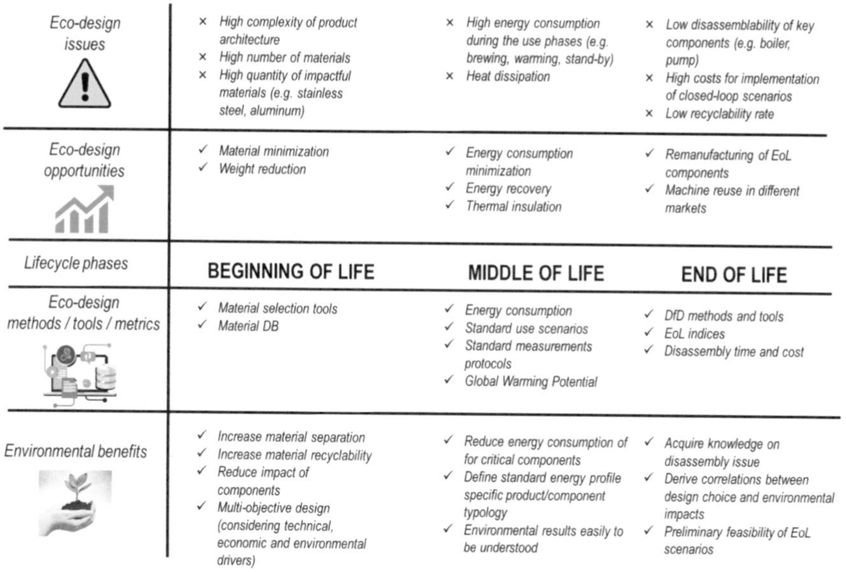

| Eco-design issues ⚠ | × High complexity of product architecture<br>× High number of materials<br>× High quantity of impactful materials (e.g. stainless steel, aluminum) | × High energy consumption during the use phases (e.g. brewing, warming, stand-by)<br>× Heat dissipation | × Low disassemblability of key components (e.g. boiler, pump)<br>× High costs for implementation of closed-loop scenarios<br>× Low recyclability rate |
|---|---|---|---|
| Eco-design opportunities 📈 | ✓ Material minimization<br>✓ Weight reduction | ✓ Energy consumption minimization<br>✓ Energy recovery<br>✓ Thermal insulation | ✓ Remanufacturing of EoL components<br>✓ Machine reuse in different markets |
| Lifecycle phases | **BEGINNING OF LIFE** | **MIDDLE OF LIFE** | **END OF LIFE** |
| Eco-design methods / tools / metrics | ✓ Material selection tools<br>✓ Material DB | ✓ Energy consumption<br>✓ Standard use scenarios<br>✓ Standard measurements protocols<br>✓ Global Warming Potential | ✓ DfD methods and tools<br>✓ EoL indices<br>✓ Disassembly time and cost |
| Environmental benefits | ✓ Increase material separation<br>✓ Increase material recyclability<br>✓ Reduce impact of components<br>✓ Multi-objective design (considering technical, economic and environmental drivers) | ✓ Reduce energy consumption of for critical components<br>✓ Define standard energy profile specific product/component typology<br>✓ Environmental results easily to be understood | ✓ Acquire knowledge on disassembly issue<br>✓ Derive correlations between design choice and environmental impacts<br>✓ Preliminary feasibility of EoL scenarios |

**Fig. 1.** Correlations among lifecycle phases, eco-design issues/potentialities and most suitable supporting methods/tools/metrics.

# 3 Case Studies

In this section, few examples of good practices oriented to eco-design are presented. Most of them refers to the product EoL that is one of the most critical phases to consider in this century when virgin resources are becoming scarce, and the management of domestic and industrial wastes is a key issue for environment.

The first example proposed within this study is the reduction of disassembly time for target components (Fig. 2). Reduction of disassembly time is directly correlated to the reduction of disassembly operations (i.e. decrease the number of components to disassemble) and the reduction of time for each operation (i.e. faster assembly/disassembly connectors). In Fig. 2 the target component is the "plastic bowl" used to contain water. Bowl requires to be disassembled during the product use for water refilling and because this component is subjected to damage. The redesign action included the integration of different component in a single one (made by only one type of plastic) and the possibility to easy disassemble it with the adoption of cylindrical plastic snap-fits, instead of the original threaded joints.

Connected with the disassembly phase, EoL business models oriented to CE (i.e. component remanufacturing, product refurbishment) are new important paradigms to pursue for companies that want to increase their business volumes. To this aim, key components with high material and manufacturing costs require a specific design. This is the case of the "main boiler" which requires to be manufactured with copper-based materials (i.e. copper or brass) due to their thermal performances. The design of main boiler changed to adapt different connectors for other espresso machine models (Fig. 3). Indeed, this component can be remanufactured with the aim to be reused in a large number

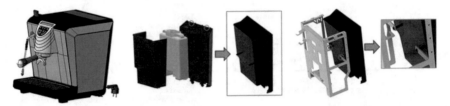

**Fig. 2.** Examples of design modifications to decrease disassembly time of plastic parts.

of models. The most important design change was related to the flange/cap that was adapted to several configurations and attachments with other components. Environmental analysis should be always coupled with cost analysis. Even remanufacturing option requires additional costs, they are lower than the economic profits provided by the implementation of this new business model.

A                                                    B

**Fig. 3.** Examples of design modifications for the main boiler (remanufacturing purpose): Original design (A) vs New design (B).

## 4   Conclusions

This paper investigates eco-design methods in the specific context of espresso coffee machine product development. The specificity of this product deals with the adoption of dedicated eco-design actions and metrics to implement CE business models. Few examples were discussed providing some models that can be exploited also in other contexts: (i) disassembly, (ii) recycling, and (iii) remanufacturing. While disassembly is a key performance indicator in the analysis of EoL options for target components, recycling and remanufacturing are new business models that can be implemented after a technical, economic and environmental analyses. Within the case study the re-design of target components allowed to implement emerging business models (i.e. product remanufacturing) encouraging the circular economy paradigm. Future work will be focused on the definition of key performance indicators (EoL indices) dedicated to the espresso coffee machine sector, able to consider the three pillars of sustainability: (i) economic, (ii) environmental, and (iii) social. These indicators can be used as decision-making tool in the assessment of design alternatives as well as in the definition of eco-design actions.

# References

1. Bey, N., Hauschild, M.Z., McAloone, T.C.: Drivers and barriers for implementation of environmental strategies in manufacturing companies. CIRP Ann. **62**(1), 43–46 (2013)
2. Dekoninck, E.A., Domingo, L., O'Hare, J.A., Pigosso, D.C.A., Reyes, T., Troussier, N.: Defining the challenges for ecodesign implementation in companies: development and consolidation of a framework. J. Clean. Prod. **135**, 410–425 (2016)
3. Reike, D., Vermeulen, W.J.V., Witjes, S.: The circular economy: new or refurbished as CE 3.0? - exploring controversies in the conceptualization of the circular economy through a focus on history and resource value retention options. Resour. Conserv. Recycl. **135**, 246–264 (2018)
4. Pigosso, D., Zanette, E.T., Filho, A.G., Ometto, A.R., Rozenfeld, H.: Ecodesign methods focused on remanufacturing. J. Clean. Prod. **18**, 21–31 (2010)
5. Vanegas, P., Peeters, J.R., Cattrysse, D., Tecchio, P., Ardente, F., Mathieux, F., Dewulf, W., Duflou, J.R.: Ease of disassembly of products to support circular economy strategies. Resour. Conserv. Recycl. **135**, 323–334 (2018)
6. Duflou, J.R., Seliger, G., Kara, S., Umeda, Y., Ometto, A., Willems, B.: Efficiency and feasibility of product disassembly: a case-based study. CIRP Ann. **57**(2), 583–600 (2008)
7. Kara, S., Pornprasitpol, P., Kaebernick, H.: A selective disassembly methodology for end-of-life products. Assem. Autom. **25**(2), 124–134 (2005)
8. Favi, C., Marconi, M., Germani, M., Mandolini, M.: A design for disassembly tool oriented to mechatronic product demanufacturing and recycling. Adv. Eng. Inform. **39**, 62–79 (2019)
9. Hicks, A.L., Halvorsen, H.: Environmental impact of evolving coffee technologies. Int. J. Life Cycle Assess. **24**, 1396–1408 (2019)
10. Favi, C., Germani, M., Luzi, A., Mandolini, M., Marconi, M.: A design for EoL approach and metrics to favour closed-loop scenarios for products. Int. J. Sustain. Eng. **10**(3), 136–146 (2019)

# A Novel Procedure to Design a Positionable and Stable Drilling Template for Spine Surgery

Nicola Cappetti, Carmen Brancaccio, Francesca De Sio, and Carlotta Fontana[✉]

University of Salerno, Fisciano, Italy
cfontana@unisa.it

**Abstract.** Spine surgery is based, nowadays, on the use of cutting-edge instruments that optimize the intervention processes in the operating room, with advantages that affect the patient himself. Among these, rapid prototyping is configured as a first-rate tool, thanks to its ability to detail the diagnostic treatment according to the specific pathological case under examination. An example of this technology is represented by the generation of a drilling template, to assist the surgeon in identifying the optimal direction of insertion of the pedicle screws, capable of significantly reduce intervention times, in addition to the inevitable exposure of the patient to ionizing radiation, to which he is subjected during a normal arthrodesis intervention procedure. The design of a drilling guide requires, however, a particular attention in identifying the undercuts present on the vertebral surface, those areas of the spinous process which, reported inside the cavity of the template, involve complications at the time of extraction. In parallel, it is vitally important to carry out an evaluation of its stability during its use. In this article, starting from the analysis of the interferences present during the insertion of the template, a semi-automatic correction model is proposed for the generation of a new profile of the same, which facilitates its extraction without causing injury to the vertebral regions involved from the contact with the mask.

**Keywords:** Computer-assisted surgery · Pedicle screw fixation · Surgical template insertion optimization · Spine · Undercuts

## 1 Introduction

The vertebral motor unit is, as for every part of the body, exposed to the possibility of degenerative pathologies, a natural process of global aging of the osteo-ligamentous structures of the vertebral column, and/or traumatic and congenital pathologies, such as to cause a structural disorder. In cases of pathologies in which results are not obtained using corrective aids or with physiotherapy, an intervention, known as pedicle arthrodesis, is indispensable. In particular, arthrodesis of the lumbar vertebrae is a surgical technique that allows to join bones in the lumbar spine to stabilize it in order to reduce pain or deformity. Experience shows that the clinical aid that best allows to guide the screws in the spinal fusion surgery with extreme precision is the use of a surgical mask [1–4], a vertebral drilling template designed for delicate pedicle arthrodesis operations. The

L. Roucoules et al. (Eds.): JCM 2020, LNME, pp. 200–205, 2021.
https://doi.org/10.1007/978-3-030-70566-4_32

surgical device allows to implant the screws in the vertebral tracts of interest with extreme safety and precision; the result is a substantial appreciable decrease in operating times (about 40%) and a massive reduction in the harmful ionizing radiation (about 80%) due to the X-ray control so far essential to perform the intervention [5, 6].

The templates, customized for individual patients, are designed according to the most modern CAD technologies and made thanks to 3D printing in bio-compatible material [7]. A normal pedicle arthrodesis operation requires that the screws must be inserted by previously making two suitably positioned holes on the vertebra which, as a rule, pass through the pedicles and are fixed in the vertebra body. Vertebra fixation is the most complex phase for this type of operation; it can be compromised by human error, morphology and cleaning of the vertebra, as extremely delicate tissues and structures are present in the intervention area which, if compromised, also cause irreversible damage to the patient. The solution to this problem consists in the design and construction of a "custom made" template for vertebral surgery, customized and optimized for the individual patient, aimed at the directional drilling of the vertebrae with a structure characterized by two hollow cylindrical geometry guides that have the function of guide the tip of the surgical drilling tool so that the hole is drilled in the position chosen in the pre-operative phase by the surgeon. Although the area to be operated is "clean", as the surgical team is preparing for a skeletal phase of the affected section, blood spills are inevitable throughout the intervention phase. This causes instability due to the positioning of the drilling mask on the vertebral surface due to the presence of superficial and deep muscle tissues and bundles affecting the area to be operated. The solution to this problem could be to design a template whose geometric profile perfectly follows the shape of the vertebral body on which it concerns. The presence of undercuts that can be found in the anchorage areas of the template can cause, however, problems during the insertion and extraction phase of the same. This type of problem is very similar to that encountered in the manufacture of molds in industrial processes. The mold must generally be designed without cavities that can limit the extraction of the piece: in particular, undercuts must be avoided, i.e. angles less than 90°, which, in fact, make the molded element indivisible from the mold. The parts of the model that during the extraction would ruin the shape are said to be undercut. The extraction of the model is possible only in the absence of undercuts, i.e. all those areas of the model that are in the shadow of the direction of extraction of the model itself. Multiple systems of automatic undercut recognition are now consolidated, according to the geometry to be obtained by means of the mold, starting from three-dimensional CAD models [8–13].

None of these methods, however, has ever been used so far for the design of such surgical devices, leading over time to neglect the problem of undercuts in vertebral drilling templates, focusing on design requirements, such as the direction of screw centering. The aim of this work, instead, is to dwell on the problems resulting from the use of the device, during the insertion and extraction phases, ensuring precision and maintenance of the positioning at the same time. This is possible through the implementation of an algorithm capable of eliminating the undercuts on the vertebra-template interaction surface, evaluating the uniqueness of positioning and the stability. The vertebra-template coupling is even more stable as the contact surface is extended. Thinking of creating a profile of the template that fully matches that of the spinous process would, however, lead to evident problems in the phase of fixing the template on the vertebra as well as in the phase of extraction of the same. Furthermore, carefully analysing the morphology of a lumbar vertebra, it can be noticed that the major surface irregularities are those affecting the spinous process. Imagining, therefore, to position the drilling mask on the vertebra, the undercut problems that are encountered are precisely those affecting the spinous process and the cavity obtained inside the central body of the template. In this regard, it is not immediate to define the correct extraction directions for the model, given that the undercuts present could prevent more than one. The design and subsequent construction of a template is, therefore, bound by the extraction direction established during the design phase. The problem is, however, that by bypassing the undercuts, contact surfaces are eliminated, reducing stability. It is important to evaluate the optimal extraction direction taking also into account the stability.

The aim of this work was to implement a corrective semi-automatic model that, after identifying the presence of any undercuts, is able to modify the anchoring profile of the template according to the extraction direction, to determine its final contact surface and its stability. The profile of the template, therefore, will be that of a 'perfectly matching' geometry with the spinous process, subtracted, however, from the relevant undercuts. In this way, it will be possible to position and extract the model from the vertebra without difficulty, without affecting any effect on stability, due to a smaller contact surface.

## 2   Methodology

For the generation of the template, a specific methodology has been implemented, aimed at optimizing the conditions of insertability, stability and uniqueness of positioning, which maximize the use of the device during the operating phase. The pursued methodology provided for the succession of a series of procedural steps, listed below:

- Using a medical image processing software (3DSlicer, slicer.org), a three dimensional surface, in.STL format, related to the spinous process of a lumbar vertebra has been obtained, and the part of the surface of the spinous process on which the template rests is duplicated.
- One of the infinite possible extraction directions has been identified. As the extraction direction changes, undercuts located in different positions will occur. Note how the insertion and extraction of the template take place in the same direction, only the orientation changes, so in the two operations the same shaded areas are found.
- To identify the undercuts obtained for an assigned direction, the scalar product was calculated between the normal vectors and the extraction direction chosen, according to the traditional *convexity approach method*. The outlines of the undercut areas have been grouped (Fig. 1).

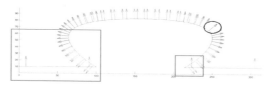

**Fig. 1.** Identification of undercut areas, highlighted in the square regions, according to the chosen extraction direction (highlighted in the circle).

- A new reference system was then adopted, in order to align the $z$ axis with the extraction direction and orientation, for all the vertices of the undercut areas.
- We proceeded with the analysis of the logics used to modify the vertices of all the triangles representative of the shadow areas found during the extraction. This modification takes place in correspondence with the control points, which, delimiting the previously highlighted regions, are not subject to change in the implementation of the algorithm. In particular, starting from the analysis of vertices with the highest value of $z$ coordinate, the following cases may occur:

  - Case 1: the node belongs to a triangle where the other two nodes are on the contour. It will be necessary to move this node perpendicular to the segment joining the two nodes of the contour;
  - Case 2: the node belongs to two triangles that have two nodes on the contour. If the triangles are adjacent, the node must be aligned with the common one;
  - Case 3: the node belongs to a triangle with a node on the contour, proceed as in Case 1 but you have to move the segment holding the node still;
  - Case 4: the node belongs to two adjacent triangles with a node on the contour, the node must be aligned with the node on the contour (same x-y coordinates in the new system);

- The algorithm outputs a cloud of points representing the new profile of the template that facilitates its extraction and that at the same time guarantees its insertion. It represents the first mandatory condition for the realization of the template and is verified respecting the criterion of minimum interference.
- The stability condition is assessed from the transverse contact surface. By the calculation of the contact surface between the template and the vertebra, starting from an extraction direction, its perpendicular is considered and the inclination of the normal at each triangle is evaluated with respect to the extraction direction itself. The contact surface is then projected onto a plane containing the extraction direction, choosing the criterion of maximum stability as that which minimizes the projection area. The process is repeated for all possible extraction directions, covering interactions of 360°, comparing the average stability assessed between each degree interval and another. The extraction direction will therefore be the one for which the minimum resistance condition occurs.
- The ideal condition of the uniqueness of positioning is that for which the maximum contact surface is guaranteed. Therefore, the extraction direction will be chosen such that, due to the removal of the undercuts, the minimum subtraction of the contact area

between the dima and vertebra is obtained. The final execution of this iterative method leads to the generation of a template model (Fig. 2), optimized by maximizing the objective function:

$$\mathbf{F.O.} = \mathbf{max(contact\ surface + min(transverse\ projection\ surface))}.$$

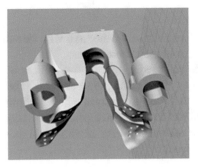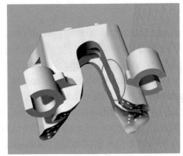

**Fig. 2.** Three-dimensional template model before the undercut removal method (to the left) and after its application (to the right). Regions of the modified undercuts are highlighted in the red areas.

## 3 Conclusion

In this work, a solution was found to the problem of interference, represented by the so-called *shaded areas*, present when inserting a template for vertebral drilling along a generic direction; the problems encountered concern the positioning of the template on the vertebral surface. These limits are mainly related to the presence of undercuts, i.e. those areas of the spinous process which, brought back into the cavity of the model body, entail complications at the time of extraction. The problem of undercuts has been limited in the present work by reducing the contact surface of the prototype with that of the spinous process. The procedure here presented allows to modify the geometric profile of the vertebral template according to the extraction direction chosen, and to compare different directions with each other, in order to identify the best solution that can, on the one hand, overcome the impediments of extraction caused by the undercuts and, on the other hand, guarantee the maximum possible adherence, compensating for the problem of stability and for the identification of a unique positioning.

## References

1. Lu, S., et al.: A novel computer-assisted drill guide template for lumbar pedicle screw placement: a cadaveric and clinical study. Int. J. Med. Robot. Comput. Assist. Surg. **5**(2), 184–191 (2009)

2. Merc, M., Drstvensek, I., Vogrin, M., Brajlih, T., Recnik, G.: A multi-level rapid prototyping drill guide template reduces the perforation risk of pedicle screw placement in the lumbar and sacral spine. Arch. Orthop. Trauma Surg. **133**(7), 893–899 (2013)
3. Farshad, M., Betz, M., Farshad-Amacker, N.A., Moser, M.: Accuracy of patient-specific template-guided vs. free-hand fluoroscopically controlled pedicle screw placement in the thoracic and lumbar spine: A randomized cadaveric study. Eur. Spine J. **26**(3), 738–749 (2017)
4. Birnbaum, K., Schkommodau, E., Decker, N., Prescher, A., Klapper, U., Radermacher, K.: Computer-assisted orthopedic surgery with individual templates and comparison to conventional operation method, Spine (Phila. Pa. 1976), vol. 26, no. 4, pp. 365–370 (2001)
5. Naddeo, F., Fontana, C., Naddeo, A., Cataldo, E., Cappetti, N., Narciso, N.: Novel design for a customized, 3D-printed surgical template for thoracic spinal arthrodesis. Int. J. Med. Robot. Comput. Assist. Surg. **15**, 1–10 (2019)
6. Naddeo, F., Cataldo, E., Naddeo, A., Cappetti, N., Narciso, N.: An automatic and patient-specific algorithm to design the optimal insertion direction of pedicle screws for spine surgery templates. Med. Biol. Eng. Comput. **55**(9), 14 (2017)
7. Cavas-Martinez, F., et al.: Early keratoconus detection by patient-specific 3D modelling and geometric parameters análisis | Detección del queratocono temprano mediante modelado 3D personalizado y análisis de sus parámetros geométricos. Dyna **94**(3), 175–181 (2019)
8. Md Yusof, M., Salman Abu Mansor, M.: Automatic core and cavity generation for 3D CAD model using normal vector and scanning ray approaches **12**(14) (2017)
9. Yusof, M.M., Abu Mansor, M.S.: Undercut feature recognition for core and cavity generation. IOP Conf. Ser. Mater. Sci. Eng. **290**(1), 012070 (2018)
10. Kumar, R., Singh, R., Madan, J.: Automated identification of complex undercut features for side-core design for die-casting parts. Proc. Inst. Mech. Eng. Part B J. Eng. Manuf. **228**(9), 1138–1152 (2014)
11. Kumar, N., Ranjan, R., Tiwari, M.K.: Recognition of undercut features and parting surface of moulded parts using polyhedron face adjacency graph. Int. J. Adv. Manuf. Technol. **34**(1–2), 47–55 (2007)
12. Ye, X.G., Fuh, J.Y.H., Lee, K.S.: A hybrid method for recognition of undercut features from moulded parts. CAD Comput. Aided Des. **33**(14), 1023–1034 (2001)
13. Shao, J., Shen, G.: Research on graph-based recognition of undercut features from molded part. In: 2nd International Conference Information Science and Engineering ICISE2010 - Proceedings, pp. 1468–1471 (2010)

# Feature-Based Modelling of Laryngoscope Blades for Customized Applications

Michele Calì[1]([⊠]) [iD], Giulia Pascoletti[2] [iD], Alessandra Aldieri[3] [iD], Mara Terzini[3] [iD], Gerardo Catapano[4] [iD], and Elisabetta M. Zanetti[2] [iD]

[1] Department of Electric, Electronics and Computer Engineering, University of Catania, Catania, Italy
michele.cali@dieei.unict.it
[2] Department of Engineering, University of Perugia, Perugia, Italy
[3] PolitoBIOMed Lab, DIMEAS, Politecnico di Torino, Turin, Italy
[4] Department of Mechanical Energy and Management Engineering, University of Calabria, Rende, Italy

**Abstract.** Laryngoscopes are used as diagnostic devices for throat inspection or as an aid to intubation. Their blade must be geometrically compatible with patients' anatomy to provide a good view to doctors with minimal discomfort to patients. For this reason, this paper was aimed to investigate the feasibility of producing customized blades.

The customizable blade model was developed following a feature-based approach with eight morphological parameters. The thickness of such a blade was determined through numerical simulations of ISO certification tests, where the finite element mesh was obtained by morphing a 'standard' mesh.

The following procedure was applied: the model was built from the selected parameters; the blade was tested *in silico*; finally, the blade was produced by additive manufacturing with an innovative biodegradable material (Hemp Bio-Plastic® -HBP-) claimed to feature superior mechanical properties. The procedure evidenced that the mechanical properties of current biodegradable materials are unsuitable for the application unless the certification norm is revised, as it is expected.

**Keywords:** Parametric drawing · Additive manufacturing · Laryngoscope blades · Patient-specific design · Biodegradable materials · Feature-based modeling · Mesh-morphing

## 1 Introduction

Laryngoscopy is the medical procedure finalized to the visualization of larynx structures such as the glottis or the vocal cords, or to facilitate tracheal intubation to perform general anesthesia, artificial ventilation, etc. [1]. The medical instrument to perform such procedure is the laryngoscope (Fig. 2) [2]. Generally, it consists of a handle and an interchangeable blade. The handle is used by the operator to apply the force needed to insert the laryngoscope into the throat. The blade has the double function of bearing the

L. Roucoules et al. (Eds.): JCM 2020, LNME, pp. 206–211, 2021.
https://doi.org/10.1007/978-3-030-70566-4_33

light source and of keeping the tongue and the soft tissues apart to provide for a free field of view. The blades can be straight or curved (Fig. 2), depending on the laryngoscope model, and are usually provided in 3–4 sizes [3]. Laryngoscopy might result challenging in the case of patients with malformations, obese patients [4], or patients younger than 4 years [5]. In all these cases, the field of view may be very narrow or teeth [6] and soft tissues [7] may be damaged, which is associated with serious morbidity and mortality. Laryngoscopy may often cause discomfort or minor lesions also to patients with no apparent anatomical airway abnormalities. A possible solution could come from the production of patient-specific blades, taking advantage of the possibilities granted by the recent advances of 3D printers. This would require performing CT scans which might be available only in very critical cases. Herein, an alternative solution is investigated in which laryngoscope blades are designed in parametric fashion and are produced with materials that may be recycled to reduce the environmental impact of disposable blades [8].

**Fig. 1.** Workflow of the proposed procedure from the customization of the parametric model, according to patient anatomy, to laryngoscope manufacturing

## 2 Materials and Methods

### 2.1 Parametrization

Figure 1 summarizes the implemented workflow for the design and manufacturing of customized laryngoscope blades. The starting step was parametrization of the design, which was based on the physical measures with the highest impact on the intubation procedure, according to the Mallampati's score [9], i.e.: the thyro-mental distance; the interincisor gap; the possible presence of tracheal deviation. Further parameters were defined with respect to the differences between pediatric and adult blades.

Two different commercial blade shapes (one for adult and the other for pediatric use, Fig. 2) were digitized with the CNC MODELA MDX-40A equipped with the 3D ZSC-1 active piezo-sensor by Roland (0.25 mm in-plane step; 0.01 mm accuracy along the Z-axis). Multiple scans were performed for different blade orientations and the respective point clouds were aligned and merged (Dr. PICZA3 and 3DEditor software). Finally, the digital representations were exported to the Geomagic Studio® software and 3D tessellated surfaces were generated for features extraction and for the subsequent parametrization.

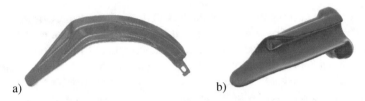

**Fig. 2.** Scanned Blades: a) adult blade; b) pediatric blade.

## 2.2 Parametric Modelling

Feature curves describing the 'tongue' were shared by both scanned laryngoscopes and were used as guide curves for surface definition through extrusions and loft. Two more principal feature curves were defined on the adult laryngoscope to reproduce its light pipe (Table 1, Fig. 3). Parameters defining the feature curves were varied at given boundary conditions and mutual relationships to generate customized blade models lying in-between the adult and the pediatric blades. Table 1 reports the parameters for the two reference models and for a customized model.

## 2.3 Mechanical Tests

The ISO 7376 standard describes what is required by laryngoscope blades in terms of mechanical strength and stiffness. The stiffness of the laryngoscope blade is a key parameter because high highly deformable blades could make vocal cord visualization more difficult [8]. According to the ISO 7376:2009 standard, when a cantilever load of 65 N is applied to the blade tip vertically downwards (Fig. 3) the respective deflection should be lower than 10 mm. In addition, the blade must bear a static load equal to 150 N without evident failures.

Mechanical tests were digitally performed by the Finite Element Method (FEM), along the following steps:

- the customized blade was generated through patient-specific modelling
- the 'standard mesh' was morphed on the patient-specific blade
- numerical simulations were performed for loads of both 65 N and 150 N, accounting for the non-linearity caused by large displacements.

Since it was decided to test the performance of biodegradable blades, Hemp Bio-Plastic® (HBP, see the next section) the material properties were set at: 3833 MPa the elastic modulus and 0.32 the Poisson ratio. The tongue thickness (P5) was optimized in agreement with the results of the finite element simulations.

**Table 1.** Sets of parameter values for three blades

| Ref. plane | Tongue | | | | | Light pipe | | |
|---|---|---|---|---|---|---|---|---|
| | Curvature | | Length | Width | Thickness | Inclination | Curvature | Thickness |
| | XZ P1 [mm] | XY P2 [mm] | XZ P3 [mm] | XY P4 [mm] | XZ P5 [mm] | XZ P6 [mm] | XY P7 [mm] | XY P8 [mm] |
| Adult blade | 69.2 | 180 | 77.3 | 21 | 2 | 125 | 12.5 | 1.6 |
| Pediatric blade | 500 | 13 | 77.3 | 22 | 1.5 | 176 | 6 | 1.5 |
| Customized blade | 150 | 17 | 77.3 | 21 | 2.5 | 160 | 7 | 1.5 |

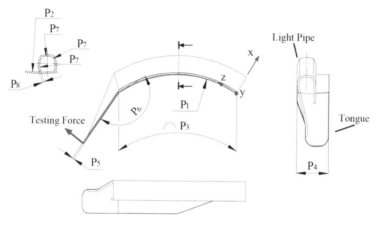

**Fig. 3.** Blade parameters and orthogonal coordinate system.

## 2.4   Additive Manufacturing of a Laryngoscope Prototype

The prototype model of the customized laryngoscope was manufactured with the D300 Technology® Printer for Fused Deposition Modeling (FDM). The main process parameters were set equal to 0.1 mm for the Z-axis resolution, and 0.05 mm for the X, Y axes accuracy (X and Y axes identify the plane containing the print bed and the Z-axis represents the vertical direction, perpendicular to the print bed). The printer was equipped with a 0.6 mm nozzle. Filaments were used made of Hemp Bio-Plastic® (HBP), a material patented by MICA s.r.l. (Ragusa, Italy), made of polylactic acid (PLA) and hemp shives. Its main advantages compared to PLA are better mechanical properties, improved superficial finish and antibacterial properties. Figure 4 shows the prototype of the custom laryngoscope blade.

## 3   Results and Discussion

Figure 1 shows the workflow that was followed. A set of parameters was chosen based on the measurements taken on a patient (Table 1), and the standard mesh was morphed on this geometry. Numerical simulations led to a minimal tongue thickness of 4 mm, as shown in Fig. 4, which would ultimately prevent from using the biodegradable blades. Alternatively, the blade thickness could be decreased close to the blade tip since the 4 mm thickness is strictly necessary only at the tongue base. It should be stressed that the ISO 7376 standard sets very strict limits both for the ultimate strength and the minimum stiffness. However, a very stiff blade could be harmful, as being less adaptable to the patient anatomy, which is an aspect that should also be taken into account.

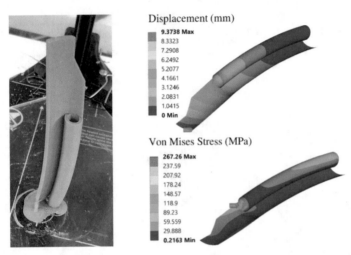

**Fig. 4.** Left: 3D printed customized blade. Right: Results of the *in silico* tests: a) displacement map for a 65 N load; b) Von Mises stress map for a 150 N

The importance of using biodegradable materials stems from the consideration that wasted plastic blades can indeed have a significant impact on the environment [8]. A recent study reported that more than 20,000 reusable laryngoscopes have been used in UK hospitals in one year [10].

In addition to the required mechanical tests, it would be interesting to study the interactions between the laryngoscope and the mouth-larynx complex with experimental models [9] and numerical methods [10, 11, 12].

Finally, biocompatibility tests should be performed to certify the absence of microbial or potentially hazardous substances after sterilization and that the sterilization process would not change neither the structure, nor the chemical nature of the laryngoscope. The customization procedure has the potential to address the specific needs of pediatric patients, for whom it would be of great help. In fact, the size of current blade is unevenly distributed with just a few sizes (one size only for some manufacturers) for pediatric patients, and at least 3–4 sizes for adult patients.

**Acknowledgments.** The research work herein reported was funded by Università degli Studi di Catania within the CRUI-CARE Agreement (research path PIA.CE.RI. 2020-2022 Linea 4).

# References

1. Berkow, L.C., Morey, T.E., Urdaneta, F.: The technology of video laryngoscopy. Anesth. Analg. **126**, 1527–1534 (2018)
2. Rossignol, F., Perrin, R., Desbrosse, F., et al.: In vitro comparison of two techniques for suture prosthesis placement in the muscular process of the equine arytenoid cartilage. Vet. Surg. **35**, 49–54 (2006)
3. Levitan, R., Ochroch, E.A.: Airway management and direct laryngoscopy: a review and update. Crit. Care Clin. **16**, 373–388. W.B. Saunders
4. Maldini, B., Hodžović, I., Goranović, T., et al.: Challenges in the use of video laryngoscopes. Acta Clin. Croat. **55**(Suppl 1), 41–50 (2016)
5. Balaban, O., Tobias, J.D.: Videolaryngoscopy in neonates, infants, and children. Pediatr. Crit. Care Med. **18**, 477–485 (2017). Lippincott Williams and Wilkins
6. Schieren, M., Kleinschmidt, J., Schmutz, A., et al.: Comparison of forces acting on maxillary incisors during tracheal intubation with different laryngoscopy techniques: a blinded manikin study. Anaesthesia **74**, 1563–1571 (2019)
7. Mourão, J., Moreira, J., Barbosa, J., et al.: Soft tissue injuries after direct laryngoscopy. J. Clin. Anesth. **27**, 668–671 (2015)
8. Sherman, J.: Reusable vs. disposable laryngoscopes - anesthesia patient safety foundation. APSF Newsl. **33**, 91 (2019)
9. Gálvez, J.A., Simpao, A.F., Dori, Y., et al.: Not just a pretty face: three-dimensional printed custom airway management devices. 3D Print. Addit. Manuf. **3**,160–165 (2016)
10. Zanetti, E.M., Ciaramella, S., Calì, M., Pascoletti, G., Martorelli, M., Asero, R., Watts, D.C.: Modal analysis for implant stability assessment: sensitivity of this methodology for different implant designs. Dent. Mater. **34**(8), 1235–1245 (2018)
11. Ambu, R., Motta, A., Calì, M.: Design of a customized neck orthosis for FDM manufacturing with a new sustainable bio-composite. In: International Conference on Design, Simulation, Manufacturing: The Innovation Exchange, pps. 707–718. Springer, Cham (2019)
12. Calì, M., Oliveri, S.M., Cella, U., Martorelli, M., Gloria, A., Speranza, D.: Mechanical characterization and modeling of downwind sailcloth in fluid-structure interaction analysis. Ocean Eng. **165**, 488–504 (2018)

# CAD and Interaction

# Magnitude Finger Forces Analysis During Simulating Pseudo-Haptic Spring

Jingtao Chen[1], Peter Mitrouchev[2(✉)], Sabine Coquillart[3], and Franck Quaine[4]

[1] Health Simulation Center SimUSanté®, Amiens University Hospital, Amiens, France
[2] Univ. Grenoble Alpes, CNRS, Grenoble, INP, G-SCOP, 38000 Grenoble, France
Peter.Mitrouchev@grenoble-inp.fr
[3] Univ. Grenoble Alpes, INRIA, CNRS, Grenoble INP, LIG, 38000 Grenoble, France
[4] Univ. Grenoble Alpes, CNRS, Grenoble INP, GIPSA-Lab, 38000 Grenoble, France

**Abstract.** This paper focuses on finger force magnitude analysis during stiffness discrimination task. In the frame of their *Study and research work* MS students from the Université Grenoble Alpes specially designed an experimental bench allowing to simulate a pseudo-haptic spring. Then, a series of stiffness discrimination tests between reals springs and a pseudo-haptic spring were performed. Finger pressing forces and students' (subjects') perception of spring stiffness were recorded and analyzed. The analysis of *psychometric curves* indicates that subjects underestimate the simulated stiffness of the pseudo-haptic spring. The results also indicate that the peak of finger force applied on pseudo-haptic spring increases as the simulated stiffness increases. Moreover, it was found that the relationships between the logarithm of stiffness and the finger force were linear for the real springs and the pseudo-haptic spring. Pseudo-haptics effect being provided by specially designed isometric force feedback device, the results of this study may be useful for computer-based rehabilitation tasks designed for motor disorder patients with muscle deficiency associated with limited joint movement range or for injured athletes in the process of rehabilitation.

**Keywords:** Muscle force · Pseudo-haptic feedback · Stiffness discrimination

## 1 Introduction

In some simulation such as: medical, sportive, remote surgery…, it is necessary to have stiffness information as it provides useful feedback to discriminate the stiffness, or the compliance, properties of different tissues. Pseudo-haptics is a technique that can provide the illusion of different physical properties such as stiffness, for instance. According to Lécuyer *et al.* [1] pseudo-haptics is defined as "*the generation, augmentation or deformation of haptic sensations by information coming from other sensory modalities*", hereinafter referred to as "*disruptive modalities*".

Besides studies about the influence of user's perception of stiffness, some studies have investigated the hand motor strategy, such as adjustment of peak force intensity or pressing duration, while pressing a real and a pseudo-haptic spring during stiffness

© The Author(s) 2021
L. Roucoules et al. (Eds.): JCM 2020, LNME, pp. 215–220, 2021.
https://doi.org/10.1007/978-3-030-70566-4_34

discrimination tasks. Tan *et al.* [2] found that higher force is needed for achieving the same magnitude of work cues for less compliant objects during discrimination compared to high compliant objects. The highest peak force is applied to the stiffest objects. This conclusion has been verified by series of studies, no matter whether participants used pinch or press movement [3] to discriminate the stiffness of a real spring.

Endo [4] investigated the exploration of movement, applied force and displacement during stiffness discrimination task with real springs. He has shown that the forces applied on the explored specimen increased with the stiffness until it reached a stabilization for the highest stiffness values. Paljic et al. [5] used pseudo-haptic feedback to simulate a pseudo-haptic torsion spring. They reported that the torque applied on the spring did not vary with its simulated stiffness. However, the comparison between the torques applied on real and pseudo-haptic torsion spring was not addressed. Freyberger et al. [6] investigated the relationship between granularity perception and vibro-tactile coding during stiffness discrimination task.

In [7] Kaim and Drewing investigated to what extent people strategically execute movements that are tuned for softness discrimination of objects with deformable surfaces also during stiffness discrimination tasks. However, a question remains: what is the force-stiffness relationship when pressing a pseudo-haptic spring compared to a real one in stiffness discrimination tasks? The answer to this question may provide a reference model for designing computer-based hand rehabilitation applications, using pseudo-haptic feedback. As far as we know, there is no work focusing on this issue.

In this context, the aim of the paper is to investigate the influence of pseudo-haptic feedback on users' finger force. For this purpose, an experimental bench was specifically designed allowing to simulate the pseudo-haptic spring. Its stiffness varies over eleven different percentages, called stiffness percentages ($-40\%$, $-30\%$, $-20\%$, $-10\%$, $0\%$, $+10\%$, $+20\%$, $+30\%$, $+40\%$, $+50\%$ and $+60\%$) from the targeted stiffness of the compared real spring. After testing both springs, participants had to identify which was stiffer: the real spring or the pseudo-haptic one. Thus, the forces applied on the real springs and pseudo-haptic spring were recorded and analyzed for investigating the trend of change in finger force depending on the change of stiffness of the springs.

## 2   Experimental Protocol and Participants

The experimental protocol presented in [8], where only the displacement of the pseudo-haptic spring was visually displayed on a computer screen, was replicated in the performed experiments. Here, the visual stimuli of the two compared springs (real and pseudo-haptic) and their respective displacements were displayed for the subjects on a computer screen for each trial. The experiment includes two parts: *i*). Maximal voluntary contraction (MVC) for force and electromyography (EMG) data calibration, *ii*). Stiffness discrimination tests. MVC consists of three items: *i*). Maximal flexion force, *ii*). Maximal extension force, *iii*). Maximal co-contraction. Each item is performed consecutively three times and consists in: performing the co-contraction during 3 s. followed by 5 s. break. After each item, subject had 5 min to rest before starting the stiffness discrimination task. For the maximal flexion and extension forces, subject had to continuously apply the maximum force by vertically pulling and pushing a ring attached with a force sensor towards down and towards up. The maximal co-contraction consists in keeping the palm and forearm horizontal, and co-contracting all the forearm muscles. During performing each item, the forces and EMG signal are recorded. During the tests, subjects, in sitting position, had to use the index finger of their dominant hands while applying the forces.

Figure 1 partially shows the test bench for the experiments carried out. Inside a box (Fig. 1a), there are four springs, labeled *1, 2, 3* and *4* with physical stiffness of *202* N/m, *304* N/m, *608* N/m and *2500* N/m respectively. The real spring is one among springs *1, 2* and *3*. The pseudo-haptic spring is the stiffest one, namely spring *4*. It has been shown [9], and our own experience confirms, that pseudo-haptic spring simulation is more relevant when using stiffer pseudo-haptic spring. Using stiffness in a similar range for pseudo-haptic and real springs precludes stiffness discrimination task, probably because subjects do not understand the task they must perform. Each spring is situated in a metallic tube and is loaded by the finger force via a button. Beneath each spring, a force sensor (*KISTLER 9017B*) records the force with a *2000* Hz sampling frequency during each trial.

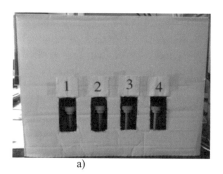
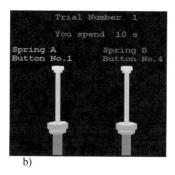

a)                                          b)

**Fig. 1.** Experimental setup: (a). Four springs inside the box; (b). a real spring (spring A, left) and the pseudo-haptic spring (spring B, right).

For the real springs no mismatch existed between the visual feedback and its displacement. However, for the pseudo-haptic one a difference existed according to the simulated targeted stiffness value, in order to lure the participants. Previous tests had shown that below −40% and above +60% the subjects' perception results were no longer influenced by the visual feedback. Thus, the experiments were performed by varying the stiffness of the pseudo-haptic spring randomly over eleven different stiffness percentages from the stiffness of the compared real spring.

During each trial of stiffness discrimination, one of the three real springs was randomly compared to the pseudo-haptic spring whose simulated stiffness varied, as previously said, from the stiffness of the compared real spring by eleven different percentages. The button's displacements of the two compared springs were displayed on the screen labeled "*Spring A*" and "*Spring B*". For each spring, the *spring number (1 to 4)* of the corresponding button was indicated just below "Spring A/B" (Fig. 1b). Subjects started a trial by pressing spring *B* first (i.e. pseudo-haptic spring) and then spring *A*, the randomized compared real springs. After they could switch back and forth between the two springs as they wished. For subject's information, in the top part of the screen, the trials' number and the time spent for this trial were displayed (Fig. 1b). Subjects had to focus on the computer screen and were not allowed to observe the current movements of their fingers while pressing the springs. For each trial, after exploring the springs, they had to press the *S* button on the keyboard, with their non-dominant hand, to *Stop* the trial. Then the question: *Which spring is stiffer?* was displayed on the computer screen. The subjects pushed key *A* or *B* to answer the question. The answer was then automatically recorded for further analysis. Fifteen unpaid voluntary and healthy subjects (twelve males and three females), aged from 23 to 25 participated in the experiments. All of them reported no: *i).* Visual impairment; *ii).* Impairments of haptic sensitivity; *iii).* Diseases or symptoms inducing hand movement disorder.

## 3  Results and Discussion

Figure 2 presents the comparison between the force on real springs and the virtual spring under identical stiffness (0% of change). A significant effect was noted for the spring type ($F = 118.32$, *p*-value $< 0.0001$) and for the stiffness ($F = 94.034$, *p*-value $< 0.0001$) meaning that the force applied on the real springs are lower than this applied on the virtual one, while the force increases from low to higher stiffness values. A significant interaction between the spring type and 'stiffness scale' ($F = 8.907$, *p*-value $< 0.001$) was also observed. The forces applied on real springs increase from *5.946 N* to *11.6 N* and from *9.5 N* to *12.79 N* on virtual spring. This means that the forces applied on the real springs were statistically lower than those applied on the virtual spring when the stiffness of real springs and the virtual spring are the same.

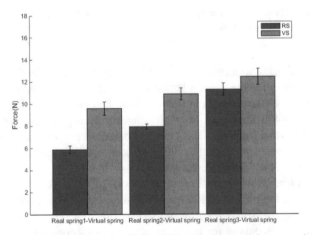

**Fig. 2.** All paired comparisons have significant differences between force on virtual spring and real springs (*p*-value $< 0.0001$).

The significant effect of *stiffness scale* on the force applied on the real springs confirms that higher stiffness induces higher force. This observation also is in agreement with the conclusions in [4]. As no significant effect of *stiffness percentage* on the force applied on the real springs was found, it can be concluded that the forces applied on the real springs were not significantly influenced by the simulated change in stiffness of the compared pseudo-haptic spring. Since the force-stiffness relationship for both, real and pseudo-haptic spring, fits the logarithmic function, it can be observed: *i)*. That subject's behavior when pressing the pseudo-haptic spring has the same pattern as when using real one, but with a shift; *ii)*. a similarity is observed between the tendency in the change of force applied on real springs and pseudo-haptic spring when stiffness increases. It appears that for the same stiffness, the force applied on the pseudo-haptic spring is higher than that applied on the real springs. This phenomenon is natural as the physical stiffness of the pseudo-haptic spring is much larger than the stiffness of the real springs. Regardless of the nature of the spring, real or pseudo-haptic, the force and the gap between them (Fig. 2) decreases with the increase in stiffness. In other words, if the simulated stiffness of the pseudo-haptic feedback is higher, less difference exists in the force's behavior pattern between the real springs and the pseudo-haptic spring. That explains the conclusion reported in [4]: the forces reach the same magnitude when the stiffness increases.

## 4  Conclusion

The aim of this study was to investigate the relationship between subjects' finger force applied on springs and spring stiffness through experimental stiffness discrimination tests (comparing the stiffness of real and pseudo-haptic spring). Subjects' perception of spring stiffness and index finger pressing force on the real springs and the pseudo-haptic spring was recorded and analyzed.

Analysis of stiffness discrimination results shows that subjects have underestimated the simulated stiffness of the pseudo-haptic spring as reported in (8). Their perception underestimation on simulated stiffness of pseudo-haptic spring implies that pseudo-haptic feedback should be better employed to simulate spring stiffness or biological tissues with higher levels of stiffness.

These findings may be useful in the design and the development of user-centered computer-based rehabilitation applications in general, and for patients with hand motor disorder (limited finger or hand movement ranges) in particular as the isometric-device simulating the pseudo-haptic springs need no displacement.

**Acknowledgements.** This work was supported by LabEx PERSYVAL-Lab (ANR-11-LABX-0025) (https://www.persyval-lab.org/index.html). We would like to thank Mr. Charles-Henry Dufetel and the other MS students for their assistance in the development of the test bench and participation in the experiments.

# References

1. Lécuyer, A., Burkhardt, J.M., Coquillart, S., Coiffet, P.: "Boundary of illusion": an experiment of sensory integration with a pseudo-haptic system. Proc. IEEE Virtual Reality **2001**, 115–122 (2001). https://doi.org/10.1109/vr.2001.913777
2. Tan, H.Z., Durlach, N.I., Beauregard, G.L., Srinivasan, M.A.: Manual discrimination of compliance using active pinch grasp: the roles of force and work cues. Percept. Psychophys. **57**(4), 495–510 (1995). https://doi.org/10.3758/bf03213075
3. Freyberger, F.K.B., Färber, B.: Compliance discrimination of deformable objects by squeezing with one and two fingers. In: Proceedings of EuroHaptics, pp. 271–276 (2006a)
4. Endo, H.: Pressing movements and perceived force and displacement are influenced by object stiffness. Physiol. Behav. **163**, 203–210 (2016)
5. Paljic, A., Burkhardtt, J.-M., Coquillart, S.: Evaluation of pseudo-haptic feedback for simulating torque: a comparison between isometric and elastic input devices. In: 12th International Symposium on Haptic Interfaces for Virtual Environment and Teleoperator Systems, Proceedings, HAPTICS 2004, pp. 216–223 (2004). https://doi.org/10.1109/haptic.2004.1287199
6. Freyberger, F.K.B., Farber, B.: Psychophysics and perceiving granularity. In: 2006 14th Symposium on Haptic Interfaces for Virtual Environment and Teleoperator Systems, pp. 387–393 (2006b). https://doi.org/10.1109/HAPTIC.2006.1627105
7. Kaim, L., Drewing, K.: Exploratory strategies in haptic softness discrimination are tuned to achieve high levels of task performance. IEEE Trans. Haptics **4**(4), 242–252 (2011). https://doi.org/10.1109/toh.2011.19
8. Lécuyer, A., Coquillart, S., Kheddar, A., Richard, P., Coiffet, P.: Pseudo-haptic feedback: can isometric input devices simulate force feedback? In: Feiner, S., Thalmann, D. (eds.) Proceedings IEEE Virtual Reality (2000). https://doi.org/10.1109/vr.2000.840369
9. Wu, W., Basdogan, C., Srinivasan, M.A., Basdogan, C.: Visual, haptic, and bimodal perception of size and stiffness in virtual environments. In: Proceeding of the ASME Dynamic Systems and Control Division Bd, p. 67 (1999)

# Parametric Hull Design with Rational Bézier Curves

Antonio Mancuso[1], Antonio Saporito[1(✉)], and Davide Tumino[2]

[1] Dipartimento di Ingegneria, Università degli Studi di Palermo, Viale delle Scienze, 90127 Palermo, Italy
antonio.saporito@unipa.it
[2] Facoltà di Ingegneria e Architettura, Università degli Studi di Enna Kore, Cittadella Universitaria, 94100 Enna, Italy

**Abstract.** In this paper, a tool able to support the sailing yacht designer during the early stage of the design process has been developed. Quadratic and cubic Rational Bézier curves have been selected to describe the main curves defining the hull of a sailing yacht. The adopted approach is based upon the definition of a set of parameters, say the length of water line, the beam of the waterline, canoe body draft and some dimensionless coefficients according to the traditional way of the yacht designer. Some geometrical constraints imposed on the curves (e.g. continuity, endpoint angles) have been conceived aimed to avoid unreasonable shapes. These curves can be imported in any commercial CAD software and used as a frame to fit with a surface. The algorithm and the related Graphical User Interface (GUI) have been written in Visual Basic for Excel. To test the usability and the precision of the tool, two sailboats with different characteristics have been replicated. The rebuilt version of the hulls is very close to the original ones both in terms of shape and dimensionless coefficients.

**Keywords:** Rational Bézier curves · Sailing yacht design · CAD

## 1 Introduction

In a work of any engineer, the design is often the central and more important part of the entire process. In a wide range of industries, such as automobile, aircraft and shipbuilding [1], the first step of the process consists of finding an existing well designed geometry to be used as a benchmark for the new model. As far as the maritime field, it is interesting to notice that the design approach is mostly based on the traditional design techniques of trial-and-error. Consequently, the obtained results are highly dependent on the experience of the designer [2, 3]. To facilitate the design of hulls, naval engineers are investigating the possibility to define the so-called Hull Equation [4]. This equation should be able to describe, from a mathematical point of view the hull of a sailboat, a motorboat or a ship. Although an intensive effort in this sense, nowadays is not possible to describe the hull with a single one equation because the geometry of a hull depends on several parameters and most of them are related one each other [5]. Several works where authors

© The Author(s) 2021
L. Roucoules et al. (Eds.): JCM 2020, LNME, pp. 221–227, 2021.
https://doi.org/10.1007/978-3-030-70566-4_36

present methods to generate a quick but detailed preliminary design or, on the other hand, approaches to optimize the geometry can be found in the literature. For instance, in [6] a design tool is developed using cubic polynomial expressions to define the control curves of a hull. In [7] cubic Bézier curves and the curve-plane intersection method are selected to properly design a submarine hull. Also [8] and [2] propose a new design framework to generate the parametric design and modification of yacht hulls. In particular, the hull is split into three regions to assure better design flexibility. Splitting the whole hull domain into sub-domains is a common practice as can be observed in [9] where the hull has two domains, one below the chine and one above the chine. Other authors were more focused on the optimization phase as in [10] where a novel simultaneous engineering design approach has been proposed or in [2] where an interactive design approach for hull forms optimization is developed. In this work, a numerical tool able to help the yacht designer to properly define and modify the hull form of a sailboat has been written.

The approach to the problem has been described in Sect. 2. It consists of the definition of a curves frame, modeled with Rational Bézier curves depending on yacht measurements (e.g. length, beam, draft) and some dimensionless coefficients. To avoid unreasonable shapes, constraints to the control points have been imposed. The efficiency of the tool is demonstrated in Sect. 3, where two different hull forms have been rebuilt and the obtained results compared with the original ones.

## 2  Design Approach

The shape of the hull is defined by three sections (fore, mid and aft), three longitudinal curves (sheer, chine and keel) and the right ahead. In this way, the whole domain is divided into three parts, as can be seen in Fig. 1. The curves have been modeled with rational Bézier curves of second and third-degree. The design variables of the problem are the Cartesian coordinates of the control points and the tangency of the curves at their ends. Rational Bézier curves [11] are defined by the following Eq. (1):

$$\frac{\sum_{i=0}^{n} w_i B_{i,n}(t) P_i}{\sum_{i=0}^{n} w_i B_{i,n}(t)} \tag{1}$$

where $B_{i,n}$ are the Bernstein polynomials while $P_i$ and $w_i$ are the control points and the weights respectively.

Rational Bézier curves differ from the Bézier curves because it is possible to increase or decrease the effect of each control point to define the shape of the curve. In this way, the designer can modify the fullness of each curve, according to the related dimensionless coefficient, without losing the condition of continuity because the tangency at the ending points of the curve is not influenced by the value of the weights.

Sections are defined with rational Bézier curves of second degree and each of them is composed of two curves: one starting from the keel to the chine and the second one starting from the chine to the sheer. In this way, it is possible to generate a wide range of shapes including sections with or without a hard chine.

To assure G0 continuity of the section (in case of hard chine hull) the position of the control points of the two curves at the chine shall be coincident. The G1 continuity (round bilge hull) is assured controlling the tangency of the curves in the common point

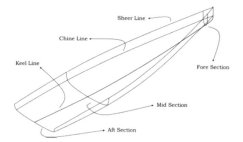

**Fig. 1.** Curves frame used to define the hull surface.

(always lying on the chine). The algorithm and the related Graphical User Interface (GUI) have been written in Visual Basic for Excel.

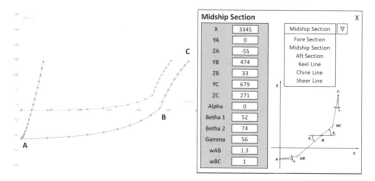

**Fig. 2.** Graphical User Interface of the tool developed in Visual Basic for Excel. Plane YZ.

The designer can modify the shape of each curve, by changing the values of the Cartesian coordinates ($Y_A$, $Z_A$, $Y_B$, $Z_B$, $Y_C$, $Z_C$), the angles at the ends of the two curves ($\alpha$, $\beta_1$, $\beta_2$ and $\gamma$) and, if necessary, the weights w, as shown in Fig. 2.

The keel line is also defined with rational Bézier curves of second degree with the same methodology but the value of $\beta_1$, $\beta_2$ are always the same to assure the G1 continuity in every part of the curve.

To properly represent chine and sheer lines rational Bézier curves of third-degree are needed. This is due to the nature of these curves that, differently from sections and keel line, are not lying on a plane. Using high order curves, in this case, helps the designer to modify the shape of the curves in both the projection planes XY and XZ as shown in Fig. 3.

A "*.pts" file, consisting of three-column list of the points' spatial coordinates (X Y Z), for each curve is automatically generated and imported in PTC Creo 4.0. An internal macro firstly generate the surface over the imported curves frame and then calculate areas and mass properties. If the design is not as good as expected the designer can modify the curves in Excel to obtain a new configuration without redesigning the entire model but just updating the "*.pts" files.

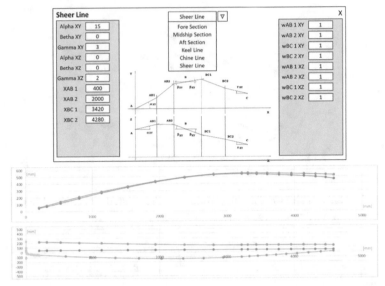

**Fig. 3.** Graphical User Interface of the tool developed in Visual Basic for Excel. Plane XY and XZ.

## 3  Case Study

To test the tool, the authors replicated two sailing dinghies participating to the 1001VELAcup competition (a sailing dinghy race held yearly in Italy; more details can be found at www.1001velacup.eu). The original CAD models of the two hulls (a round bilge hull and a hard chine hull) were defined by means of cubic B-spline surfaces over a set of about 20 × 10 control points. Starting from these models, the curves frame as defined in Fig. 1 have been obtained and placed as background images for the rebuilding process, following the procedure described in the previous section. Although the rebuilt curves have been generated starting from few information (a set of 9 points and 9 angles at all), a good match has been found.

**Fig. 4.** Original curves (black) and rebuilt curves (red): round bilge (left) and hard chine (right)

Particularly, Fig. 4 shows the overlap of the original and rebuilt curves, while Fig. 5 shows the cut-off of the rebuilt surfaces in CREO with transversal planes to show the sections (red curves), with horizontal planes to show the waterlines (blue curves) and with longitudinal planes to show the buttocks (green curves). A well faired curves frame has been obtained without undesired changes in slope or curvature confirming the goodness of the proposed approach.

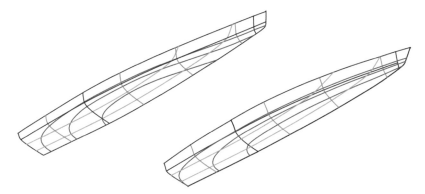

**Fig. 5.** Rebuilt hull surfaces: round bilge (left) and hard chine (right)

Once the geometry of the hulls is defined, the main characteristics of the two sailboats are automatically calculated. Table 1 shows the comparison of the main hull characteristics whose definitions can be found in [5].

**Table 1.** Main characteristic comparison between original and rebuilt hulls.

| Characteristics | Symbol | Unit | Round bilge | | Hard chine | |
|---|---|---|---|---|---|---|
| | | | Original | Rebuilt | Original | Rebuilt |
| Displacement | $\nabla$ | $m^3$ | 0.238 | 0.233 | 0.262 | 0.263 |
| Length overall | $L_{OA}$ | m | 4.60 | 4.60 | 4.60 | 4.60 |
| Length water line | $L_{WL}$ | m | 4.40 | 4.42 | 4.49 | 4.49 |
| Max beam water line | $B_{WL}$ | m | 1.05 | 1.05 | 0.95 | 0.95 |
| Wetted surface | $S_W$ | $m^2$ | 3.38 | 3.40 | 3.46 | 3.50 |
| Water plane area | $A_W$ | $m^2$ | 3.14 | 3.17 | 3.10 | 3.14 |
| Max transversal area | $A_X$ | $m^2$ | 0.107 | 0.103 | 0.094 | 0.093 |
| Long. centre of buoyancy | LCB | m | 2.48 | 2.52 | 2.25 | 2.26 |
| Long. centre of flotation | LCF | m | 2.68 | 2.67 | 2.60 | 2.60 |
| Max draught | $T_c$ | m | 0.14 | 0.14 | 0.17 | 0.17 |

Concerning length, areas and volume coefficients, a maximum deviations close to 1% has been achieved between original and rebuilt hulls coefficients. This value is lower than the sensitivity of the most common numerical application like, for instance, the bare hull resistance estimation [12].

## 4   Conclusion

In this work, a tool and the relative methodology to design hulls of sailing boats is presented. The algorithm and the related Graphical User Interface (GUI) have been written in Visual Basic for Excel. A total of seven Bézier curves of the second and third-degree are selected to define the geometry of the hull. To prove the validity of the tool and the applied approach, two existing sailboat hulls have been successfully replicated.

## References

1. Nam, J., Bang, N.S.: A curve based hull form variation with geometric constraints of area and centroid. Ocean Eng. **133**, 1–8 (2017)
2. Khan, S., Gunpinar, E., Sener, B.: GenYacht: an interactive generative design system for computer-aided yacht hull design. Ocean Eng. **191**, 106462 (2019)
3. Cirello, A., Cucinotta, F., Ingrassia, T., Nigrelli, V., Sfravara, F.: Fluid-structure interaction of downwind sails: a new computational method. J. Mar. Sci. Technol. **24**, 86–97 (2019)
4. Mancuso, A.: Parametric design of sailing hull shapes. Ocean Eng. **33**(2), 234–246 (2006)
5. Larsson, L., Eliasson, R.E., Orych, M.: Principles of Yacht Design. McGraw-Hill Education, New York (2014)
6. Calkins, D.E., Schachter, R.D., Oliveira, L.T.: An automated computational method for planing hull form definition in concept design. Ocean Eng. **28**(3), 297–327 (2001)
7. Chrismianto, D., Zakki, A.F., Arswendo, B., Kim, D.J.: Development of cubic Bezier curve and curve-plane intersection method for parametric submarine hull form design to optimize hull resistance using CFD. Mar. Sci. Appl. **14**, 399–405 (2015)
8. Khan, S., Gunpinar, E., Dogan, K.M.: A novel design framework for generation and parametric modification of yacht hull surfaces. Ocean Eng. **136**, 243–259 (2017)
9. Pérez-Arribas, F.: Parametric generation of planing hulls. Ocean Eng. **81**, 89–104 (2014)
10. Ingrassia, T., Mancuso, A., Nigrelli, V., Tumino, D.: A multi-technique simultaneous approach for the design of a sailing yacht. Int. J. Interact. Des. Manuf. **11**, 19–30 (2017)
11. Sederberg, T.W.: Computer Aided Geometric Design Course Notes. Computer Aided Geometric Design. BYU Scholars Archive (2012)
12. Keuning, J.A., Katgert, M.: A bare hull resistance prediction method derived from the results of the delft systematic yacht hull series extended to higher speeds. In: International Conference on Innovation in High Performance Sailing Yachts, Lorient, France (2008)

# A Framework for Curvature-Based CAD Mesh Partitioning

Yifan Qie[1]([⊠]), Lihong Qiao[2], and Nabil Anwer[1]

[1] LURPA, ENS Cachan, Université Paris-Saclay, 94235 Cachan, France
`yifan.qie@ens-paris-saclay.fr`
[2] School of Mechanical Engineering and Automation, Beihang University, Haidian District,
37 Xueyuan Road, Beijing 100191, China

**Abstract.** In ISO Geometrical Product Specifications and Verification Standards (GPS) [1], partition is one of the fundamental operations used to obtain ideal or non-ideal features of a product. The operation of partition produces independent geometrical features by decomposing the object. A curvature-based CAD mesh partitioning framework is proposed in this paper. The framework combines several key steps including curvature-based attribute calculation, local shape type refinement, region growing, slippage analysis and statistical modeling. The partitioned features are classified into seven invariance classes of surface in the context of ISO GPS. A case study shows that not only appropriate partitioning but also accurate invariance class recognition for GPS are achieved by the proposed framework.

**Keywords:** Partitioning · ISO GPS · Digital geometry processing · Curvature

## 1 Introduction

In ISO GPS, feature operations such as partition, extraction, filtration and association are used to obtain ideal and non-ideal features. Among these operations, partition is used to identify ideal or non-ideal features. The default partition segments the part into maximal surface portions that correspond to one of the seven invariance classes of surfaces, i.e. plane, sphere, cylinder, surface of revolution, prismatic, helix, and complex surface. The partitioned feature could be used to facilitate other operations such as association and filtration.

In this paper, existing partition methods are reviewed and a comprehensive classification is proposed. A curvature-based partitioning framework in the context of ISO GPS is developed. The framework contains two main stages, initial region partitioning and region refinement. A case study on a CAD mesh is presented to demonstrate the performance of the framework.

## 2 Related Work

Partition or Segmentation is an important problem in geometry processing and is addressed by different topics and fields such as computer vision, civil engineering and

© The Author(s) 2021
L. Roucoules et al. (Eds.): JCM 2020, LNME, pp. 228–234, 2021.
https://doi.org/10.1007/978-3-030-70566-4_37

mechanical engineering in the context of reverse engineering. Anwer et al. [2] put forward a classification of the existing partitioning approaches to provide scientific fundamentals for ISO GPS partitioning.

A more comprehensive classification is proposed as show in Fig. 1. The existing point cloud (PC, in short) and mesh partitioning methods are classified into six categories based on their methodologies. The attributes of the PC or the mesh such as curvature, geodesic distance, normal, dihedral angles, slippage, etc. are calculated in order to provide criteria for different methods [3].

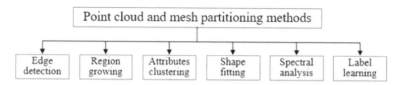

**Fig. 1.** Classification of point cloud and mesh partitioning methods.

Edge detection methods, or boundary-based methods, attempt to locate the edges of the object then group the points inside the boundaries [4]. Partitioning by region growing decomposes a mesh or a point cloud into regions that share similar properties [5]. A merging step is always needed to reduce the number of the obtained regions. The core of attributes clustering methods is to address the attributes for each point based on the geometric characteristics and then calculate the distance of each node to the specific region [6, 7]. Here we divide the clustering method into iterative clustering [6] and hierarchical clustering (or "direct clustering methods") [7] according to the clustering process. Shape fitting methods try to fit primitive shapes such as planes, spheres, cylinders, cones or tori to the point cloud. Points are labeled as one partitioned region if they belong to the same primitive shape [8]. Spectral analysis methods try to use the eigenvalues and eigenvectors of a properly structured square matrix, which is informative of local geometric attributes of a mesh [9]. Recent advances in machine learning, especially deep learning technologies, provide novel solutions for the 3D point cloud and mesh segmentation. Attributes are used as the label features to train the learning model and the well-turned model achieves a significant improvement in results [10].

The key techniques and the attributes used for each type of methods introduced above are highlighted in Table 1. The output types of each reference are classified into semantic (e.g. a shaft) and geometric (e.g. a cylinder).

It can be seen that existing methods are able to segment the object into different regions. However, it is critical to identify the segmented regions as seven invariance classes in the context of ISO GPS. Meanwhile, most of the methods use post-processing steps to ensure the segmentation results. Therefore, a framework that combined a variety of techniques is preferable to be implemented for obtaining robust results.

## 3   Curvature-Based Framework for ISO GPS Partitioning

Curvature is an important tool for analyzing the geometry of surfaces since it specifies the second-order properties of a surface and is independent of the used parametrization.

**Table 1.** Summary and comparison of existing partitioning methods

| References | Category | Input | Technique | Attributes used | Output |
|---|---|---|---|---|---|
| [4] | Edge detection | Mesh | Random walk | Shortest graph path | Semantic |
| [5] | Region growing | PC | An octree-based region growing method | Normal vector and residual value | Geometry |
| [6] | Attributes clustering | Mesh | Heat mapping | Heat mean signature | Semantic |
| [7] | Attributes clustering | Mesh | Hough transformation, mean shift | Mean curvature field | Geometry |
| [8] | Shape fitting | PC | Random sample consensus | Normal | Geometry |
| [9] | Spectral analysis | PC | Heat walks and a curvature-aware segmentation | Symmetric point cloud Laplacian | Semantic |
| [10] | Label learning | PC | Conditional random field | Unary features | Semantic |

Based on the curvature estimation of points on a given mesh, a framework of partitioning in the context of ISO GPS is presented as shown in Fig. 2.

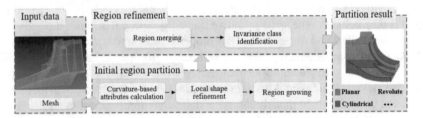

**Fig. 2.** The framework of the proposed partitioning method.

Since a point-cloud does not provide much point-set topological information, a mesh representation should be firstly reconstructed according to some specific criteria as the input of the proposed framework [11]. The output are the partitioned features in accordance with the seven invariance classes defined in ISO GPS. Initial region partition and region refinement are conducted as two main steps in the framework in order to obtain robust partition results.

### 3.1 Initial Region Partition

Firstly, the principal curvatures of each vertex ($k_1$ and $k_2$, i.e. maximum and minimum principal curvature) in the mesh is calculated. Secondly, shape index and curvedness [12, 13] in the neighborhood of each vertex are calculated based on the principal curvatures. Shape index specifies the local shape type and it is independent from size and the assignment of principal directions. Curvedness specifies the size and it is the amount of the surface curvatures in the neighborhood of a vertex. The shape index and the curvedness of a vertex are defined as (1) and (2).

$$s(\mathbf{p}_i) = \frac{-2}{\pi} \arctan\left(\frac{\kappa_1(\mathbf{p}_i) + \kappa_2(\mathbf{p}_i)}{\kappa_1(\mathbf{p}_i) - \kappa_2(\mathbf{p}_i)}\right) \tag{1}$$

$$c(x_i) = \sqrt{\frac{\kappa_1^2(x_i) + \kappa_2^2(x_i)}{2}} \tag{2}$$

Each vertex on the mesh will be clustered into sharp edges and several shape types by evaluating their shape index and curvedness values. Vertices belonging to sharp edges are detected at first in the framework.

Due to the error of point data during the acquisition process in practice [14], the vertices in a relatively small region do not share the same value of curvedness and shape index. Therefore, an iterative voting method [15] is conducted in order to refine the local shape type. The 11 types of local shape are the label calculated for each vertex according to its shape index value.

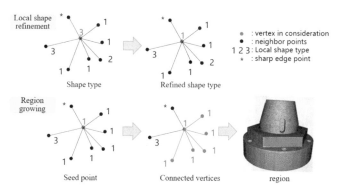

**Fig. 3.** Illustration of local shape refinement and region growing in the framework.

After the local shape type refinement, labelled points on the mesh are connected to each other by a region growing method, as illustrated in Fig. 3. In each iteration, the point which is not labeled will be chosen as the initial seed for region growing process. All the neighbor vertices of the seed will be checked if they share the same local shape type index. The neighbor points will be set as the new seed of growing process if they share the same local shape type index. The full region growing process stops when all the vertices are labeled by a group number.

## 3.2   Region Refinement

The obtained regions from the region growing process become the initial patches as the input of region refinement. A slippage signature-based algorithm [16] is used to merge the regions. In the method, a geometrical feature called slippable motion are calculated for each initial patch. Adjacent region pairs are evaluated by the slippage signature and are merged into one region if they share the same value. Further post-processing steps are conducted after the slippage merging of the surface in order to reduce the impact by noise data.

Merged regions by slippage analysis are not classified in the context of ISO GPS seven invariance classes. Therefore, vertices which are grouped from the input mesh should be registered according to local properties. A statistical modeling method proposed by [17] is utilized to provide a statistical description of each region with strong mathematical basis. After invariance class recognition, quick and accurate identification of features is addressed by labeling the regions.

## 4   Case Study

A test part is used as a case study to show the effectiveness of the proposed framework for CAD mesh partitioning. The part is designed by CAD system and manufactured as in Fig. 4.

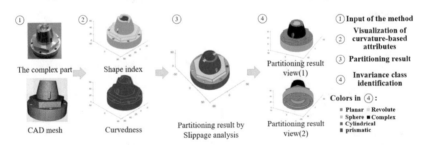

**Fig. 4.** A case study of the proposed framework of mesh partitioning.

The 3D nominal shape of the part is represented by a mesh model. Shape index and curvedness of each vertex on the mesh are calculated. Then local shape type refinement and region growing methods are implemented to stabilize the indicator of each point in the mesh.

Slippage analysis is used to merge small patches into a large region. The independent 38 regions of the mesh (in random color) are the partitioning result by slippage analysis. An invariance class recognition method by statistical modeling are then implemented to label these regions. The results are recognized and visualized according to seven invariance classes.

# 5 Conclusion

A curvature-based mesh partitioning method framework is proposed based on a comprehensive literature review. The proposed framework is able to partition 3D CAD mesh into regions in the context of ISO GPS. The result of implementation and testing shows that the proposed method can partition the tessellated nominal models appropriately. Future work will investigate the applicability of the approach in the context of measurement for ISO GPS verification.

**Acknowledgments.** This research has benefitted from the financial support of China Scholarship Council (first author).

# References

1. ISO 17450-1:2011, Geometrical product specifications (GPS)—General concepts—Part 1: Model for geometrical specification and verification. Geneva, International Organization for Standardization (2011)
2. Anwer, N., Scott, P.J., Srinivasan, V.: Toward a classification of partitioning operations for standardization of geometrical product specifications and verification. Procedia CIRP **75**, 325–330 (2018)
3. Shamir, A.: A survey on mesh segmentation techniques. Comput. Graph. Forum **27**(6), 1539–1556 (2008)
4. Golovinskiy, A., Funkhouser, T.: Randomized cuts for 3D mesh analysis. ACM Trans. Graph. **27**(5), 1 (2008)
5. Chen, J., Chen, B.: Architectural modeling from sparsely scanned range data. Int. J. Comput. Vis. **78**(2–3), 223–236 (2008)
6. Fang, Y., Sun, M., Kim, M., et al.: Heat-mapping: a robust approach toward perceptually consistent mesh segmentation. In: CVPR 2011, pp. 2145–2152. IEEE (2011)
7. Xiao, D., Lin, H., Xian, C., et al.: CAD mesh model segmentation by clustering. Comput. Graph. **35**(3), 685–691 (2011)
8. Schnabel, R., Wahl, R., Klein, R.: Efficient RANSAC for point-cloud shape detection. Comput. Graph. Forum **26**(2), 214–226 (2007)
9. Williams, R.M., Ilieş, H.T.: Practical shape analysis and segmentation methods for point cloud models. Comput. Aided Geom. Des. **67**, 97–120 (2018)
10. Kalogerakis, E., Hertzmann, A., Singh, K.: Learning 3D mesh segmentation and labeling. In: ACM Transactions on Graphics (TOG), vol. 29, no. 4, p. 102. ACM (2010)
11. Dey, T.K.: Curve and Surface Reconstruction: Algorithms with Mathematical Analysis. Cambridge University press, Cambridge (2006). (Cambridge Monographs on Applied and Computational Mathematics)
12. Koenderink, J.J., Van Doorn, A.J.: Surface shape and curvature scales. Image Vis. Comput. **10**(8), 557–564 (1992)
13. Anwer, N., Mathieu, L.: From reverse engineering to shape engineering in mechanical design. CIRP Ann. **65**(1), 165–168 (2016)
14. Weißgerber, M., Ebermann, M., Gröger, S., et al.: Requirements for datum systems in computer aided tolerancing and the verification process. Procedia CIRP **43**, 238–243 (2016)

15. Cai, N., Anwer, N., Scott, P.J., et al.: A new partitioning process for geometrical product specifications and verification. Precis. Eng. **62**, 282–295 (2020)
16. Gelfand, N., Guibas, L.J.: Shape segmentation using local slippage analysis. In: Proceedings of the Eurographics/ACM SIGGRAPH Symposium on Geometry Processing, pp. 214–223 (2004)
17. Chiabert, P., Costa, M.: Statistical modelling of nominal and measured mechanical surfaces. J. Comput. Inf. Sci. Eng. **3**(1), 87–94 (2003)

# A Topology Optimization Method for Stochastic Lattice Structures

Filippo Cucinotta, Marcello Raffaele$^{(\boxtimes)}$, and Fabio Salmeri

Department of Engineering, University of Messina, Contrada Di Dio (S. Agata),
98166 Messina, Italy
fsalmeri@unime.it

**Abstract.** Stochastic lattice structures are very powerful solutions for filling three-dimensional spaces using a generative algorithm. They are suitable for 3D printing and are well appropriate to structural optimization and mass distribution, allowing for high-performance and low-weight structures. The paper shows a method, developed in the Rhino-Grasshopper environment, to distribute lattice structures until a goal is achieved, e.g. the reduction of the weight, the harmonization of the stresses or the limitation of the strain. As case study, a cantilever beam made of Titan alloy, by means of SLS technology has been optimized. The results of the work show the potentiality of the methodology, with a very performing structure and low computational efforts.

**Keywords:** Topology Optimization · Additive Manufacturing · Mechanical design · Lattice structures

## 1 Introduction

Topology Optimization (TO) is a very powerful tool [1] and it allows to obtain lightweight structures, optimizing one or more parameters [2]. Recently, thanks to the rapid development of the Additive Manufacturing (AM) technologies, also TO has had a rapid increase in interest and application [3]. The main fields range from biomedical [4, 5] to sports [6], automotive [7] and aerospace [8].

In 2019, Cucinotta et al. proposed a 2D TO method based on a new algorithm developed in the Rhinoceros-Grasshopper environment (grasshopper3d.com) [9, 10] based on a Voronoi tessellation. In this work, the authors present an important evolution of that method, with a 3D-volume application and the use of a reticular structure, instead of the hollowing. In this case the generated structure is lattice, based on points, called seeds, distributed in a stochastic way or with a distribution law.

The method is oriented for a Selective Laser Sintering (SLS) AM technology, or similar. These techniques consist in the sintering of metal powders with the energy supply coming from a laser light [6, 10].

© The Author(s) 2021
L. Roucoules et al. (Eds.): JCM 2020, LNME, pp. 235–240, 2021.
https://doi.org/10.1007/978-3-030-70566-4_38

## 2    Materials and Methods

### 2.1    The Algorithm

The algorithm has been developed in the Rhinoceros – Grasshopper environment. The Grasshopper is a visual programming language that allows to parametrize the drawing tools of the Rhinoceros CAD (Fig. 1). The first step of the algorithm is the filling of the regions that are to be optimized with the seeds. Each seed is the point where the lattices will connect to each other. It is possible to decide the number and the distribution of the seeds, that can be uniform or not. In the second step, the seeds are interconnected by linear prismatic elements. In this phase it is possible to decide a limit of a number of connections for each node and the maximum and minimum length limits of the connectors. Obviously, by increasing the number and the limits, the number of connectors that make up the lattice structure increases. At the end of each stage, the algorithm exports a ready-to-use universal .STP file.

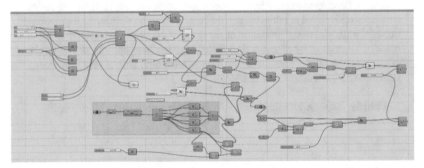

**Fig. 1.**  The algorithm on the Grasshopper canvas

### 2.2    The Case Study

The case study is a monolithic cantilever beam, with a fixed constrain in a side and the load on the opposite side, that is optimized only in the core volume, fixing the upper and a lower face with a constant thickness. In this way, the result, is a sandwich panel [12] with all the potentiality in terms of weight and stress of this kind of structures [13, 14]. The objective of the TO has been set as the reduction of the global weight of the system, until a set stress threshold is reached (in this case set as 70% of the yield stress, i.e. about 600 MPa). The component is a 100 (length) $\times$ 50 (width) $\times$ 13 (thickness) mm cantilever beam, with a fixed constrain in a side, and a vertical force (2 kN), distributed on the edge, on the opposite side. At the iteration 0, the beam was monolithic. After that iteration, at each subsequent step, the algorithm populates with seeds the volume between the upper and lower face, realizing a sandwich. The thickness of the two skin was fixed (in the case study 3 mm). The seeds have been increasingly populated, keeping the points that reach a stress threshold, set at 70% of the yield point of the material, plus a random quota. The random quota can be distributed uniformly, as in the case of this work, or non-uniformly with an ad hoc distribution law. For each seed, the algorithm connects the nearest 5 seeds with a beam, provided they are neither longer nor shorter than certain limits. Each beam was imposed with a circular section and a fixed diameter of 0.5 mm.

For this work the Ti-6Al-4V ELI alloy has been used hypothesizing an AM process in SLS technology [11]. This type of material is used in various industrial sectors, in relation to its good mechanical properties and in the biomedical field in consequence of its biocompatibility [14].

The Finite Element Analysis (FEA) was performed using the Nastran solver within Siemens NX 1859. The choice of mesh was a structured hexahedral grid with size of 1mm, in the monolithic component and in the skins, and beams elements for the lattice structure. A linear static analysis was performed using a Nastran SOL 101. To be sure that the boundary conditions were always the same in each simulation, fixed selection boxes have been created.

## 3   Results and Discussion

The Fig. 2 shows the FEA results at the last step (with 1000 seeds). It is possible to observe that, if the skins are linearly loaded, as expected, the beams are more or less uniformly loaded, because the number of seeds change in function of the stress.

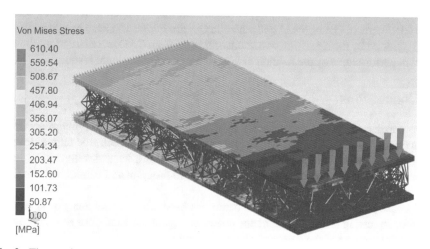

**Fig. 2.** The maximum von Mises equivalent stress on the FEA at the last step. The distributed force is of 2 kN. The constrain is fixed on the opposite side of the force.

As the number of seeds increases, the number of beams increases and consequently the structure becomes more robust but also weightier (Fig. 3). To test the effectiveness of the algorithm, many population cycles were carried out until the maximum stress value stabilized without producing significant advantages. At that point, the weight/stress ratio grows.

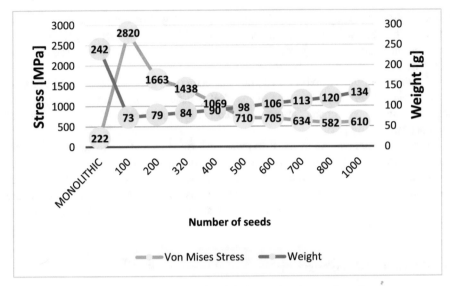

**Fig. 3.** The trend of stress and weight increasing the number of seeds

At the end of the optimization, the component reached the fixed allowable stress (600 MPa) with a weight reduction of 50% with 800 seeds and about 2700 beams.

The workstation used for running the algorithm was equipped with an Intel Core I7-8700 CPU with 16 GB of RAM and a nVidia Quadro P5000 16 GB. The computational time requested by the algorithm was about of 60 s for each CAD reconstruction, showing the high performance of the method also in terms of computational resources.

## 4 Conclusions

In the paper a novel 3D optimization method, based on a mix of random and high-stress points, have been presented. The method is based on a stochastic lattice structure inside two, or more, monolithic walls. The algorithm has been developed inside the Rhinoceros-Grasshopper environment that allows to program and parametrise the CAD commands.

As case study a simple cantilever beam has been choose. The analysis shows the different effects, in terms of weight and stress, changing the number of seeds necessary for the construction of the stochastic lattice structure.

The method showed high potentialities with a drastic weight reduction (up to 50%) and very good stress performance, with a very low computational effort. Furthermore, the stochastic structure ensures good performance even when real loads differ significantly from design loads.

## References

1. Sigmund, O., Maute, K.: Topology optimization approaches. Struct. Multidiscip. Optim. **48**, 1031–1055 (2013). https://doi.org/10.1007/s00158-013-0978-6

2. Barone, S., Cucinotta, F., Sfravara, F.: A comparative life cycle assessment of utility poles manufactured with different materials and dimensions (2017). https://doi.org/10.1007/978-3-319-45781-9_10

3. Gardan, J.: Additive manufacturing technologies: state of the art and trends. Int. J. Prod. Res. **54**, 3118–3132 (2016). https://doi.org/10.1080/00207543.2015.1115909

4. Cucinotta, F., Guglielmino, E., Longo, G., Risitano, G., Santonocito, D., Sfravara, F.: Topology optimization additive manufacturing-oriented for a biomedical application. Adv. Mech. Des. Eng. Manuf. II (2019). https://doi.org/10.1007/978-3-030-12346-8_18

5. Singh, S., Ramakrishna, S.: Biomedical applications of additive manufacturing: present and future. Curr. Opin. Biomed. Eng. **2**, 105–115 (2017). https://doi.org/10.1016/j.cobme.2017.05.006

6. Graziosi, S., Rosa, F., Casati, R., Solarino, P., Vedani, M., Bordegoni, M.: Designing for metal additive manufacturing: a case study in the professional sports equipment field. Procedia Manuf. **11**, 1544–1551 (2017). https://doi.org/10.1016/j.promfg.2017.07.288

7. Reddy, K.S.N., Maranan, V., Simpson, T.W., Palmer, T., Dickman, C.J.: Application of topology optimization and design for additive manufacturing guidelines on an automotive component (2016). https://doi.org/10.1115/DETC2016-59719

8. Zhu, J.-H., Zhang, W.-H., Xia, L.: Topology optimization in aircraft and aerospace structures design. Arch. Comput. Methods Eng. **23**, 595–622 (2016). https://doi.org/10.1007/s11831-015-9151-2

9. Cucinotta, F., Raffaele, M., Salmeri, F.: A stress-based topology optimization method by a Voronoi tessellation additive manufacturing oriented. Int. J. Adv. Manuf. Technol. **102** (2019). https://doi.org/https://doi.org/10.1007/s00170-019-03676-4

10. Cucinotta, F., Raffaele, M., Salmeri, F.: A Topology Optimization of a Motorsport Safety Device. Springer, Heidelberg (2020). https://doi.org/10.1007/978-3-030-31154-4_34

11. Fotovvati, B., Namdari, N., Dehghanghadikolaei, A.: Fatigue performance of selective laser melted Ti6Al4V components: state of the art. Mater. Res. Express. **6**, 14 (2019). https://doi.org/10.1088/2053-1591/aae10e

12. Cucinotta, F., Sfravara, F., Neri, P., Razionale, A.: Composite sandwich impact response: experimental and numerical analysis. Frat. ed Integrita Strutt. **13**, 367–382 (2019). https://doi.org/10.3221/IGF-ESIS.47.27

13. Cucinotta, F., Paoli, A., Risitano, G., Sfravara, F.: Optical measurements and experimental investigations in repeated low-energy impacts in powerboat sandwich composites. Proc. Inst. Mech. Eng. Part M J. Eng. Marit. Environ. (2017). https://doi.org/10.1177/1475090217720619

14. Cucinotta, F., Guglielmino, E., Risitano, G., Sfravara, F.: Assessment of damage evolution in sandwich composite material subjected to repeated impacts by means optical measurements. Procedia Struct. Integr. **2**, 3660–3667 (2016). https://doi.org/10.1016/j.prostr.2016.06.455

# Generative Design for Additively Manufactured Textiles in Orthopaedic Applications

V. Ricotta[1(✉)], R. Ian Campbell[2], T. Ingrassia[1], and V. Nigrelli[1]

[1] Dipartimento di Ingegneria, Università degli Studi di Palermo, Viale delle Scienze,
90128 Palermo, Italy
vito.ricotta@unipa.it
[2] Loughborough Design School, Loughborough University, Loughborough LE11 3TU, UK

**Abstract.** The aim of this work is to implement a new process for the design and production of orthopaedic devices to realize entirely by Additive Manufacturing (AM). In particular, a generative algorithm for parametric modelling of flexible structures to use in orthopaedic devices has been developed. The developed modelling algorithm has been applied to a case study based on the design and production of a customized elbow orthosis made by Selective Laser Sintering. The results obtained have demonstrated that the developed algorithm overcomes many drawbacks typical of traditional CAD modelling approaches. FEM simulations have been also performed to validate the design of the orthosis. The new modelling algorithm allows designers to model flexible structures with no deformations or mismatches and to create parametric CAD models to use for the production of orthopaedic devices through AM technologies.

**Keywords:** Additive Manufacturing · Additively manufactured textiles · Generative algorithms · CAD modelling · Elbow orthosis

## 1 Introduction

Additive Manufacturing (AM) is a constantly evolving technology and many researchers have explored its application in different fields, like engineering and medicine [1, 2]. In this last field, AM has been largely used for orthopaedic applications to produce customised prostheses and orthoses [3, 4].

Customised orthoses can be better adapted to the patient's body than prefabricated devices, thus ensuring better performance [1, 2, 4] but, usually, they could have very complex shapes. Therefore, for this kind of application, AM is the most suitable production technology because it facilitates creating complex shapes with a large variety of materials [5]. It also has lower costs than manual fabrication [6] and very good surface finish, if the process parameters are suitably chosen [7, 8].

Moreover, in recent years the use of AM has been investigated to produce flexible materials. The basic idea is to create a flexible 3D printed textile which could be used to make, for example, clothing and accessories.

L. Roucoules et al. (Eds.): JCM 2020, LNME, pp. 241–248, 2021.
https://doi.org/10.1007/978-3-030-70566-4_39

The aim of this work is to implement a new process aimed at the design and production of orthopaedic devices that are fully manufacturable by AM. For this purpose, a new modelling approach based on generative algorithms and on the use of the Additively Manufactured Textiles (AMT) technology [9] has been proposed. A test case, based on the modelling and production of a customized elbow orthosis, has been developed.

## 2   AMT Flexible Structures

To improve the aesthetics of orthopaedic devices today, thanks to modern additive technologies, it is possible to use additively manufactured textiles to create flexible structures [10] that also allow a better fit of the devices to patients.

Of course, additively manufactured textiles are very interesting and could be widely used for different applications, but the lack of efficient approaches for their 3D modelling limits their use. In this work, two different approaches have been used to model AMT structures: in the first, a standard CAD software was used; in the second, a generative algorithm [11, 12] was used. This second approach was developed to overcome the problems encountered during the CAD modelling of the AMT structures with conventional software.

### 2.1   CAD Modelling of Flexible Structures by Commercial Software

To investigate a standard approach to model additively manufactured textiles, a test case study related to the design and modelling of a customized elbow orthosis was developed (Fig. 1).

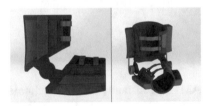

**Fig. 1.**   CAD model of the designed elbow orthosis.

The analysed elbow prosthesis was initially modelled using a classical reverse engineering approach [13] and its structural parts were dimensioned using numerical methods [14].

The flexible structures of the orthosis were modelled using the commercial software Rhinoceros. Initially, it was decided to create flexible structures consisting of networks of rings. For this purpose, the Paneling Tools plug-in was used. The 3D basic module of the network is shown in Fig. 2 (left). The structure created and adapted to the external surface of a part of the orthosis is shown in Fig. 2 (right).

**Fig. 2.** Basic module to repeat (Left) - CAD model of the flexible structure by Rhinoceros (Right).

The modelling of the analysed flexible structure using this approach showed some drawbacks mainly due to the deformations of the rings of the basic module after the panelization on the surface. As shown in Fig. 3, in many cases, wide variations of the diameter of the rings were detected. In some cases, a diameter reduction of around 40–50% was calculated. Moreover, there were also misalignments among the rings of different modules (Fig. 3).

**Fig. 3.** Deformations and misalignments of the rings of the flexible structure using a commercial software.

The above-mentioned modelling problems, besides representing a drawback from the aesthetic point of view, could generate some problems during the manufacturing phase and could affect the strength of the structure. This aspect was confirmed during the realization of a prototype by AM. One of the two parts of the orthosis (Fig. 4), in fact, reported defects due to the deformation of the rings, while the other part was not fully made due to the too small thickness of the rings.

**Fig. 4.** STL model of flexible structure (on the left) and prototype (on the right).

## 2.2  CAD Modelling of Flexible Structures by a New Approach

To overcome the drawbacks of the modelling of flexible structures with Rhinoceros, a new modelling approach based on generative algorithms was proposed. The new algorithm was developed using Grasshopper, a graphical algorithm editor that can be used as a plug-in to Rhinoceros. The developed algorithm allows the creation of flexible structures made of different basic modules (for example rings) that are perfectly mapped onto any surface with no deformation and/or misalignment. The algorithm consists of three main features that allow the:

- 3D parametric modelling of a basic module;
- definition of a grid of points on a surface;
- paneling of the basic module on the surface.

The function blocks diagram of the developed algorithm is shown in Fig. 5.

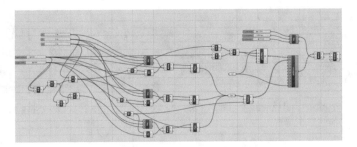

**Fig. 5.**  Function blocks diagram in Grasshopper.

The developed algorithm was tested by modelling the same flexible structures previously created with the Rhinoceros software. Initially, the basic module was modelled. A group of three connected rings (Fig. 6) was modelled. The relative positions of the three rings were parameterized depending on the main input parameters, such as the diameters of the toroid and the transversal circular section.

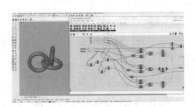

**Fig. 6.**  Basic module (on the left) and block diagram of the 3D parametric modelling function (on the right).

The developed modelling algorithm allowed the creation of a flexible structure perfectly adapted to the surface of the orthosis (Fig. 7).

**Fig. 7.** Flexible structure modelled by the algorithm.

As can be seen in Fig. 8, there is no deformation, misalignment or interference among the rings.

**Fig. 8.** Details of the flexible structure.

The final CAD model of the orthosis made through the new algorithm is shown in Fig. 9.

**Fig. 9.** CAD model of the elbow orthosis.

## 3   FEM Analysis

FEM analyses have been performed to validate the design in terms of topology and main dimensions of the structure. In fact, only the structural parts of the orthosis in PA 2200 have been analysed. In particular, the flexion and extension movements have been simulated. A very high test load, equal to 150 N, has been imposed to the orthosis to simulate extremely hard-working conditions. Figure 10 and 11 show Von Mises stress maps during extension and flexion movements. In all cases, it can be observed that all the maximum values of the Von Mises stress are lower than the ultimate tensile stress of PA 2200 ($\approx$48 MPa). Experimental tests will be useful for a better characterization of the materials and validating the numerical models [15–17] in different working conditions.

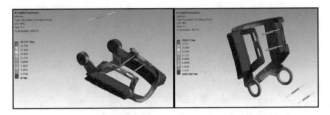

**Fig. 10.** Extension movement: Von Mises stress maps

**Fig. 11.** Flexion movement: Von Mises stress maps

## 4   Prototype Manufacturing by Additive Manufacturing

After the CAD model of the designed orthosis was completed, a prototype of the device was produced in PA 2200 using an SLS 3D printer (EOS FORMIGA P100). PA 2200 is a fine white powder based on polyamide, used for the manufacture of products with high surface quality and very good thermal and mechanical characteristics [5]. Figure 12 shows the prototype from which it is possible to observe the good surface finish obtained with this AM technology.

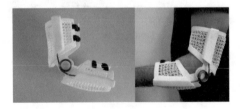

**Fig. 12.** Final prototype of the designed elbow orthosis.

## 5   Conclusions

In this work, a new process aimed at the design and production of orthopaedic devices fully manufacturable by Additive Manufacturing has been implemented.

To better customise and improve the aesthetic appearance of an elbow orthosis, it was decided to use Additively Manufactured Textiles technology. Initially, a commercial CAD software was used to model the non-structural parts of the orthosis and its flexible structures. This approach had some drawbacks mainly due to the deformations and misalignments of the AMT structures. To overcome these problems, a new modelling approach based on generative algorithms was developed. The obtained results have shown that this generative algorithm is a very good solution that allows the user to create very complex flexible structures with no drawbacks typical of traditional CAD modelling approaches. FEM analyses have been carried out considering the movements of flexion and extension of the arm. Obtained results allowed to validate the design. Subsequently, a functional prototype of the elbow orthosis has been realized by means of the SLS technology. The results present innovative elements of originality in the CAD modelling sector, which can contribute to solving problems related to modelling for Additive Manufacturing in different application fields.

# References

1. Javaid, M., Haleem, A.: Additive manufacturing applications in medical cases: a review literature based. Alexandria J. Med. **54**(4), 411–422 (2018)
2. Singh, S., Ramakrishna, S.: Biomedical applications of additive manufacturing: present and future. Curr. Opin. Biomed. Eng. **2**, 105–115 (2017)
3. Ingrassia, T., Nalbone, L., Nigrelli, V., Ricotta, V., Pisciotta, D.: Biomechanical analysis of the humeral tray positioning in reverse shoulder arthroplasty design. Int. J. Interact. Des. Manuf. **12**(2), 651–661 (2018)
4. Ingrassia, T., Nalbone, L., Nigrelli, V., Pisciotta, D., Ricotta, V.: Influence of the metaphysis positioning in a new reverse shoulder prosthesis. In: Advances on Mechanics, Design Engineering and Manufacturing, Lecture Notes in Mechanical Engineering. Springer International Publishing AG (2017). https://doi.org/10.1007/978-3-319-45781-9_47
5. Singh, S., Ramakrishna, S., Singh, R.: Material issues in additive manufacturing: a review. J. Manuf. Process. **25**, 185–200 (2017)
6. Gao, W., Zhang, Y., Ramanujan, D., Ramania, K., Chen, Y., Williams, C.B., Wang, C.C.L., Shin, Y.C., Zhang, S., Zavattieri, P.D.: The status, challenges, and future of additive manufacturing in engineering. Comput. Aided Des. **69**, 65–89 (2015)
7. Mancuso, A., Pitarresi, G., Saporito, A., Tumino, D.: Topological optimization of a structural naval component manufactured in FDM. In: Lecture Notes in Mechanical Engineering, pp. 451–462 (2019). https://doi.org/10.1007/978-3-030-12346-8_44
8. Ingrassia, T., Nigrelli, V., Ricotta, V., Tartamella, C.: Process parameters influence in additive manufacturing. In: Advances on Mechanics, Design Engineering and Manufacturing, Lecture Notes in Mechanical Engineering, pp. 261–270. Springer International Publishing AG (2017). https://doi.org/10.1007/978-3-319-45781-9_27
9. Lussenburg, K., Van der Velden, N., Doubrovski, Z., Geraedts, J., Karana, E.: Designing with 3D printed textiles. In: Proceedings of 5th International Conference on Additive Technologies, Wien, AT, pp. 74–81, 16–17 October 2014 (2014)
10. Bloomfield, M. and Borstrock, S.: Modeclix. The additively manufactured adaptable textile. Mater. Today Commun. **16**, 212–216 (2018)
11. Krish, S.: A practical generative design method. Comput. Aided Des. **43**(1), 88–100 (2011)
12. Khabazi, Z.: Generative Algorithms Using Grasshopper (2010)
13. Ricotta, V., Campbell, R.I., Ingrassia, T., Nigrelli, V.: Additively manufactured textiles and parametric modelling by generative algorithms in orthopaedic applications. Rapid Prototyping J. (2020). https://doi.org/10.1108/RPJ-05-2019-0140
14. Mancuso, A., Pitarresi, G., Tumino, D.: Using FEM simulation to predict structural performances of a sailing dinghy. Int. J. Interact. Des. Manuf. **12**(3), 811–822 (2018)

15. Tumino, D., Ingrassia, T., Nigrelli, V., Pitarresi, G., Urso Miano, V.: Mechanical behavior of a sandwich with corrugated GRP core: numerical modeling and experimental validation. Frattura ed Integrita Strutturale **30**, 317–326 (2014)
16. Ingrassia, T., Lombardo, B., Nigrelli, V., Ricotta, V., Nalbone, L., D'Arienzo, A., D'Arienzo, M., Porcellini, G.: Influence of sutures configuration on the strength of tendon-patch joints for rotator cuff tears treatment. Injury **50**, S18–S23 (2019)
17. Ingrassia, T., Nigrelli, V., Ricotta, V., Nalbone, L., D'Arienzo, A., D'Arienzo, M., Por cellini, G.: A new method to evaluate the influence of the glenosphere positioning on stability and range of motion of a reverse shoulder prosthesis. Injury **50**, S12–S17 (2019)

# Modeling Symmetric Minimal Surfaces by Mesh Subdivision

Stefano Rosso[1(✉)], Andrea Curtarello[2], Federico Basana[2], Luca Grigolato[2], Roberto Meneghello[1], Gianmaria Concheri[2], and Gianpaolo Savio[2]

[1] Department of Management and Engineering, University of Padova, Padua, Italy
stefano.rosso.3@phd.unipd.it
[2] Department of Civil, Environmental and Architectural Engineering, University of Padova, Padua, Italy

**Abstract.** Thanks to the great diffusion of additive manufacturing technologies, the interest in lattice structures is growing. Among them, minimal surfaces are characterized by zero mean curvature, allowing enhanced properties such as mechanical response and fluidynamic behavior. Recent works showed a method for geometric modeling triply periodic minimal surfaces (TPMS) based on subdivision surface. In this paper, the deviation between the subdivided TPMS and the implicit defined ones is investigated together with mechanical properties computed by numerical methods. As a result, a model of mechanical properties as a function of the TPMS thickness and relative density is proposed.

**Keywords:** Lattice structures · Additive manufacturing · Triply periodic minimal surfaces · Design for additive manufacturing

## 1 Introduction

The spreading of additive manufacturing (AM) technologies makes it possible to produce parts with unprecedented complexity, such as biomimicry products, organic shapes and lightweight components. Nevertheless, anisotropic materials, surface finish, costs and security are just some of the open issues [1]. Among the lightweight structures, cellular solids or lattice structures are characterized by superior properties: they have high specific stiffness and strength, and they are good heat exchangers, energy absorbers and acoustic insulators [2, 3, 4]. Lattice structures, indeed, can find application from microscale to macroscale, from tissue engineering [5], to construction [6]. Minimal surfaces are a kind of lattice structures defined as surfaces with zero mean curvature or surfaces that minimize the surface area for given boundary conditions [7]. Minimal surfaces repeating themselves in three dimensions are called triply periodic minimal surfaces (TPMS) [8]. These surfaces have received huge attention in the research community due to their intrinsic properties [9], especially in the biomedical field, where it is mandatory to use porous scaffolds designed to allow fluid exchange and tissue regrowth [10, 11]. Furthermore, due to their curvature continuity, the stress concentration at nodal points is eliminated and fatigue life is improved compared to beam-like lattice structures [12].

L. Roucoules et al. (Eds.): JCM 2020, LNME, pp. 249–254, 2021.
https://doi.org/10.1007/978-3-030-70566-4_40

TPMS can be modeled adopting different methods: boundary representation (BRep), volume representation (VRep) and constructive solid geometry (CSG) [2, 13]. As emerged in recent studies, current mesh or parametric models are not ideal for modeling lattice structures, due to the significant computational resources required, processing times, robustness, rendering and visualization issues [14, 15]. To overcome these limitations, a novel method for geometric modeling variable thickness triply periodic surfaces based on a subdivision surface algorithm has been recently proposed [13]. Nevertheless, the geometrical accuracy of the modeling method and the mechanical characterization of the thickened structure still need to be investigated.

As an extension of the aforementioned work, in this study, the geometrical deviation of the subdivided TPMS P-Surface from the implicit defined one is investigated, then the mechanical properties of the unit cell are computed by finite element analyses. As a result, a model of mechanical properties as a function of the cell relative density is proposed and compared to the ones available in literature. The findings show the accuracy of the proposed modeling method; more, the numerical model allows to relate the thickness of the part to the mechanical properties by simulating a single cell inside the lattice structure, saving computational time, and giving directions for tailored applications with lattices that present variable properties in the design volume. Furthermore, knowing the properties of a single cell is a key feature when integrating topology optimization in the design workflow and also for applying the homogenization method [16] during the mechanical characterization of lattice structures.

## 2    Methods

A 1 mm unit cell of P-Surface type was modeled as described in [13]. A coarse mesh was used to model the unit cell of a P-surface; the Catmull-Clark subdivision surface algorithm was then adopted to achieve a smooth geometric model; finally, a thickness is assigned adopting a differential offset algorithm.

The deviation of the subdivided mesh at the third iteration of the Catmull-Clark subdivision scheme from the reference minimal surface was then computed in Rhinoceros 6 by "Mesh-Mesh Deviation" tool from "Rhino Open Projects" [17, 18]. The reference minimal surface was built in Netfabb introducing the minimal surface implicit equation in the mathematical part library (accuracy at 0.05 mm).

The finite element analysis software ANSYS R18.1 was then used to obtain the mechanical properties of the TPMS cells: Young's modulus (E), Poisson's ratio ($v$), and Shear modulus (G). The used material was a titanium alloy, Ti6Al4V ELI (Grade 23), with the following bulk properties: $E_0 = 113800$ MPa, $v_0 = 0.342$, $G_0 = 42400$ MPa [19]. The stl file of the cell after three iterations of the subdivision algorithm was imported in the software and a second order tetrahedral elements mesh was used. Then, a set of constraints was imposed to the single cell. Even if a single cell is studied, the boundary conditions have been imposed in order to simulate an entire lattice structure, so the cell must deform accordingly. As Fig. 1 shows, for E and $v$ determination, a 0.05 mm displacement along Y axis is imposed to the upper face, while X and Z directions are free; the bottom face is fixed along Y axis (i.e., a 0 mm displacement is imposed), and X and Z direction displacements are free. Remote points are used to control the degrees of freedom of the cell faces; this technique allows to relate a point (the remote point) to a node, a face, or to the entire body, and to control the behavior of the connected part by directly imposing loads and/or constraints to the remote point. A remote point with a deformable behavior, i.e. the geometry is free to deform, is assigned to the bottom face; remote points are also connected to the lateral faces of the unit cell, along the normal direction of each face, with a coupled behavior. To take into account the presence of the adjacent cells, coupling equations are assigned to the nodes of opposite faces [16], so each node of a face moves with the same displacement of the opposite node. To obtain E and $v$, the reaction force of the bottom face and the contraction of lateral faces are evaluated. Similarly, another set is imposed to the P-Surface cell to determine G. A displacement along the X direction is imposed to the upper face, and reaction force of the bottom face are used to calculate G; the remote displacement and coupling equations of opposite faces are also assigned to ensure a displacement compatible with adjacent cells. The simulations were run four times, targeting four different thickness of the P-Surface cell, 0.1 mm, 0.2 mm, 0.3 mm and 0.4 mm, corresponding to a relative density of 0.23, 0.45, 0.64, 0.80, respectively.

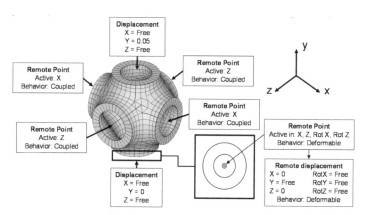

**Fig. 1.** Set of constraints for E and $v$ determination.

## 3 Results

Figure 2 shows the deviation map between the subdivided and the implicit minimal surface on a 1 mm unit cell. Subdivided minimal surfaces slightly differ from the ones defined by implicit equations, less than 1.3% of the cell dimension.

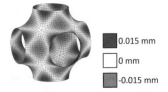

0.015 mm

0 mm

-0.015 mm

**Fig. 2.** Deviation map between subdivided and implicit minimal surfaces on a 1 mm P-surface unit cella. Max range: ±0.0127 mm.

Figure 3 shows the results for the elastic modulus E. The ratio between the obtained modulus E and the modulus of the bulk material $E_0$ is plotted against relative density ($\rho$). A power law well fits the results (Standard Deviation = 0.0146). The experimental data are also described by a quadratic equation so, as a first approximation, the model can be simplified (Standard Deviation = 0.0301).

In the same graph, a comparison with the results by Bobbert et al. [9] and Lee et al. [20] is presented. Lee's results are close to the ones obtained in this study; the trend found by Bobbert et al., instead, slightly differs but their results were experimentally obtained from compression tests, while the ones of the present study refer to numerical analyses of tensile tests. The proposed numerical model also extends in a wider relative density range, 0.2–0.8, if compared to Bobbert's, 0.3–0.5, and Lee's, 0.02–0.3. Graphs in Fig. 4 show the results for the determination of $\nu$ and G.

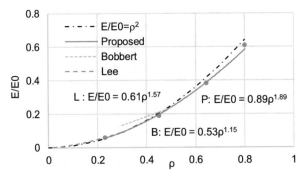

**Fig. 3.** Elastic modulus simulation results for P-Surface cell, and comparison with the literature.

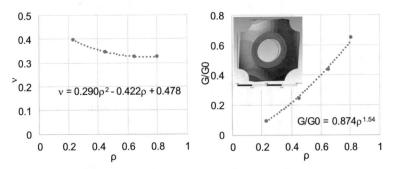

**Fig. 4.** Poisson's ratio (left) and shear modulus (right) simulation results for P-Surface cell. The figure in the shear plot highlights the effects of the deformable and coupled conditions.

# 4 Conclusions

In this work, a TPMS P-Surface modeled with a subdivision surface method was geometrically and mechanically characterized, and a model of the mechanical properties as a function of the relative density was obtained. The numerical method that was used only requires a single unit cell resulting in accurate solutions, and reducing computational time, since computational time and cost increase cubically as the number of cells increases.

The results show that Young's and shear modulus increase with relative density. The results for Young's modulus, if compared with data available in the literature, present a good agreement and extend for a wider relative density range.

Mechanical characterization of subdivided TPMS opens new possibilities for the implementation of the topology optimization in the modeling workflow of lattices with variable properties and allows to simulate this type of structures using the homogenization method, reducing time and computational costs.

**Acknowledgments.** This work was partially funded by Fondazione Cassa Di Risparmio di Padova e Rovigo (CARIPARO) and grant "BIRD 190850" by Department of Civil, Environmental and Architectural Engineering, University of Padova.

# References

1. Bacciaglia, A., Ceruti, A., Liverani, A.: Additive manufacturing challenges and future developments in the next ten years, pp. 891–902 (2020). https://doi.org/10.1007/978-3-030-31154-4_76
2. Savio, G., Rosso, S., Meneghello, R., Concheri, G.: Geometric modeling of cellular materials for additive manufacturing in biomedical field: a review. Appl. Bionics Biomech., 1–14 (2018). https://doi.org/10.1155/2018/1654782
3. Tamburrino, F., Graziosi, S., Bordegoni, M.: The design process of additively manufactured mesoscale lattice structures: a review. J. Comput. Inf. Sci. Eng. **18**(4), 1–6 (2018). https://doi.org/10.1115/1.4040131
4. Plocher, J., Panesar, A.: Review on design and structural optimisation in additive manufacturing: towards next-generation lightweight structures. Mater. Des. **183** (2019). https://doi.org/10.1016/j.matdes.2019.108164
5. Lanzotti, A., Martorelli, M., Russo, T., Gloria, A.: Design of additively manufactured lattice structures for tissue regeneration. Mater. Sci. Forum MSF **941**, 2154–2159 (2018). https://doi.org/10.4028/www.scientific.net/MSF.941.2154
6. Paolini, A., Kollmannsberger, S., Rank, E.: Additive manufacturing in construction: a review on processes, applications, and digital planning methods. Addit. Manuf. **30**(July), 100894 (2019). https://doi.org/10.1016/j.addma.2019.100894
7. Lord, E.A., Mackay, A.L.: Periodic minimal surfaces of cubic symmetry. Curr. Sci. **85**(3), 346–362 (2003)
8. Schoen, A.H.: Infinite periodic minimal surfaces without self-intersections (1970)
9. Bobbert, F.S.L., et al.: Additively manufactured metallic porous biomaterials based on minimal surfaces: a unique combination of topological, mechanical, and mass transport properties. Acta Biomater. **53**, 572–584 (2017). https://doi.org/10.1016/j.actbio.2017.02.024

10. Kapfer, S.C., Hyde, S.T., Mecke, K., Arns, C.H., Schröder-Turk, G.E.: Minimal surface scaffold designs for tissue engineering. Biomaterials **32**(29), 6875–6882 (2011). https://doi.org/10.1016/J.BIOMATERIALS.2011.06.012

11. Ambu, R., Morabito, A.E.: Modeling, assessment, and design of porous cells based on schwartz primitive surface for bone scaffolds. Sci. World J. **2019** (2019). https://doi.org/10.1155/2019/7060847

12. Speirs, M., Van Hooreweder, B., Van Humbeeck, J., Kruth, J.-P.: Fatigue behaviour of NiTi shape memory alloy scaffolds produced by SLM, a unit cell design comparison. J. Mech. Behav. Biomed. Mater. **70**, 53–59 (2017). https://doi.org/10.1016/j.jmbbm.2017.01.016

13. Savio, G., Meneghello, R., Concheri, G.: Design of variable thickness triply periodic surfaces for additive manufacturing. Prog. Addit. Manuf. **4**(3), 281–290 (2019). https://doi.org/10.1007/s40964-019-00073-x

14. Feng, J., Fu, J., Lin, Z., Shang, C., Li, B.: A review of the design methods of complex topology structures for 3D printing. Vis. Comput. Ind. Biomed. Art **1**(1), 5 (2018). https://doi.org/10.1186/s42492-018-0004-3

15. Pasko, A., Fryazinov, O., Vilbrandt, T., Fayolle, P., Adzhiev, V.: Procedural function-based modelling of volumetric microstructures. Graph. Models **73**(5), 165–181 (2011). https://doi.org/10.1016/j.gmod.2011.03.001

16. Savio, G., Curtarello, A., Rosso, S., Meneghello, R., Concheri, G.: Homogenization driven design of lightweight structures for additive manufacturing. Int. J. Interact. Des. Manuf. (2019). https://doi.org/10.1007/s12008-019-00543-0

17. Savio, G.: Rhino Open Projects. https://www.food4rhino.com/app/rhino-open-projects. Accessed 06 Feb 2020

18. Savio, G., Meneghello, R., Concheri, G.: Optical properties of spectacle lenses computed by surfaces differential quantities. Adv. Sci. Lett. **19**(2), 595–600 (2013). https://doi.org/10.1166/asl.2013.4724

19. AZoM, Grade 23 Ti 6Al 4V ELI Alloy (UNS R56401). AZO Materials (2013). https://www.azom.com/article.aspx?ArticleID=9365. Accessed 07 Feb 2020

20. Lee, D.-W., Khan, K.A., Abu Al-Rub, R.K.: Stiffness and yield strength of architectured foams based on the Schwarz Primitive triply periodic minimal surface. Int. J. Plast. **95**, 1–20 (2017). https://doi.org/10.1016/J.IJPLAS.2017.03.005

# Analysis of the Accuracy of Reconstruction of a Human Cornea by Two Geometric Modelling Techniques: A Comparative Study

F. J. F. Cañavate[1], F. Cavas[1], J. S. Velázquez[1(✉)], J. M. Bolarín[2], and J. L. Alió[3]

[1] Department of Structures, Construction and Graphical Expression, Technical University of Cartagena, 30202 Cartagena, Spain
jose.velazquez@upct.es

[2] Technology Centre for IT and Communications (CENTIC), Scientific Park of Murcia, 30100 Murcia, Spain

[3] Division of Ophthalmology, Miguel Hernández University, 03690 Alicante, Spain

**Abstract.** This study compares two techniques of reconstruction and representation of the anterior and posterior surfaces of the human cornea, both for healthy and for keratoconus eyes, in a CAD software environment using raw tomographic data, by two different methods: a mesh model (obtained from a grid of points) and a NURBS surface model. After reconstruction, a morpho-geometric analysis was made, and several parameters were defined and measured in a set of 100 healthy eyes and 61 keratoconus eyes, detecting the statistical analysis significant differences between methods for both groups. Mesh method proved to be more accurate, but less eyes could be modelled. Volumetric and surface measurement values showed a total equivalence between both methods, so these would be the ones to use when modelling eyes for keratoconus detection and characterization purposes.

**Keywords:** NURBS · Scheimpflug tomography · Computer-aided design · Point cloud · Cornea

## 1 Introduction

Geometric modelling allows the representation of a physical shape from a discrete cloud of scanned points [1], being currently used for a wide variety of engineering applications, such as reconstruction of 3D urban models [2], geo-morphometry assessment [3], design parametrization in reverse engineering [4], prosthesis production [5] or rare disease diagnosis [6], among others.

In addition, the use of CAD tools in biological structure modelling is a well-stablished technique that allows creating both virtual and printed 3D models [7]. However, not much attention has been paid to which technique fits better the geometrical reconstruction, and how this affects clinical diagnosis.

Many commercial tools with auto-surfacing capabilities can easily convert point clouds into surface models [4], but the reconstruction methods are not unique, and the accuracy of the results may vary depending on which method is selected. Usual

© The Author(s) 2021
L. Roucoules et al. (Eds.): JCM 2020, LNME, pp. 255–260, 2021.
https://doi.org/10.1007/978-3-030-70566-4_41

alternatives are using a "mesh" model (obtained from a grid of points) or a "patch" model based in non-uniform rational basis splines (NURBS).

Meshes have no topological constraints and have a simple definition [8], but they are discrete and therefore performing operations with them can be difficult [9]. NURBS are a mathematical model commonly used in CAD environments for generating and representing curves and surfaces [10]. They are parametrically defined, and therefore, they are smooth no matter the scale, and operations such as intersection or trimming can be simpler because coordinates can be calculated anywhere on the surface. However, maths involved are difficult and therefore data exchange with other 3D applications is almost impossible [9].

This study presents a comparative analysis of two reconstruction and representation techniques applied to the anterior and posterior surfaces of the human cornea, for both healthy eyes and eyes with keratoconus (KC), to assess the strengths and weaknesses of each representation in KC disease detection.

## 2 Materials and Methods

### 2.1 Geometrical Reconstruction of the Human Cornea

In this research, we have reconstructed both anterior and posterior corneal surfaces from the raw tomographic data (unprocessed by the device software, which avoids possible biases) generated by the corneal tomographer Sirius® (CSO, Italy) in the form of two clouds of points. To do so, we used the surface-handling CAD software Rhinoceros® V 5.0 (MCNeel & Associates, Seattle, USA), which offered two alternative reconstruction methods, between which we wanted to compare:

- "Patch" surface generation function. Uses a NURB surface to reconstruct both corneal surfaces, minimizing the nominal distance between the 3D point cloud and the solution surface. A cubic NURB surface was used with a segmentation of 256 in each parametric direction. The surface so obtained, approximates the point cloud.
- "Mesh" model to generate a grid. In this case, the points lie on each surface, and then surfaces are obtained by interpolation. A rectangular grid of 21 rows and 256 columns was selected, being then deformed to minimise the nominal distance between the spatial points and the grid surface.

Rhinoceros® "PointDeviation" function was used to estimate the mean distance error between the generated surface and the cloud of points used for its generation. For both surface generation methods, and for both anterior and posterior surfaces, a threshold limit of $10^{-3}$ was fixed to consider the points as "not valid", while a value of $10^{-4}$ was considered a "good" point. Eyes with a mean value of error for all their points lower than $5 \times 10^{-3}$ were discarded from the original set.

The procedure followed to generate full 3D customized corneal models consisted in two stages (Fig. 1): a first one of virtual 3D modelling of the cornea, in which corneal tomography system acquires the altimetry data of the corneal surfaces, that is later exported to Rhinoceros® to reconstruct the surface by the two methods previously exposed; and a second one of morpho-geometric analysis, in which the customized

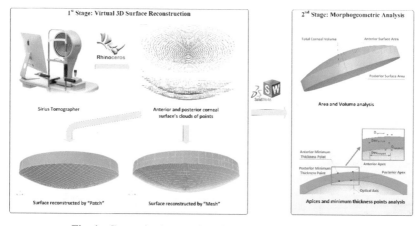

**Fig. 1.** Customized corneal model reconstruction procedure

3D corneal model obtained in Rhinoceros® was transferred to the CAD/CAM program SolidWorks® V 2015 (Dassault Systèmes, Vélizy-Villacoublay, France), that allowed performing a thorough analysis of the custom model and therefore characterizing its morpho-geometry.

The final output after this stage, was the set of thirteen parameters related with area and volume shown in Table 1. These parameters, which have been described in detail in a precedent study [11], were finally studied in a specific set of eyes, in order to determine which reconstruction method characterized better KC.

## 2.2  Practical Application

This comparative study included 161 eyes of 161 patients selected at random from the official database "Iberia" of KC cases (RETICS-OFATARED).

The sample was later divided into two groups: a "healthy" group formed by 100 healthy eyes of 100 individuals, and a "keratoconus" group that included 61 KC eyes of 61 patients. The presence and degree of the disease was stablished according to state-of-the-art evaluation techniques and the Amsler-Krumeich classification [12], creating subgroups for the Grades I (27 subclinical KC eyes) and II + III + IV (34 clinical KC eyes).

The research was developed at Vissum Corporation Clinic in Alicante, Spain, and was ratified by the institution's Ethical Board, according to the tenets the Declaration of Helsinki (7th revision, October 2013, Fortaleza, Brazil). An agreed consent form was signed by each participant.

## 2.3  Statistical Analysis

The software used to perform all statistical analyses was SPSS V23.0 (IBM, Armonk, EEUU). To check the normality of data, Kolmogorov-Smirnov test and Shapiro-Wilks test were used. To determine if the measured values of the morpho-geometric parameters depended on the function used when reconstructing the corneal surface ("mesh" or "patch"), Student and Mann-Whitney tests for unpaired data were used. A significance level of 0.05 was fixed for p-values in all statistical tests. Bland-Altman diagrams were used with comparison purposes.

**Table 1.** Difference in value and statistical significance of "patch" vs. "mesh" functions comparison for each morpho-geometrical parameter. Stu.: Student's test p-value; Wilc.; Wilcoxon's test p-value; Abs.: Absolute difference in value; %: Relative difference in percentage.

| Parameter (unit) | Difference | | Healthy (n = 89) | | Grade I (n = 24) | | Grade II + III + IV (n = 18) | |
|---|---|---|---|---|---|---|---|---|
| | Abs | %. | Stu. | Wilc. | Stu. | Wilc. | Stu. | Wilc. |
| Total volume (mm³) | 0.04 | 0.16 | 0.049 | 0.043 | 0.233 | 0.383 | 0.873 | 1.000 |
| Ant. Apex area (mm²) | 0.05 | 0.08 | <0.001 | <0.001 | 0.025 | 0.041 | 0.178 | 0.183 |
| Post. Apex area (mm²) | 0.02 | 0.03 | <0.001 | <0.001 | 0.026 | 0.003 | 0.152 | 0.197 |
| Total area (mm²) | 0.04 | 0.03 | 0.015 | 0.061 | 0.443 | 0.603 | 0.553 | 0.875 |
| Sag. plane apex area (mm²) | 0.04 | 0.93 | 0.373 | 0.419 | 0.875 | 0.692 | 0.029 | 0.018 |
| Sag. plane area at MTP (mm²) | 0.02 | 0.39 | 0.033 | 0.161 | 0.655 | 0.534 | 0.862 | 0.409 |
| Ant. apex dev. (mm) | 0.03 | 482.7 | 0.008 | 0.011 | 0.086 | 0.005 | 0.001 | 0.001 |
| Post. apex dev. (mm) | 0.10 | 92.37 | 0.786 | < 0.001 | 0.063 | 0.064 | 0.464 | 0.084 |
| Centre of mass X (mm) | 0.00 | 8.69 | 0.006 | 0.009 | 0.417 | 0.405 | 0.773 | 0.671 |
| Centre of mass Y (mm) | 0.00 | 12.37 | 0.013 | 0.005 | 0.271 | 0.3 | 0.11 | 0.127 |
| Centre of mass Z (mm) | 0.00 | 7.47 | 0.841 | 0.85 | 0.05 | 0.052 | 0.856 | 0.829 |
| Ant. MTP deviation (mm) | 0.16 | 16.62 | 0.077 | 0.045 | 0.729 | 0.966 | 0.184 | 0.245 |
| Post. MTP deviation (mm) | 0.16 | 17.81 | 0.125 | 0.014 | 0.796 | 0.853 | 0.374 | 0.889 |

## 3   Results

Table 1 shows the result of the comparison between both methods. For each case, it was studied if the difference between the values measured by both methods was different from zero (meaning zero total equivalence between them). As shown, significant differences were found between them when considering the total number of eyes, and also for each grade of the disease.

It is important to highlight that not all the corneas (161) could be reconstructed from the cloud of points. Some of them did not lead to any result at all when treated with Rhinoceros®, and some others showed values of the mean deviation error from the cloud of points to the final surface higher than $10^{-3}$, which invalidated them for the study. Table 2 summarizes how each function performed.

**Table 2.** Distribution and total number of cases successfully modelled by each function

|               | Healthy   | Keratoconus | Total     |
|---------------|-----------|-------------|-----------|
| Total raw data | 100       | 61          | 161       |
| Total modelled |           |             |           |
| Grid          | 89 (89%)  | 42 (69%)    | 131 (81%) |
| Patch         | 96 (96%)  | 56 (92%)    | 152 (94%) |

## 4   Conclusions

Results in Table 1 indicate that both methods are not equivalent, especially when measuring the anterior/posterior apices and anterior/posterior minimum thickness points deviations. The lack of concordance between methods for posterior apex deviation (92% difference) is of paramount importance, as this is considered the critical parameter for early KC detection [13]. The difference for anterior apex deviation is even higher (482.73%), but as anterior apex starts to displace in most advanced stages of the disease [14] and disease can be diagnosed by other parameters at this stage, this difference is less important.

According to Table 2, the use of "patch" function allows modelling a higher amount of corneal surfaces not only for healthy eyes but also for KC ones. Its main drawback is the tolerance or approximation to the cloud of points, which makes it less sensible for modelling KC at a local level in its early stages than the "mesh" function. Conversely, the degree of precision with "mesh" function is higher, and therefore the approximation to the cloud of points is better, but its main downside is that a great number of KC corneas could not be modelled. When the surface is very abrupt, or highly steeped (advanced stage of the KC pathology) such a sensible method of surface reconstruction like "mesh" presents local reconstruction problems that lead to non-convergence, which could be used as a clue of presence of the disease.

**Funding.** This publication forms part of the Thematic Network for Co-Operative Research in Health (RETICS), reference number RD16/0008/0012, financed by the Carlos III Health Institute - General Subdirection of Networks and Cooperative Investigation Centers (R&D&I National Plan 2013–2016) and European Regional Development Funds (FEDER), and the Results Valorisation Program (PROVALOR-UPCT) financed by the Technical University of Cartagena.

## References

1. Berger, M., Tagliasacchi, A., Seversky, L., Alliez, P., Levine, J., Sharf, A., Silva, C.: State of the art in surface reconstruction from point clouds. In: Eurographics 2014 - State of the Art Reports, Strasbourg, France, vol. 1, pp. 161–185, 07 April 2014. https://doi.org/10.2312/egst.20141040.https://hal.inria.fr/hal-01017700/document
2. Wang, R., Peethambaran, J., Chen, D.: LiDAR point clouds to 3-D urban models : a review. IEEE J. Sel. Topics Appl. Earth Obs. Remote Sens. **11**(2), 606–627 (2018). https://doi.org/10.1109/JSTARS.2017.2781132

3. Eltner, A., Kaiser, A., Castillo, C., Rock, G., Neugirg, F., Abellán, A.: Image-based surface reconstruction in geomorphometry-merits, limits and developments. Earth Surf. Dyn. **4**(2), 359–389 (2016). https://doi.org/10.5194/esurf-4-359-2016

4. Chang, K.-H., Chen, C.: 3D shape engineering and design parameterization. Comput. Aid. Des. Appl. **8**(5), 681–692 (2011). https://doi.org/10.3722/cadaps.2011.681-692

5. Vukašinović, N., Kolšek, T., Duhovnik, J.: Case study – surface reconstruction from point clouds for prosthesis production. J. Eng. Des. **18**(5), 475–488 (2007). https://doi.org/10.1080/09544820701403805

6. Cavas-Martínez, F., Fernández-Pacheco, D.G., Cañavate, F.J.F., Velázquez-Blázquez, J.S., Bolarín, J.M., Alió, J.L.: Study of morpho-geometric variables to improve the diagnosis in keratoconus with mild visual limitation. Symmetry **10**, 306 (2018). https://doi.org/10.3390/sym10080306

7. Lohfeld, S., Barron, V., McHugh, P.E.: Biomodels of bone: a review. Ann. Biomed. Eng. **33**(10), 1295–1311 (2005). https://doi.org/10.1007/s10439-005-5873-x

8. Wittek, A., Grosland, N.M., Joldes, G.R., Magnotta, V.: Miller KJAobe from finite element meshes to clouds of points: a review of methods for generation of computational biomechanics models for patient-specific applications. Ann. Biomed. Eng. **44**(1), 3–15 (2016)

9. Rutten, D.: Mesh or surface, understanding the difference. Available via McNeel Rhinoceros Forums. https://discourse.mcneel.com/t/mesh-or-surface-understanding-the-difference/75151. Accessed 13 Jan 2020

10. Dimas, E., Briassoulis, D.: 3D geometric modelling based on NURBS: a review. Adv. Eng. Softw. **30**(9), 741–751 (1999). https://doi.org/10.1016/S0965-9978(98)00110-0

11. Bolarín, J.M., Cavas, F., Velázquez, J.S., Alió, J.L.: A machine-learning model based on morphogeometric parameters for RETICS disease classification and GUI development. Appl. Sci. **10**, 1874 (2020). https://doi.org/10.3390/app10051874

12. Krumeich, J.H., Daniel, J., Knulle, A.: Live-epikeratophakia for keratoconus. J. Cataract Refract. Surg. **24**(4), 456–463 (1998). https://doi.org/10.1016/s0886-3350(98)80284-8

13. Cavas-Martinez, F., Fernandez-Pacheco, D.G., Canavate, F.J.F., Velazquez-Blazquez, J.S., Bolarin, J.M., Tiveron, M., Alio, J.L.: Early keratoconus detection by patient-specific 3D modelling and geometric parameters analysis. Dyna **94**, 171–175 (2019). https://doi.org/10.6036/8895

14. Velázquez, J.S., Cavas, F., Alió del Barrio, J., Fernández-Pacheco, D.G., Alió, J.: Assessment of the association between In Vivo corneal morphogeometrical changes and keratoconus eyes with severe visual limitation. J. Ophthalmol. **2019**, 7 (2019). https://doi.org/10.1155/2019/8731626

# Customization of Kayak Paddle Grips by Using Reverse Engineering, Computer Aided Design and Additive Manufacturing Tools

Eneko Solaberrieta[1(✉)], Xabier Amezua[1], Xabier Garikano[1], Mikel Iturrate[2], Jose Antonio Oriozabala[1], and Iñaki Martin[1]

[1] Department of Graphic Design and Engineering Projects, Gipuzkoa Faculty of Engineering, University of the Basque Country UPV/EHU, Donostia-Eibar, Spain
`eneko.solaberrieta@ehu.eus`
[2] Department of Business Management, Gipuzkoa Faculty of Engineering, University of the Basque Country UPV/EHU, Donostia, Spain

**Abstract.** In this paper, due to the importance of maintaining a secure grip with the control hand in kayaking, a simple three phase process is presented for the massive development of personalized grips which allow the improvement of this handgrip. This process consists of obtaining the 3D geometry of the paddler's handgrip by using Reverse Engineering (RE) tools, designing the grip from the obtained 3D geometry by using Computer Aided Design (CAD) tools and manufacturing the grip by using Additive Manufacturing (AM) tools. Therefore, this paper shows that the RE, CAD and AM tools available today allow the customization of products for many applications.

**Keywords:** Reverse engineering · Computer aided design · Additive manufacturing · Customization · Kayak paddle grip

## 1 Introduction

In the world of canoeing and kayaking, the term "kayak" denotes a closed-cockpit boat that is controlled by a paddle with a blade at either end by one or more people who are sitting [1]. What could be considered the first kayaks were developed by the Inuit people of the Arctic as a means of hunting and transportation and were designed to handle the implacable environmental conditions of the Arctic [2, 3]. Today there is a wide variety of kayaks and paddles, so guidelines have been developed to help users make the correct selection [3, 4].

In any case, to achieve an optimum performance, among other points, it is important to grip the paddle correctly. For this reason, there are rules for holding the paddle [3]. Among these rules it should be noted that during any type of stroke (foward stroke, backward stroke, draw stroke, etc.) one of the hands, which is called the control hand (usually the right hand for right-handed people and the left hand for left-handed people), should never allow the paddle to rotate [3]. This way the control hand maintains the

© The Author(s) 2021
L. Roucoules et al. (Eds.): JCM 2020, LNME, pp. 261–267, 2021.
https://doi.org/10.1007/978-3-030-70566-4_42

control of the paddle and the opposite hand allows any rotation of the paddle that is necessary. With this gripping technique, it is possible to maintain the effectiveness of each stroke and reduce the risk of injury [3].

Due to the importance of maintaining a secure grip with the control hand during navigation, in this paper is presented a procedure for the development of customized kayak paddle grips that allow to improve the secure grip of the control hand in a more ergonomic way for the users. The ability to produce individualized products with the same efficiency and cost as production is known as Customization [5].

The simple three-phase customization process presented in this paper is outlined in Fig. 1. Hence, this process consists of obtaining the 3D geometry of the handgrip (also called 3D anthropometric data) by using Reverse Engineering (RE) tools, designing the grip from the obtained 3D geometry by using Computer Aided Design (CAD) tools and manufacturing the grip by means of Additive Manufacturing (AM) tools. This technology is an emerging technology and there are many possibilities nowadays [6–8].

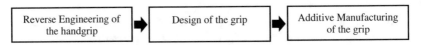

**Fig. 1.** The simple three-phase process for the customization of kayak paddle grips.

This process has been carried out in the DEHI laboratory (www.ehu.eus/dehi) of the Faculty of Engineering of Gipuzkoa of the University of the Basque Country UPV/EHU. As far as the authors know, although 3D anthropometry data has been used for the development of several customized products in different sectors (including sports) [9, 10], this is the first time it is used for the customization of kayak paddle grips.

## 2   Materials and Methods

The material resources used to carry out the three-phase procedure described in Fig. 1 are the following ones: a cylindrical sample of the same diameter as the paddle shaft, a malleable impression material (modelling clay, Jovi), a 3D digitalization system (the portable white light Go!SCAN20 3D scanner and its software VX elements 6.3 SR1, Creaform), a CAD software (Solid Edge ST10, Siemens) and an AM machine (uPrint SE FDM 3D printer, Stratasys). Now, a detailed description of how the procedure was carried out is given.

### 2.1   Phase 1: Reverse Engineering of the Handgrip

This first phase consists of obtaining the 3D geometry of the paddle user's handgrip. To carry out this phase, the following steps were followed:

1. Obtain an impression of the paddle user's handgrip (Fig. 2a). To perform this task, a thick layer of modelling clay has been placed on the paddle shaft sample and then the paddle user has gripped it.

2.  Prepare the impression for the digitalization (Fig. 2b). As the portable 3D scanner used employs for its positioning both the features and texture of the object to be digitalized and positioning targets, to obtain a more accurate result, several of these positioning targets were placed around the impression.

3.  Digitalize the impression (Fig. 2c). To perform this task, keeping the relative position between the positioning targets fixed and changing the relative position between the 3D scanner and the impression, different captures were taken until a point cloud that completely defined the geometry of the impression was obtained.

4.  Process the point cloud to obtain the polygon model (triangle mesh) of the impression (Fig. 2d). This process involves principally reducing the noise in the data collected, merging the point clouds collected in different captures and creating a unique polygon model. This process was carried out in the 3D scanner's software and was performed automatically after adjusting certain parameters such as: resolution (0,4 mm), optimize scan mesh (level 0 of 100), decimate scan mesh (level 0 of 100), automatically fill holes (level 5 of 100) and remove isolated patches (level 0 of 100).

5.  Clean the polygon model (Fig. 2e). This process consists of removing from the polygon model all the unnecessary data for the design. This task was carried out in the 3D scanner's software.

6.  Define the global alignment of the polygon model (Fig. 2f). To perform this task, certain geometric features must previously be created: a cylinder that best fits the paddle shaft and a plane perpendicular to it located at any point along the paddle shaft. To define the global coordinate system, the z-axis of the reference system must coincide with the axis of the created cylinder, and the x-y plane of the reference system must coincide with the created plane. The position of the x-y plane along the cylinder axis and the orientation of its axes are not important in this case.

7.  Improve the polygon model to close holes, clean the edges, increase resolution on high curvature areas and minimize scanner noise (Fig. 2g) This process consists of editing the triangle mesh by using tools such as: close holes, decimate the mesh, smooth the mesh, etc. This process was carried out in the 3D scanner's software and was performed by using all the necessary tools to obtain a clean and watertight mesh by varying the values of its parameters for different regions of the polygon model depending on its geometrical characteristics.

8.  Generate an organized patch layout surface model from the polygon model (Fig. 2h). This process was carried out in the 3D scanner's software and as it was an organic geometry, it was performed by using the auto surface tool.

9.  Export the surface model to the CAD software in STL format.

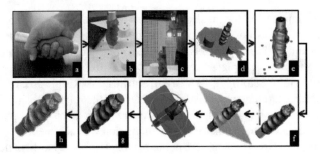

**Fig. 2.** Phase 1: reverse engineering of the grip.

## 2.2   Phase 2: Design of the Grip

This second phase consists of designing the grip from the 3D geometry obtained in the preceding phase. To carry out this phase, the following steps were followed in the CAD software:

1. Transform the surface model into a solid model (Fig. 3a).
2. Generate the paddle shaft's hole (Fig. 3b). To perform this task, a circular passing hole with the same diameter and direction as the paddle shaft was created.
3. Generate paddle shaft's sleeve (Fig. 3c). To perform this task, a thin cylinder with the same inner diameter and axis as the paddle shaft was created.
4. Design the joining system and divide the resulting solid into two parts (Fig. 3d). As a kayak paddle has a blade at either end, to enable its assembly on the paddle shaft, it must be composed of at least two parts. Also, since the grip transmits the force to the paddle, for applications where high paddle forces are expected, it may be necessary to include geometric modifications in its design to avoid both sliding and rotation onto the paddle shaft (Fig. 3d).
5. Export the grip design in STL format.

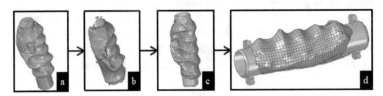

**Fig. 3.** Phase 2: design of the grip.

## 2.3   Phase 3: Additive Manufacturing of the Grip

This phase consists of manufacturing the final design of the kayak paddle grip by using an AM machine. Today there is a wide variety of AM technologies available, which allow manufacturing parts with different materials [11]. But in any case AM allows a direct translation from the design to the part, without having to program (the STL file of the design is enough). As an example, one of the designs was manufactured with the uPrint SE FDM 3D printer (Stratasys) (Fig. 4c).

## 3   Results and Discussion

The following two designs were generated: the design for activities that involve a small paddling force (e.g., touring kayaking) (Fig. 4a) and the design for activities that involve a high paddling force (e.g., whitewater kayaking) (Fig. 4b). The difference between the two designs is that the first one restricts the sliding and rotation of the grip onto the paddle shaft by the pressurized assembly of its two parts while the second one restricts this movement by the modification of the geometry of both the sleeve and the paddle shaft.

Therefore, the first design would be able to avoid the movement between the paddle shaft and the grip at low paddling forces and the second design would be able to avoid it at higher paddling forces. However, taking into account that the position of the hands on the paddle is different for each user [3, 4], the first design would allow sharing the same paddle between different users, while the second one, as it requires modifications in the paddle geometry, would restrict this possibility. Therefore, before selecting the design, it is necessary to determine the specifications that the grip must satisfy. In any case, as reflected in this paper, both designs could be developed in a massive way.

As for the duration of the development of these personalized grips, this depends on factors such as the hardware and software used in each phase, the experience of the developers and the geometric characteristics of both the paddler's hand and the final design. In the case of the prototype developed in this study, it was as follows: 0.5 h for phase 1, 0.75 h for phase 2 and 12 h for phase 3. The high duration of phase 3 is due to the printing time required by the 3D printer used. Finally, in order to calculate the cost of the development process, it would have to take into account, among other things, the labor costs, the raw material costs (in this case 15 €/kg) and the amortizations, which are completely dependent on the development company.

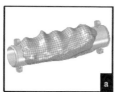

**Fig. 4.**   The two designs generated (a,b) and the result of the AM of the first design (c).

# 4  Conclusions

This paper shows that the RE, CAD and AM tools available today allow the customization of kayak paddle grips. Although some designs involve modifications in the paddle's geometry, the development process of the customized grip is still massive. The process presented in this document can be easily adapted to any other similar product from any other sector.

**Acknowledgments.** The authors of this paper thank the Faculty of Engineering of Gipuzkoa forlocating the DEHI laboratory in their facilities and the study was supported by the Gipuzkoa Country Council (grant number 70/19) and MINECO Ministry of Economy and Competitiveness (grant number PID2019-108975RA-I00).

# References

1. Poynter, S.: Instant canoeing and kayaking: tips and techniques for perfect paddling. Infinite Ideas Limited (2012)
2. Hutchinson, D.C.: The Complete Book of Sea Kayaking, 6th ed. Falcon (2018)
3. Parsons, S.: How to Paddle a Kayak: The 90 Minute Guide to Master Kayaking and Learn to Paddle Like a Pro (2012)
4. Foster, N.: The Art of Kayaking. Everything You Need to Know about Paddling. Falcon (2017)
5. Raja, V., Fernandes, K.J.: Reverse Engineering: An Industrial Perspective. Springer, Berlin (2008)
6. Zindani, D., Kumar, K.: Integrated manufacturing system for complex geometries: towards zero waste in additive manufacturing. In: Handbook of Research on Green Engineering Techniques for Modern Manufacturing, pp. 14–23 (2019)
7. Keshavamurthy, R.: Additive manufacturing process and their applications for green technology. In: Handbook of Research on Green Engineering Techniques for Modern Manufacturing, pp. 262–281 (2019)
8. Miar, S., et al.: Additive manufacturing for tissue engineering. 3D printing and biofabrication. In: Reference Series in Biomedical Engineering, pp: 3–54 (2018)
9. Jones, P.R.M., Rioux, M.: Three-dimensional surface anthropometry: applications to the human body. Opt. Lasers Eng. **28**(2), 89–117 (1997)
10. Pang, T.Y., et al.: Fit, stability and comfort assessment of custom-fitted bicycle helmet inner liner designs, based on 3D anthropometric data. Appl. Ergon. **68**, 240–248 (2018)
11. Tofail, S.A.M., et al.: Additive manufacturing: scientific and technological challenges, market uptake and opportunities. Mater. Today **21**(1), 22–37 (2018)

# A Fast and Reliable Optical 3D Scanning System for Human Arm

Francesco Buonamici, Monica Carfagni, Luca Puggelli, Michaela Servi,
and Yary Volpe[✉]

Department of Industrial Engineering of Florence – DIEF, Via di S. Marta 3, 50139 Firenze, Italy
yary.volpe@unifi.it

**Abstract.** The article discusses the design of an acquisition system for the 3D surface of human arms. The system is composed by a 3D optical scanner implementing stereoscopic depth sensors and by an acquisition software responsible for the processing of the raw data. The 3D data acquired by the scanner is used as starting point for the manufacturing of custom-made 3D printed casts. Specifically, the article discusses the choices made in the development of an improved version of an existing system presented in [1] and presents the results achieved by the devised system.

**Keywords:** 3D scanning · Arm scanner · 3D printing · Orthoses · Body scanner

## 1 Introduction

The introduction of personalized medicine procedures in the traditional clinical practice calls for the development of innovative measurement systems capable of acquiring different kind of 3D patient-specific data. Traditional diagnostic tools remain an important data source for the design and production of custom-made medical devices, but some limiting aspects (e.g. CT and MRI administration of radiations to patients) can be identified. Accordingly, the development of optical 3D scanners based on passive technologies or non-invasive active ones, could foster the application of high-quality personalized medicine procedures in new areas (e.g. dentistry [2], monitoring of congenital malformations [3]).

One of the areas that could benefit the most from the introduction of optical non-invasive scanners is orthopedics [2, 5], where they could be applied for the design of different classes of devices. Orthopedic appliances rely on customized geometries that fit the patient's anatomy to be effective. Accordingly, the development of 3D scanners with specially designed features for the acquisition of specific anatomies could improve the treatment of several injuries and pathologies.

In this context, the authors designed a system for the acquisition of the geometry of the arm-wrist-hand district, called *Oplà* 1.0, deputed to the fabrication of 3D printed casts for the treatment of wrist fractures. The system is composed of a 3D optical scanner and an acquisition software that is responsible for collecting the reference data and processing it. For the interested reader, all the details pertaining to the development and

© The Author(s) 2021
L. Roucoules et al. (Eds.): JCM 2020, LNME, pp. 268–273, 2021.
https://doi.org/10.1007/978-3-030-70566-4_43

testing of the first version of the system can be found in [1, 5, 6]. The main features sought in the development of the system were: i) fast acquisition speed, required in order to minimize the incidence of human movements, ii) relative low-cost, iii) a maximum acquisition error of 1 mm. The scanner was successfully applied in a clinical trial that led to the production and application of ten 3D printed orthoses, generated using the CAD procedure detailed in [6].

While the overall results achieved with *Oplà* 1.0 were satisfactory, several features to be enhanced were identified: 1) a faster acquisition, in order to further reduce the effects of micro-movements on the quality of the result; 2) reduce the variability observed in the quality of the acquisition depending on the type of surface; 3) reduce the incidence of environment light condition; 4) increase the comfort for the patience and 5) improve the easiness of use of the entire system, allowing for a future full-autonomous use of this technology for the medical staff. A new version of the system, *Oplà* 2.0, has been designed from this specification. The present article discusses the design process and performances of the new device. In detail, Sect. 2 presents the renewed hardware, Sect. 3 discusses the revisions made to the software procedure; finally, Sect. 4 presents a first look on the results provided by *Oplà* 2.0 and the work's conclusions. 3D scanner design.

The analysis of the limitations observed in the first version of the device and the imposed medical requirements led to some major changes in the hardware of the 3D scanner. The design process started from the identification of new 3D sensors capable of addressing the major flaws of the previous system.

**Table 1.** Depth cameras specifications

| Sensor | Intel SR300 | Intel D415 |
|---|---|---|
| Depth technology | Coded light | Active stereoscopic |
| Operating range | 0.3 m–2 m | ~0.16 m–10 m |
| Resolution | VGA | 1280 × 720 |
| Framerate | Up to 60fps | Up to 90fps |
| Field of view FOV | H73°, V59°, D90° | H69°, V43°, D77° ($\pm3°$) |

As previously discussed, the accuracy of *Oplà* 1.0 was considered sufficient; on the other hand, a further reduction of the scanning time guaranteed by the device was necessary. While the SR300 RGB-D Intel® Realsense™ sensors used in *Oplà* 1.0 allow a fast acquisition rate (up to 30fps), the integration of eight cameras observing a shared area had imposed a sequential activation in order to avoid interference between the sensors. As discussed in [1], the total time required for the sequential activation of each sensor adds to circa 2.5 s. Accordingly, a new model of low-cost 3D camera, i.e. Intel® RealSense™ Depth Camera D415 [7], has been selected to replace the predecessor. A full metrological characterization of the sensor was performed to evaluate its suitability [8]. As reported in Table 1, which summarizes most notable differences between the SR300 and the D415, the new sensor relies on a different technology for the acquisition of 3D data. Specifically, the 3D information is extrapolated from the comparison between two RGB images acquired by a stereo pair of cameras. An infrared laser projector is also used to enhance the differences between the two frames and ease the computation of corresponding points. Accordingly, the new system makes use of a new procedure where all the sensors are triggered together to acquire a single "image" of the scene. The entire acquisition procedure requires 0.5 s to be performed.

The new sensor was also tested in the acquisition of the different kind of surfaces typically encountered in the considered application. Specifically, it was positively tested the behavior of the sensor w.r.t. human skin, which proved to be source of errors for the SR300 [6].

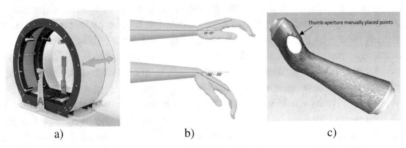

a)                              b)                              c)

**Fig. 1.** a) *Oplà* 2.0: the ring-shaped structure holding the sensors can be moved to account for different dimensions of the arm. A is the elbow support, B the hand support; b) Arm configuration to be acquired; c) selection of points defining the thumb opening.

A new disposition was studied to account for the different FOV and operating range characterizing the D415. *Oplà* 2.0 maintains the same structure of the first version of the scanner, with two circular arrays of 4 sensors each observing the arm. The radius of the new circle was determined according to the operating range of the D415 and was set to 250 mm. The new setting was computed exploiting a custom optimization procedure [9] tuned to identify the set of positions and orientations for the sensors that: i) maximizes the arm surface observed by all sensors; ii) reduces and possibly removes the portion of the patient's arm that remain hidden from all sensors; iii) guarantees a sufficient amount of overlapping between the point clouds of the sensors for the registration of the data. This was achieved by means of a visibility analysis able to assess the visibility of each point of a digital model of a target (arm) from the optical centers of the sensors. A set of six digital models of arms of different dimensions and features was used to introduce the desired variability in the analysis.

Finally, the structure of the device was redesigned to maximize the comfort of the patient during the acquisition; to this purpose, a relative movement between the structure supporting the sensors and the elbow support point was introduced (Fig. 1a). This way, by means of a manual adjustment, the distance between the two supports (and the position of the sensors) can be tuned to maximize the patient's comfort and to assure that the region of interest is fully covered by the sensors. In addition, the new medical requirements foresee the possibility to scan the arm in two different configurations (Fig. 1b). Accordingly, some changes in the general structure of the old system are introduced to address this aspect. The shape of the supports (A and B in Fig. 1a) was edited to maximize the patient's comfort as well as to allow the rotation of the wrist while maintaining a pivot point in B such to maintain the arm principal axis approximately aligned with the central axis of the system; at the same time, the wrist can assume the indicated position.

## 2 Software Design

The use of the *Oplà* scanning system is facilitated thanks to the development of a simple and intuitive Graphical User Interface (GUI). In this work, a significant effort has been devoted to the design of a renewed GUI to improve the user experience. The main steps of the procedure are confirmed in 2.0 version of the system, but the graphic design was restyled mainly on the basis of the results of usability tests [5]. As illustrated in Fig. 2, software tools are now divided into two main tabs in the main window, one dedicated to the scanning phase and the other to custom orthosis modelling. This separation is intended to give the user some guidance on the order of operations and reduce the possibility of human errors. Moreover, 3D navigation and interaction tools have been moved in the GUI upon the 3D viewer, following the typical convention of 3D modelling environments. In the new setup, the steps were the human intervention is required are clearly presented to the user; these are: 1) selection and removal of all the points of the arm that will not be covered by the orthosis; 2) provide indications on the position and shape of the thumb opening; 3) indicate where the orthosis closing mechanism features are placed [6]. These operations are now grouped under the "orthosis modelling" tab and presented in the correct order to the user.

The core steps of the software have been revised following two principal requirements: 1) patient's arm must be acquired in two different positions; the modelling mechanism of the orthosis has been modified accordingly considering the two positions, while remaining transparent to the user, who only needs to select whether the acquisition refers to a right or left arm, as shown in Fig. 2. 2) According to the usability tests performed on *Oplà* 1.0, the procedure for the selection of the points defining the opening for the thumb on the cast was perceived as the most challenging by the medical staff. Previously, the user was asked to define the entire curve for the aperture; such operation was manually difficult for a person not accustomed to CAD software. The new procedure, depicted in Fig. 1c, asks for the independent selection of a number of points that lay on the desired opening contour. The selected points are then automatically processed to generate the profile required by the procedure.

**Fig. 2.** New GUI of the Oplà 2.0 software. Left: arm scan, Right: orthosis modelling

## 3 Discussion and Conclusions

The scanner has been validated by performing an acquisition of a rigid object (i.e. cylinder of Fig. 3). Ground truth data has been obtained using a Romer RS1 3D scanner mounted on a 7520-SI absolute arm by Hexagon Metrology. Except some local errors, *Oplà* 2.0 performed well within the limits imposed by the accuracy requirements; all errors measured in the reconstruction were in the range [−2.9, 1.5] mm, the mean error of the signed distance is −0.49 mm with a standard deviation of 0.64 mm. The reference diameter of the cylinder was 59.2 mm, while the reconstructed diameter is 58.8 mm. These values have been considered acceptable for the application as restrictive orthopaedical devices are always characterized by clearance between skin and orthosis that can go up to 2 mm.

**Fig. 3.** Deviation map obtained acquiring a solid cylinder.

The system was tested with positive results in the acquisition of 20 arms of potential patients. The composition of the panel group has allowed the validation of the acquisition system on significantly different hand-wrist-arm anatomies. Figure 4 shows a selection of the obtained results. Positive results were obtained throughout the test, with all the acquisition performed successfully, both with a gauze applied on the arm and without it (Fig. 4). As the acquired anatomy varied, the data produced by the sensors was in any circumstance able to produce a full reconstruction of the arm. The results achieved on this aspect were positive, as all the perturbation of the surface that were present in the SR300 acquisition are not present on the acquisitions performed with the new sensor.

**Fig. 4.** Results: left) right arm with gauze; right) left arm without gauze, different position.

## References

1. Carfagni, M., Furferi, R., Governi, L., Servi, M., Uccheddu, F., Volpe, Y., Mcgreevy, K.: Fast and low cost acquisition and reconstruction system for human hand-wrist-arm anatomy. Procedia Manuf. **11**, 1600–1608 (2017)

2. Haleem, A., Javaid, M.: 3D scanning applications in medical field: a literature-based review. Clin. Epidemiol. Glob. Heal. **7**(2), 199–210 (2019)
3. Lain, A., Garcia, L., Gine, C., Tiffet, O., Lopez, M.: New Methods for imaging evaluation of chest wall deformities. Front. Pediatr. **5**, 257 (2017)
4. Cha, Y.H., Lee, K.H., Ryu, H.J., Joo, I.W., Seo, A., Kim, D.-H., Kim, S.J.: Ankle-foot orthosis made by 3D printing technique and automated design software. Appl. Bionics Biomech. **2017**, 1–6 (2017)
5. Servi, M., Volpe, Y., Uccheddu, F., Furferi, R., Governi, L., Lazzeri, S.: A preliminary usability assessment of a 3D printable orthosis design system. In: Proceedings of the Communications in Computer and Information Science. Springer Verlag, vol. 850, pp. 273–280 (2018)
6. Buonamici, F., Furferi, R., Governi, L., Lazzeri, S., McGreevy, K.S., Servi, M., Talanti, E., Uccheddu, F., Volpe, Y.: A practical methodology for computer aided design of custom 3D printable casts for wrist fractures. Vis. Comput. 1–16 (2019)
7. Stereo Depth – Intel® RealSense™ Depth and Tracking Cameras Available online: https://www.intelrealsense.com/stereo-depth/. Accessed 29 Jan 2020
8. Carfagni, M., Furferi, R., Governi, L., Santarelli, C., Servi, M., Uccheddu, F., Volpe, Y.: Metrological and critical characterization of the intel D415 stereo depth camera. Sensors (Switzerland) **19**(3), 489 (2019)
9. Buonamici, F., Furferi, R., Governi, L., Marzola, A., Volpe, Y.: Scene acquisition with multiple 2D and 3D optical sensors: a PSO-based visibility optimization. Sensors (Switzerland) **20**(6), 1726 (2020)

# Sensitivity Analysis for Simulated Annealing-Based Fitting of CAD Models to Point Clouds of Digitized Parts

Ghazanfar Ali Shah[1,2,3], Arnaud Polette[1], Jean-Philippe Pernot[1(✉)], Franca Giannini[2], and Marina Monti[2]

[1] Arts et Métiers, LISPEN, EA 7515, HeSam, Aix-en-Provence, France
jean-philippe.pernot@ensam.eu
[2] Istituto di Matematica Applicata e Tecnologie Informatiche "Enrico Magenes", CNR, Via De Marini 6, 16149 Genova, Italy
[3] DIME-Dipartimento di Ingegneria Meccanica, Energetica, Gestionale e dei Trasporti, Università degli Studi di Genova, Genova, Italy

**Abstract.** This paper addresses the way a simulated annealing-based fitting strategy can be enhanced by leveraging a sensitivity analysis able to characterize the impact of the variations in the parameters of a CAD model on the evolution of the deviation between the CAD model itself and the point cloud of the digitized part to be fitted. The principles underpinning the adopted fitting algorithm are briefly recalled. The applied sensitivity analysis is described together with the comparison of the resulting sensitivity evolution curves with the changes in the CAD model parameters imposed by the simulated annealing algorithm. This analysis suggests several possible improvements that are discussed. The overall approach is illustrated on the fitting of single mechanical parts but it can be directly extended to the fitting of parts' assemblies. It is particularly interesting in the context of the Industry 4.0 to update digital twins of physical products and systems.

**Keywords:** Sensitivity analysis · CAD model parameters · Simulated annealing · Reverse engineering · Digital twins

## 1 Introduction

Today, being able to reconstruct or update 3D representations of existing products and systems has become mainstream to support the creation and exploitation of digital twins in the context of Industry 4.0 [1]. Clearly, the relative ease of access to more or less sophisticated 3D acquisition devices has certainly accelerated the demand while extending the range of possible applications. However, the point cloud treatment and the reverse engineering of CAD models are not yet fully automatized. Most of the existing reverse engineering techniques follow a time-consuming patch-by-patch reverse engineering strategy that does not satisfy the Industry 4.0 requirements [2]. Indeed, in most applications, the structure of the products or systems does not change, and only the position and orientation of its constitutive parts is to be updated. Thus, the update problem can be

L. Roucoules et al. (Eds.): JCM 2020, LNME, pp. 274–280, 2021.
https://doi.org/10.1007/978-3-030-70566-4_44

transformed in a fitting problem for which the optimal parameter values of CAD models have to be found to minimize the deviation between the digitized point clouds and the digital twins.

To support the fitting of parts and assemblies through the reverse engineering process, a simulated annealing (SA) based fitting algorithm has been developed [3]. It considers as input the digitized point cloud, the CAD models to be fitted and the set of its parameters to be modified by the SA algorithm to minimize the deviation. This metaheuristic algorithm has a stochastic behavior [4], with few knowledge on how the CAD parameters' values are to be changed throughout the optimization. A possible way to understand the influence between the output and parameters' evolution is sensitivity analysis [5, 6]. It has been used in various contexts, as, for example, for the parameter choices of SA for continuous network design [7], and various indices have been specified [8–10]. This paper studies the possibility of coupling a sensitivity analysis to the SA algorithm. The sensitivity analysis allows the evolution of the influence of the CAD model parameters to be monitored throughout the fitting process. Comparing the obtained sensitivity evolution curves to the changes of the parameters directly imposed by the SA algorithm provides hints to improve the fitting process.

The paper is organized as follows. Section 2 briefly reviews the previously developed fitting framework, and the new sensitivity analysis is introduced in Sect. 3. Section 4 discusses the results on a L-like shape and how such an analysis could be intelligently exploited throughout the fitting process. The last section concludes the paper and discusses the future works.

## 2 Simulated Annealing-Based Fitting of CAD Models

Recently, a breakthrough has been achieved in the definition of part-by-part reconstruction or update of editable CAD models fitting the point cloud of a digitized mechanical part or assembly avoiding the patch-by-patch reconstruction [3]. The idea is to work directly at the level of the part whose parameters are modified by a SA algorithm until a good-quality fitting is obtained. The inputs are the point cloud PC to which a parameterized CAD model $\mathcal{M}_0$ is to be fitted. The point cloud is composed of $N$ points PC[$i$], with $i \in [1..N]$, and the CAD model is parameterized by $N_p$ control parameters $p_k$, with $k \in [1..N_p]$. During the prearrangement step, the user locates in a coarse manner the part inside the point cloud, this initializes the parameters values $p_{k,0}$ (Fig. 1a). At each iteration $j$, the SA algorithm then tries to modify the parameters values $p_{k,j}$ until the updated CAD model perfectly fits the point cloud (Figs. 1b and 1c). The quality of the fitting is assessed thanks to an energy function which characterizes the overall deviation between the point cloud and the CAD model. The process stops when the energy function no longer decreases with respect to a given accuracy $\varepsilon_s$, or when a max number of iterations $M_{iter}$ is reached.

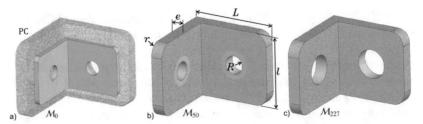

**Fig. 1.** Global fitting of a L-like shape defined by 5 controlled parameters: (a) initial configuration after the prearrangement step, (b) fitted part after 50 iterations of the SA algorithm, (c) final fitted part minimizing the deviation with the point cloud ($\varepsilon_s = 10^{-1}$ and $M_{iter} = 1000$).

At an iteration $j$ of the SA algorithm, the energy function to be minimized is evaluated as follows:

$$E\left(p_{1,j}, \cdots, p_{N_p,j}\right) = \sum_{i=1}^{N} \mathrm{d}^2\left(\mathrm{PC}[i], \mathcal{M}_j^{\triangleright}\left(p_{1,j}, \cdots, p_{N_p,j}\right)\right), \quad \text{with } j \geq 0 \qquad (1)$$

wherein $\mathcal{M}_j^{\triangleright}$ represents the tessellation of the CAD model $\mathcal{M}_j$ after its update by the CAD modeler at the jth iteration of the SA algorithm, and d(*point*, *mesh*) is the distance function that returns the closest distance between a *point* and a *mesh*.

## 3  Sensitivity Analysis

The previous section has introduced the newly developed fitting algorithm that exploits a SA algorithm to minimize the overall deviation between the point cloud and the CAD model to be fitted. Even though this process has proved to generate good-quality results when compared to the traditional patch-by-patch fitting strategy, the path followed to reach the minimum remains under the control of the SA algorithm, which works as a black box and that neither takes into account the semantics associated to the various parameters $p_k$ nor the way their variations may influence the evolution of the energy function step after step. Indeed, while acting on the variables to be optimized, the SA algorithm does not distinguish for instance the lengths, widths, thicknesses and radius that parameterize the features, and all the variables are considered in the same way. It also does not take into account the fact that the influence of a parameter variation on the energy may vary a lot depending on whether the parameter is associated to a structural feature (e.g. the length $L$ of the L-like shape in Fig. 1) or to a detail feature (e.g. the radius $r$ in Fig. 1).

To better compare the influence of the parameters, a sensitivity estimator $S_{k,j}$ is here introduced to compute at each step $j$ of the optimization process the sensitivity of the energy function with respect to each parameter $p_k$:

$$S_{k,j} = \left.\frac{\partial E}{\partial p_k}\right|_{\left(p_{1,j}, \cdots, p_{N_p,j}\right)}, \quad \text{with } j \geq 0 \qquad (2)$$

Since the energy function is computed using a nearest point algorithm, it cannot be differentiated. Thus, at the jth iteration, the sensitivities are approximated using a first-order finite difference scheme (either forward or backward) directly controlled by the step $h_k$ chosen for the corresponding parameter $p_k$:

$$S^+_{k,j} = \frac{E_j\left(p_{1,j},\cdots,p_{k,j}+h_k,\cdots,p_{N_p,j}\right) - E_j\left(p_{1,j},\cdots,p_{k,j},\cdots,p_{N_p,j}\right)}{h_k} + O(h_k), \quad \text{with } j \geq 0 \qquad (3)$$

$$S^-_{k,j} = \frac{E_j\left(p_{1,j},\cdots,p_{k,j},\cdots,p_{N_p,j}\right) - E_j\left(p_{1,j},\cdots,p_{k,j}-h_k,\cdots,p_{N_p,j}\right)}{h_k} + O(h_k), \quad \text{with } j \geq 0 \qquad (4)$$

## 4  Results and Discussion

The core of the fitting algorithm has been implemented in MATLAB®, which is able to call the built-in functions of SolidWorks® to perform the successive CAD model updates and ensure the consistency of the resulting B-Rep model during the optimization loops. All the steps $h_k$ used to compute the sensitivities are equal and only depend on the type of parameter they are associated to. Thus, step $h_k = 0.1$ mm in case $p_k$ is a length, and $h_k = 0.1°$ in case it is an angle. Those values have been chosen in coherence with the accuracy of the adopted CAD modeler.

In order to compare the sensitivity of the parameters to the way the SA algorithm modifies the parameters values step after step, the evolution of the parameters can be tracked in an absolute or relative manner, as follows:

$$\delta p_{k,j} = p_{k,j+1} - p_{k,j} \qquad (5)$$

$$\Delta p_{k,j} = \frac{\delta p_{k,j}}{p_{k,j}} \qquad (6)$$

Thus, the value $\delta p_{k,j}$ characterizes the absolute evolution of the parameter $p_k$ between iterations $j$ and $j+1$, whereas $\Delta p_{k,j}$ refers to its relative evolution. At each iteration $j$ of the SA algorithm, it is therefore possible to compare the sensitivity $S_{k,j}$ of each parameter $p_k$, to the decision taken by the SA algorithm on how to modify the parameter value for the next step $j+1$. This is illustrated in Fig. 2 for the fitting of the L-like shape introduced in Sect. 3. For this example, the initial temperature of the SA algorithm has been set up to $T_0 = 10$ following the initialization procedure suggested in [3]. For the sensitivities, the forward finite difference scheme $S^+_{k,j}$ has been used when the absolute evolution $\delta p_{k,j} > 0$, and the backward scheme $S^-_{k,j}$ otherwise.

From this experimentation, one can clearly see that the sensitivities of the five parameters defining the L-like shape are relatively different at the beginning, and that they slightly evolve as the fitting process goes on (Fig. 2 right). On the contrary, the changes imposed by the SA for each parameter at each iteration $j$ are very close. For sake of clarity, the sensitivity of the five parameters at the beginning and the average of first ten iterations are given in Table 1. This table also shows the desired values, $p_{k,D}$, and the final values $p_{k,F}$ for the object's parameters.

One can also see that the successive changes imposed by the SA algorithm to the parameters' values (Fig. 2 left) cannot be easily correlated to the evolution of the sensitivities. Thus, several ways of improvement can be sketched:

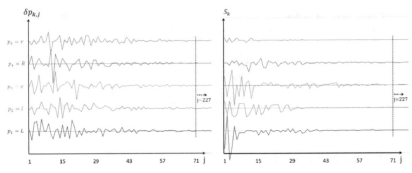

**Fig. 2.** Comparison between the evolution of the sensitivities $S_{k,j}$ (right) and the changes $\delta p_{k,j}$ imposed by the SA algorithm (left) at each iteration $j$ of the fitting process.

**Table 1.** Values of the comparison metrics for the initial iteration ($j = 0$) and average of first ten iterations ($j \in [0..9]$).

| $k$ | 1 | 2 | 3 | 4 | 5 |
|---|---|---|---|---|---|
| $p_{k,0}$ | 25.0 | 20.0 | 2.0 | 3.0 | 2.0 |
| $S_{k,0}$ | −8.0E+4 | 3.0E+4 | 4.0E+4 | 0.0E+00 | 1.0E+4 |
| $\delta p_{k,0}$ | 0.56 | −0.45 | −0.22 | 0.15 | 0.16 |
| $\Delta p_{k,0}$ | 0.02 | −0.02 | −0.11 | 0.05 | 0.09 |
| $\overline{S_{k,j \in [0..9]}}$ | 4.4E+4 | 2.9E+4 | 4.1E+4 | 6.0E+3 | 6.0E+3 |
| $\overline{\delta p_{k,j \in [0..9]}}$ | 2.55 | 1.25 | 1.76 | 1.79 | 1.60 |
| $\overline{\Delta p_{k,j \in [0..9]}}$ | 0.10 | 0.06 | 0.88 | 0.60 | 0.80 |
| $p_{k,D}$ | 35 | 30 | 5 | 6 | 5 |
| $p_{k,F}$ | 34.92 | 29.84 | 4.89 | 5.93 | 5.49 |

- the parameters $p_k$ could be grouped according to their levels of sensitivity, so as to treat them in several batches for which the sensitivity is comparable;
- the way the SA algorithm changes the parameters values could be further optimized while considering the sensitivity analysis all along the process;
- the sensitivity analysis could be performed at a lower level, while decomposing the energy function and tracking which parts of the energy function are affected by the changes to save time when computing the overall energy.

The first idea has been implemented, but due to space limitation results cannot be shown here. On the example of the L-like shape, two main categories of parameters can be distinguished when comparing their sensitivities at the beginning of the fitting process: $\{L, \ell, e\}$ and $\{R, r\}$ gather together parameters having comparable sensitivities. Such a decomposition in two groups gives rise to two successive optimization loops.

Overall, doing this way the accuracy of the fitting is improved compared to the use of a single optimization loop.

## 5 Conclusion

This paper has explored the possibility to integrate a sensitivity analysis within a SA algorithm used to fit a CAD model in the point cloud of a digitized part. Such an approach helps grouping the CAD model's parameters so as to treat batches having comparable sensitivities. Other improvements have also been sketched to be more accurate and to reduce optimization times.

## References

1. Lu, Y.: Industry 4.0: a survey on technologies, applications and open research issues. J. Ind. Inf. Integr. **6**, 1–10 (2017)
2. Falcidieno, B., Giannini, F., Léon, J.-C., Pernot, J.-P.: Processing free form objects within a product development process framework. In: Advances in Computers and Information in Engineering Research, pp. 317–344 (2014)
3. Shah, G.A., Polette, A., Pernot, J.-P., Giannini, F., Monti, M.: Simulated annealing-based fitting of CAD models to point clouds of mechanical parts' assemblies. To appear in Engineer with Computers (2020). https://doi.org/10.1007/s00366-020-00970-8
4. Kirkpatrick, S., Gelatt, C.D., Vecchi, M.P.: Optimization by simulated annealing. IBM Research Report RC 9355, Acts of PTRC Summer Annual Meeting (1982)
5. Iooss, B., Lematre, P.: A review on global sensitivity analysis methods. In: Uncertainty Management in Simulation-Optimization of Complex Systems, pp. 101–122. Springer (2015)
6. Hamby, D.M.: A review of techniques for parameter sensitivity analysis of environmental models. Environ. Monit. Assess. **32**(2), 135–154 (1994)
7. Yang, J., Xu, M., Gao, Z.: Sensitivity analysis of simulated annealing for continuous network design problems. J. Transp. Syst. Eng. Inf. Technol. **9**(3), 64–70 (2009). ISSN 1570-6672
8. Gamboa,F., Janon, J., Klein, T., Lagnoux, A.: Sensitivity analysis for multidimensional and functional outputs (2013). https://arxiv.org/abs/1311.1797
9. Spagnol, A., Le Riche, R., Da Veiga, S.: Global sensitivity analysis for optimization with variable selection. SIAM/ASA J. Uncertain. Quantif. **7**(2), 417–443 (2019)
10. Lamboni, M., Monod, H., Makowski, D.: Multivariate sensitivity analysis to measure global contribution of input factors in dynamic models. Reliab. Eng. Syst. Saf. **96**(4), 450–459 (2011). ISSN 0951-8320

# Digital Operative Guides for Helping First Responders in Public Buildings During Emergency Interventions

Antonio Ruiz-Cartiel[1]($\boxtimes$), César García-Hernández[2], and Pedro Ubieto-Artur[2]

[1] Fire Service, Zaragoza Council, Valle de Broto, 16, 50015 Zaragoza, Spain
[2] Department of Design and Manufacturing Engineering, University of Zaragoza, Campus Río Ebro, C/María de Luna, 3, 50018 Zaragoza, Spain

**Abstract.** Graphical tools for firefighters in emergency interventions have been proven to be very effective. Particularly, paper-based operational guides and digital guides with 360° images were already assessed during the drills developed, during a training program in a real, but obsolete and not in use, hotel. The obtained results were considered very interesting, being part of the study previously published by this research team. In the previous study, three different methods (two of them based on graphical contents) were compared in order to communicate the features of a building to firefighters, five minutes before starting a victim rescue during the drills of the mentioned live fire training program. These good results encouraged us to continue the development of the digital guides and this paper describes the first one created for a real building, placed in Zaragoza (Spain), which is in use nowadays. In this study, two versions of the digital guide are presented, one based on a PDF file and another one implemented with online 360° images.

**Keywords:** Guides · Firefighters · Emergency interventions · Digital tools

## 1 Introduction

One of the main factors that is very important in an emergency intervention is the time. This factor can be decisive in a rescue intervention of a victim, in which every second counts on the possibilities of rescuing the victim alive.

Any firefighter assistance tool that serves to improve intervention time, while improving firefighter safety conditions, should be taken into account [1, 2].

Within tools, it should be emphasized that they are not only working equipment, such as thermal cameras or personal protection equipment. Other tools, like protocols, practices or drills, should also be considered [3–7]. There is a graphical tool that could be among those ones, which is the base of this paper, in order to consider the information of the place where the emergency occurs. Places where there is a high-risk level have documents (e.g.: self-protection plans) with information related to the building based on text and technical drawings [8]. The problem is that, in general, these documents are too extensive and contain too much information that is not relevant for first responders

© The Author(s) 2021
L. Roucoules et al. (Eds.): JCM 2020, LNME, pp. 281–288, 2021.
https://doi.org/10.1007/978-3-030-70566-4_45

in emergency interventions. For this reason, self-protection plans are not considered useful by firefighters, according to the opinion of all the professionals interviewed for this study. The reason is that, just before the intervention, they do not have time enough to select the information that is really useful for them [9]. In addition, these documents contain many technical drawings that can be hard to interpret for firefighters that have just 5 to 10 min before arriving at the emergency area.

In this paper, we describe two methods, based on graphic content, applied to a real public building. On the one hand, self-protection plans are summarized extracting the information that is really required by firefighters before the emergency and, on the other hand, to add more graphical information, trying to make rescues easier for firefighters. The chosen public building, which is currently in use, is the Ebro Hydrographic Confederation, placed in Zaragoza, Spain.

Both methods have been tested, in a previous study, by simulating a rescue in an abandoned hotel. More than 150 firefighters were involved in this study, based on the use of graphical tools for helping firefighters in rescues, conducted by researchers from the University of Zaragoza [10]. The results of this study demonstrated that these methods significantly reduced intervention times.

## 2  Method

The two methods previously mentioned are based on graphical tools. These tools are technical documents used to describe the most critical aspects of buildings and their main features. They are based on the self-protection plans or emergency plans, which companies and other organizations make in order to prevent and control risks, to optimize the first response in emergencies. These plans comply with the regulations and help firefighters to act with the greatest diligence, speed, effectiveness and decision in their interventions. The differences between the two methods are that one includes conventional pictures, with a classical paper-based document structure, while the other one is based on interactive guides made up of 360° images, which include relevant information on them. Although the first documents could be consulted in digital format, e.g. with a tablet, or printed on paper, the second ones can only be accessed by means of a digital device.

These two types of methods were developed in a public building, built between 1933 and 1946, which has been declared a site of cultural interest [11], thanks to its unique features. This is the reason why the building has problems to comply the self-protection plans, because in some cases, modifications are not allowed and in others, their costs are too high to be all assumed at once.

This building has four floors of offices and a basement. Its theorical capacity is 746 people, but the real occupancy is 377 people. We can face a range of risks in the basement such as a boiler room, a power generator, a processing plant and a large room for storing documents.

The self-protection plan would be the tool that firefighters would use in the event of an emergency in the building. The self-protection plan, despite of being incomplete and having some sections which doesn´t comply the regulation [12, 13], has an extension of 138 text pages and 20 technical drawings in ISO-A3 format.

These self-protection plans are not helpful for firefighters, because when they are in an emergency, they only have 5 or 10 min to review them before arriving to the emergency place.

Therefore, both methods mentioned before were developed, based on self-protection plans, in order to optimize the first response in emergencies. With each method, firefighters have access to the most relevant information in just a few minutes, optimizing their interventions and significantly reducing intervention times.

## 2.1  Operational Guide

The first graphical tool developed was the Operational Guide. According to a previous research study [14] which examines the information requirements for interventional procedures, too much information could overload the receptors. In fact, too much information is usually considered as problematic as a lack of it. This study emphasizes the use of operational information which must be "accurate, relevant and timely" focused on the requirements of all the personnel in the emergency environment.

In a similar way, trying to develop a useful tool for first responders in emergencies, a group of professional firefighters, which included construction experts and occupational health and safety specialists, defined three key rules to be followed [10]:

1. They should be concise, with only the information elements considered relevant for emergency situations.
2. Graphical information is a totally relevant aspect, so the innovative tools should include appropriate maps, conventional images, 360° images. Beyond that, highlighting their most important aspects to facilitate a quick and clear communication.
3. Order and clearness of the included information must be always preserved.

The operational guide developed for the building used in this study is based on carefully chosen information, according to the previously described rules, and it has a total extension of 28 text pages.

**Fig. 1.** Route from one of fire stations and building aerial view.

The guide sections are:

– Route from the closest fire stations to the building and its placement, as shown in Fig. 1.
– Contact phones.
– Surroundings of the building (Fig. 2).

**Fig. 2.** Main entrance and roof of the building.

– Remarkable features of the building.
– Main risks.
– Preferential protection points with special danger.
– Fire protection.
– Evacuation routes and exits (Fig. 3).

**Fig. 3.** Different emergency exits of the building.

Finally, technical drawings with the most relevant information for firefighters are included such as:

– Location technical drawings of the risk elements.
– Evacuation route on technical drawings.

## 2.2   Reduced Operational Guide

It is developed from the Operational Guide, which has an extension of 28 text pages, extracted from the self-protection plan of 158 pages. The extension of the self-protection plan was clearly reduced, but this office building gave the possibility to condense the Operational Guide without losing its main function of containing the relevant information for firefighters.

The development of the Reduced Operated Guide was based on formatting the information in ordered and clear tables (including contact phones, main risks, preferential protection points with special danger and fire protection). Its main difference with the operational guide is that the information of the 13 technical drawings was reduced to 4. This was possible because, as it is an office building, risks were clear with only 4 technical drawings.

In contrast, for industrial or more complex buildings, additional technical drawings could be required. Another difference of the Reduced Operational Guide, compared with the Operational Guide, is that the graphical information of the evacuation routes, exits and the surroundings of the building were reduced. In the Reduced Operational Guide, images of the evacuation doors were not included. Finally, this guide also includes a location map and routes from the closest fire stations. The final version of the Reduced Operational Guide has an extension of 8 text pages (Fig. 4).

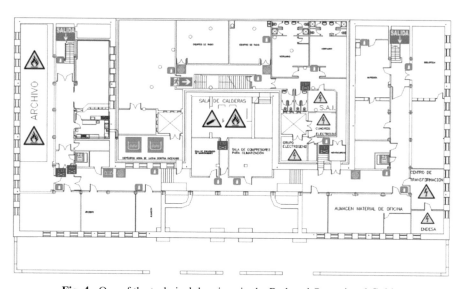

**Fig. 4.** One of the technical drawings in the Reduced Operational Guide.

## 2.3   Digital Operational Guide

It is developed from the Operational Guide. One of the most important needs for firefighters, before the first response in an emergency, is graphical information. In the Operational

Guides, graphical information can be included, but it must be limited to avoid excessive number of pages. In the case of this building, it was necessary to reduce the graphical information to decrease its number of pages, obtaining an additional version, i.e. the Reduced Operational Guide.

Due to the problems of the previously described tools, the Digital Operational Guide (DOG) was developed. The most important difference of these DOGs is the possibility to include 360° images and the intuitive interaction. Users can easily navigate through the guide, touching icons, being possible to go from a 360° image to another one, virtually visiting the building (Fig. 5).

**Fig. 5.** One of the 360° images (flattened).

To develop this guide, strategic points to make the 360° images were carefully selected, after thoroughly studying the technical drawings. The main idea was that, with the lower number of images, all the required graphical information should appear.

Additional information was added over these 360° images. This additional information was the same that the Operational Guides included. This additional information was added using small icons, as signs of different types, to indicate hazards, etc. The small size of the signs makes possible to understand the images easily. It is possible to rotate or zoom the images without difficulty, touching the digital device screen and moving fingers on it.

When one of the placed signs is clicked, additional information is displayed or access to a different location is given. The intention of this user interface is to be simple and very intuitive, to make the learning process very fast. All the contents of the Digital Operational Guides are accessible via internet connection (after previously being uploaded to a server), although they could also be stored on tablets, to ensure their functionality in places with a difficult access to online data.

## 3   Results and Conclusions

The graphic tools described show digital guides of a building in use, made with real images in 360°, adapting and improving the preliminary guides. Those initial guides were made in a much simpler building, in order to experimentally compare the use of virtual 3D images with 360° photographs.

The results show us that it is possible to adapt a building of great complexity and large size to a format of digital guides. These guides facilitate, using a tablet, the transmission of the relevant information for a real emergency. Firefighters can consult the guides before and during the emergency.

As these guides have been developed recently, they have only been shown to the members of the Occupational Risk Prevention Department and several firefighters, receiving positive feedback. We have not used them in a drill yet, to obtain quantitative data in order to test and compare them. Future drills have already been considered to be implemented in further research.

The digital operational guide can be considered a robust and flexible tool which can be used without an internet connection by storing the guides in the tablet's internal memory (or memory card). The tools help processing and using the information in a graphical way, so they could be adapted to different fields, attending their magnitudes and the characteristics of their associated risks.

## 4   Limitation and Further Research

This study will continue in order to solve its limitations. It can be expected that these digital tools will help firefighting professionals in their first response activities, although the results of their application to real emergency services must be assessed during their future implantation, as they have only been previously tested under research conditions.

## References

1. Buttussi, F., Chittaro, L.: Effects of different types of virtual reality display on presence and learning in a safety training scenario. IEEE Trans. Visual. Comput. Graph. **24**, 1063–1076 (2018). https://doi.org/10.1109/TVCG.2017.2653117
2. Ford Daniel, A.L.: A study into fire and rescue services, supporting the development of a community engagement model, p. 141 (2009)
3. Cha, M., Han, S., Lee, J., Choi, B.: A virtual reality based fire training simulator integrated with fire dynamics data. Fire Saf. **50**, 12–24 (2012). https://doi.org/10.1016/j.firesaf.2012.01.004
4. Hamp, Q., et al.: New technologies for the search of trapped victims. Ad Hoc Network. **13**, 69–82 (2014)
5. Kolmanic, S., Guid, N., Nerat, A.: SIN: multimedia-based teaching tool for computer-supported fire-fighter training. Fire Safety J. **61**, 26–35 (2013). https://doi.org/10.1016/j.firesaf.2013.08.006
6. Padgett, K.: The importance of live-fire training (2008). https://firerescuemagazine.firefighternation.com/2008/07/31/the-importance-of-live-fire-training/#gref. Accessed 2 Mar 2020

7. Xu, Z., Lu, X.Z., Guan, H., Chen, C., Ren, A.Z.: A virtual reality based fire training simulator with smoke hazard assessment capacity. Adv. Eng. Softw. **68**, 1–8 (2014). https://doi.org/10.1016/j.advengsoft.2013.10.004

8. Castro, D., Sans, J. Los planes de autoprotección como instrumento técnico y educativo. Educar **50** (2014). https://doi.org/10.5565/rev/educar.127

9. García, I., Ruiz, A.: Planes de intervención operativa. Emergencias 112: Revista Asociación Profesional de Técnicos de bomberos (2016). https://www.aptb.org/revista-112-emergencias-magazine. Accessed 2 Mar 2020

10. García-Hernández, C., Sánchez-Álvarez, E.J., Ubieto-Artur, P., Huertas-Talón, J.L.: Graphical tools for helping firefighters in victim rescues. Assessment during a live fire training program. Saf. Sci. (2019). www.sciencedirect.com/science/article/pii/S0925753518308506

11. Hydrographic Confederation of the Ebro, Bien Cultural: Patrimonio Cultural de Aragón (2008). https://www.patrimonioculturaldearagon.es/bienes-culturales/confederacion-hidrografica-del-ebro-zaragoza. Accessed 2 Mar 2020

12. Spain. Ley 31/1995, de 8 de noviembre, de prevención de Riesgos Laborales. Boletín Oficial del Estado, 269, de 10 de noviembre de 1995, pp. 2092–2095 (1995)

13. Spain. Real Decreto 393/2007, de 23 de marzo, por el que se aprueba la Norma Básica de Autoprotección. Boletín Oficial del Estado, 72, de 24 de marzo de 2007, pp. 12841–12850 (2007)

14. Chief Fire & Rescue Adviser. Fire and Rescue Service Operational Guidance – Operational Risk Information, Department for communities and local goverment, London (2012)

# Design and Optimization of a Photovoltaic Canopy for an Electric Vehicle Charging Station in Urban Environments

Dolores Parras-Burgos[1]([✉]), Jesús Melgarejo-Teruel[1], Antonio Mateo-Aroca[2], Daniel G. Fernández-Pacheco[1], Francisco J. F. Cañavate[1], and Francisco L. Sáez-Gutiérrez[1]

[1] Departamento de Estructuras, Construcción y Expresión Gráfica, Universidad Politécnica de Cartagena, Cartagena, Spain
`dolores.parras@upct.es`
[2] Departamento de Automática, Ingeniería Eléctrica y Tecnología Electrónica, Universidad Politécnica de Cartagena, Cartagena, Spain

**Abstract.** Nowadays, the use of renewable energies and electric vehicles has become particularly relevant in order to lower the high pollution levels surrounding our cities. The design of a photovoltaic canopy for charging electric vehicles is a highly promising combination that can be set up in urban areas. To favour installing them in different places, this communication provides details of the technico-functional aspects that have been considered to design and fit them, along with other aesthetic and user-centred aspects that help stimulate our society to use such infrastructures.

**Keywords:** Renewable energy · User-centered design · Aesthetics

## 1 Introduction

Canopies tend to be installed to protect users from adverse climate effects, like sun, snow or wind. They are employed in many places, such as entrances to buildings, bars or restaurants, car parks or public transport stops. The main use of canopies specifically employed for parked vehicles is to protect users from the sun and to avoid high temperatures inside vehicles. One way of putting this energy to good use is the so-called "photovoltaic canopies", whose roofing is covered by photovoltaic panels that allow solar energy to be absorbed and employed in several ways (Alghamdi et al. 2017; Bushur et al. 2019; Umer et al. 2019; Zarcone et al. 2016). One such use is charging electric vehicles, which are becoming particularly relevant for society that wishes to use less polluting energies. The design of such installations must provide urban area with places that promote this energy type. So an aesthetic user-centred design is just as important as technico-functional aspects that have been considered to build them (De Fusco 1981; Gay and Samar 2004; Macdonald 2001; Parras-Burgos et al. 2019).

This communication presents an optimised photovoltaic canopy design for charging electric vehicles. This novel proposal presents a system that not only supplies part of the

L. Roucoules et al. (Eds.): JCM 2020, LNME, pp. 289–294, 2021.
https://doi.org/10.1007/978-3-030-70566-4_46

energy required for charging electric vehicles with renewable energy, but also supplies users with an attractive and convenient rest area while they wait.

## 2   Materials and Methods

To design and set up such installations, the regulations and characteristics of each area must be taken into account. The study presented in this communication was conducted for an area in Murcia (Spain), but can be extrapolated elsewhere. The proposed photovoltaic canopy would be formed by four parking spaces with charging points for electric vehicles. Its structure has a span of 5,9 m, is 12,2 m long with two openings, and users can freely move around the rest area. Bearing in mind the location of this installation, the optimum slope of the photovoltaic panels would be 30°. However, adapting these layouts to the metal structure implies considerably concentrating stresses. This is why a slope of 15° was set by assuming loss of the photovoltaic installation's power (Fig. 1).

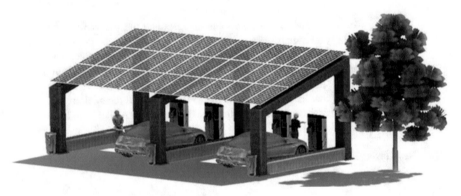

**Fig. 1.** General view of the photovoltaic canopy as an electric vehicle-charging station.

In order to provide this type of canopies distinguishing aesthetics and to extend their functionality, trapezoidal single-pitch gantries were designed with two pillars measuring 4.396 m and 2.815 m respectively, with an upper lintel on which two roof purlins rest, and a lower one to act as the rest area for waiting while charging cars. These rest areas include wooden seating to make the area welcoming and comfortable (Fig. 2). The aesthetic and user-centered design aspects that stand out from this model are its large surfaces for customer use, combining various colors and materials that give the whole a feeling of resistance and comfort.

The most economic separation among frames to meet economic criteria should not exceed 5 m or 6 m. This measure is to reduce manufacturing costs (the structure's weight, labour, assembly, etc.) and to make full use of the length of the bars forming the cross-sections that act as purlins (usually about 12 m long). The structure's sizes will meet these criteria if this aspect is taken into account.

To streamline, and to consequently lower the main metal structure's cost, standard rolled steel cross-sections are used, whose design was studied to achieve mechanical

**Fig. 2.** Details of the rest area

efficiency and economic materials. For the purlins, we chose to use formed steel given the importance of its function in photovoltaic installations.

The self-service photovoltaic installation design is made up of the following elements:

- Photovoltaic panels: they produce electric energy from solar energy and have their own electric protection.
- Support structure: a canopy that acts as a structure for the photovoltaic panels to rest on.
- Inverter: a device that transforms the direct current produced in the photovoltaic panels into suitable alternating current to be used.
- Generated energy counter: measuring equipment that controls the energy generated by the photovoltaic installation.
- Connection to the distribution network, along with its protections.
- Other supplementary fittings: wiring, protector boxes, etc.

The photovoltaic panels cover the top of the canopy structure and act as roofing. The panel model to be used is SI-ESF-M-BIPV-SM-P156–72 from the SOLAR INNOVA company. This is a BIPV (Building Integrated PhotoVoltaic) panel model so that the photovoltaic generator can be completely built into the structure. Its main mechanic and electric features are shown in Table 1.

One very important aspect when streamlining photovoltaic installations is limiting the spaces restricted by a canopy's sizes. This is why a 36-panel photovoltaic generating system with 11,7 kW power was designed in relation to charging powers, which also adapts to the available sizing to be installed.

The inverter's power is determined by the photovoltaic generator's nominal power under standard test conditions (STC) by considering the inverter's performance and

**Table 1.** The main properties of the SI-ESF-M-BIPV-SM-P156-72 photovoltaic panel from the SOLAR INNOVA company.

| Magnitude | Value |
|---|---|
| Sizes | 1.948 × 976 × 30 (mm) |
| Weight | 29 kg |
| Maximum power voltage (Vmpp) | 38,1 (V) |
| Open-circuit voltage (Voc) | 46,7 (V) |
| Maximum power current (Impp) | 8,53 (A) |
| Short-circuit current (Icc) | 9,07 (A) |
| Maximum power rating (Pmpp) | 325 (W) |

the power loss factors of the photovoltaic modules under real operating conditions. We decided to fit two inverters whose sizes allow the equipment to be built into the canopy's framing and to achieve its best aesthetic integration. The chosen inverter model is the three-phase Fronius Symo 5.0-3-M with its flexible design, whose sizes enable it to be included in the canopy (Table 2). A three-phase inverter was selected to facilitate its connection to the three-phase electric vehicle-charging installation and to avoid any type of imbalance among phases.

**Table 2.** Characteristics of the three-phase Fronius Symo 5.0–3-M inverter model's input/output.

| Input | Value | Output | Value |
|---|---|---|---|
| Maximum generated power FV | Peak 10 kW | Nominal power | 5 kw |
| Voltage range MPP | 163–800 V | Maximum temperature | 60 °C |
| Maximum input voltage | 1000 | Maximum current | 7,2 A |
| Maximum input current | 16 A | Nominal voltage | 400 V |
| Direct current | 24 A | Nominal frequency | 50/60 Hz |
| No. Inputs | 2 + 2 | Network type | TT/T |
| | | Performance | 97,9% |

The Fronius inverter includes a Fronius Smart Meter. This equipment sends the most complete data to the monitoring system and does not allow the inverter to inject energy into the electric network. In line with the features of the inverter selected as the employed panels' characteristics, 18 panels were placed in series to form a chain.

The RAPTION-22 model was chosen as the electric vehicles-charging system to allow vehicles to be quickly charged whenever required and there is not enough power. Depending on their capacity, batteries can be partially or completely charged in short times. The RAPTION-22 station operates with powers up to 22 kW, which means it can be fitted in simpler installations and saves extra costs related to special rates. This device has an interactive screen, and communications (Ethernet, 3G) that facilitate users' interaction, as well as remote management to the control centre. This selected product permits two charge types, direct current (DC) and alternating current (AC), so it adapts to the charging system of any vehicle on the market. This means that it has connectors CHAdeMO and CCS Combo 2 in DC and type 2 Mennekes in AC. It also allows three-phase and single-phase currents (Table 3).

**Table 3.** Characteristics of input in AC, output in DC and output in AC of the RAPTION 22 charging station.

| Input AC | Value | Output DC | Value | Output AC | Value |
|---|---|---|---|---|---|
| AC supply | 3F + N + PE | Maximum output current | 56 Acc | Maximum output current | 32 A |
| AC voltage CA | 400 Vac ± 10% | Maximum output power | 22 kW | Maximum output power | 22 kW |
| Current nominal input | 64 A | Range of output voltage | 150–550 Vcc | Range of output voltage | 400 Vca |
| Power factor | ≥0.98 | | | | |
| Efficiency | 94% | | | | |
| Frequency | 50/60 Hz | | | | |

The total cost of the system taking into account foundations, structures (pillars, light structures for roofs and beams) and installations (grounding, pipes, cables, protection boxes, photovoltaic solar modules and electric vehicle charging stations) amounts to 185.052 €.

## 3   Conclusions

One area where electric vehicles will be very important is urban areas because electric vehicles do not emit direct polluting emissions. It is here where not only the structure's functionality and installations are important, but also its aesthetics and design. Installing canopies in cities to quickly charge electric vehicles is becoming a very effective solution, provided the structure can be integrated into the domain it is located in. Accordingly, the proposed canopy design can be classified as a canopy with a strong visual impact and a completely built in photovoltaic system. It strikes the best balance between energy efficiency and its architectural integration into its surroundings. The main advantage of this vehicle-charging canopy is that energy is produced in the same place where it is used, which avoids resorting to lengthy energy transports and substantially enhances the system's total efficiency. For this reason, this installation fulfils the described objectives: an integrating and optimised structure that is functional, modular and user-centred that adapts to any location.

# References

Alghamdi, A.S., Bahaj, A.S., Wu, Y.: Assessment of large scale photovoltaic power generation from carport canopies. Energies **10**(5) (2017). https://doi.org/10.3390/en10050686

Bushur, A., Ward, K., Flahaven, T., Kelly, T., Jo, J.H., Aldeman, M.: Techno-economic evaluation of installing EV and PV combined infrastructure on Academic Institution's Parking Garages in Illinois, USA. AIMS Energy **7**(1), 31–45 (2019). https://doi.org/10.3934/energy.2019.1.31

De Fusco, R.: Historia de la arquitectura contemporánea: H. Blume Ediciones (1981)

Gay, A., Samar, L.: El diseño industrial en la historia. Ediciones Tec., Córdoba, Argentina (2004)

Macdonald, A.S.: Aesthetic intelligence: optimizing user-centred design. J. Eng. Des. **12**(1), 37–45 (2001). https://doi.org/10.1080/09544820010031562

Parras-Burgos, D., Hernández, J., Velázquez, J., Cavas-Martínez, F., Cañavate, F., Fernández-Pacheco, D.: Combined urban furniture designed by a bio-inspired approach. In: Advances on Mechanics, Design Engineering and Manufacturing II, pp. 564–572. Springer (2019)

Umer, F., Aslam, M.S., Rabbani, M.S., Hanif, M.J., Naeem, N., Abbas, M.T.: Design and optimization of solar carport canopies for maximum power generation and efficiency at Bahawalpur. Int. J. Photoenergy **2019** (2019). https://doi.org/10.1155/2019/6372503

Zarcone, R., Brocato, M., Bernardoni, P., Vincenzi, D.: Building Integrated Photovoltaic System for a Solar Infrastructure: liv-lib' project. Paper presented at the Energy Procedia (2016)

# Taking into Account Users' Perceptions in the Design Process: Principles to Create a Digital Design Tool

Renato Fonseca Livramento Da Silva[1], Angelina Dias Leão Costa[1], and Guillaume Thomann[2($\boxtimes$)]

[1] Universidade Federal da Paraíba, PPGAU, João Pessoa, PB 58033-455, Brazil
[2] Univ. Grenoble Alpes, CNRS, Grenoble INP, G-SCOP, 38000 Grenoble, France
`guillaume.thomann@grenoble-inp.fr`

**Abstract.** User Centered Design approach is used in many sectors and appropriated by many design teams to defend principles of products adapted to the final users. In the Architectural and Industrial Design disciplines, architects and designers defend principles that could be able to create spaces, public areas or innovated products that are closer as possible as the user behavior. The issue is still the complexity of the user perception and the variability of its interpretation of the environment. The research method used in this research is to combine Universal Design and Usability approaches to be able to extract one first list of principles. The combination of this list with the five human sensorial systems identified in the literature give the structure of a tool that can be proposed to projectists like architects and industrial designers to better consider user perception during the designing process. The result of the research is the proposition of a software coupled with a user friendly interface dedicated to architects and industrial designer. It has the aim to simplify the organization of the early phases of the design process, taking into account designers and architects design priorities and integrating the final user specific sensorial situation.

**Keywords:** Users centered design · Universal design · Usability · Human sensorial system · Digital tool design

## 1 Introduction

Architecture and design give the development of environments, products and services a culturally rooted visually-based bias. In this context, the significant growth of new technologies, especially information-oriented, has exacerbated the predominance of vision as a source of absorption of information available to metropolitan populations. That said, in a way, it is comfortable, from the point of view of the designer, to keep thinking about the project based mainly on visual experience, because culturally the Western world tends to organize itself in this way [1].

This fact reinforces the premise of a material world still largely thought to be visually lived. However, [2] points out that in relation to the challenges of the project in

L. Roucoules et al. (Eds.): JCM 2020, LNME, pp. 295–300, 2021.
https://doi.org/10.1007/978-3-030-70566-4_47

today's world, the profusion of dynamically and complexly inserted "codes" in people's daily lives also needs attention, such as the questions related to intangible attributes, aligned with aspects of human behavior, aesthetic and psychological factors. In this sense, according to the author "new creative tools are necessary to fill the gaps that the methodological models used from now are no longer able to answer the current issues" [2].

Studies related to the design process in architecture and design have multiplied considerably in recent years. In particular since the 1960s, the number of proposals for methods, techniques and tools aimed at contributing to the control and organization of the design process has increased considerably, mainly due to the technological increasing complexity of products [3–5].

However, in recent decades, discussions in the design field have paid particular attention to User Centered Design (UCD), which focuses more on taking into account behavioral, social, physical and cognitive aspects of the users [6]. It is based on the effective participation of users throughout the whole project, in order to collect information on their experiences for solutions closer to their needs [7, 8]. In this context, Universal Design (UD) and Usability contribute significantly to the UCD process, while presenting similarities in their criteria: DU aimed to establish wide access to products and Usability focused on the ease of use [9].

Due to previous observations, there is currently no tool that can help the industrial designer or the architect to take in consideration a variety of human perceptions of an environmental space occupation.

Objectives of the current research are focused on:

– From architects and industrial designers points of view, assessing the need for a tool to help the structure of ideas in the early phases of the design process,
– How Design Universal, Usability and the Human Sensorial System can contribute to such a process?
– Finally, is it possible to structure a tool to assist architects and industrial designers during these initial steps of the process?

## 2   Methods

The research method is organized in two steps.

First of all the different fields of the literature review give some orientation of the research. Thanks to the evolution of the history of the architecture domain, it is clear that the UCD approach has to be included into the current study. UCD is an approach in which user characteristics, usability goals, environment, tasks and workflow of a product, service or process are given extensive attention at each stage of the design process. In the context of designing an environmental space, the usability of this space has to be study. Moreover, this space should ideally suit any individual. Universal Design (UD) is the design of any layout, product, equipment, program or service that can be used by any person, without requiring any adaptation or special design, regardless of gender, age, location or his handicap. To be sure not to be limited to the visual perception of the user, all the human sensory perception system has to be taken into account during this study.

Secondly, data collection was carried out in two steps to better fix the requirements of specialists and being able to develop a first version of a tool. The first one took place through a questionnaire with 13 Brazilian students from the Architecture and Urbanism Post-Graduate Program in one Brazilian Federal University. All of them were regularly enrolled in the subject. The students had a degree in architecture, engineering and economics, with 4 men and 9 women between 25 and 43 years old. Questionnaire was structured in three parts: (1) participants characteristics, (2) knowledge about creativity and creativity tools and (3) personal experience related to UD, Usability and sensory-human systems.

In the second one, students' groups follow-up was carried out at a French university. 20 students divided into four groups during the development of creative design steps (common project) were observed. Students were between the ages of 18 and 27, with 13 women and 7 men; all of them have some experience in product design and/or environments. The teams were composed of students of architecture, design and product engineering of different French universities and of different periods in their respective courses. The aim of this study was (1) to explore the design methodologies known by all the members before the realization of the project and (2) how they combine and decide the use of one unique methodology during the project (steps and tool used in relation to UD, Usability and sensory-human systems).

Finally, a first version of a tool structure was proposed and a first prototype of software was developed. It has to be evaluated.

# 3   Results

## 3.1   Contributions from Literature

Universal design is known as a global design trend and has a broad and diverse scope [10]. However, its principles are relatively difficult to apply because there are few clearly defined methods for this [11]. On the other hand, [12] declares that "the theme of user-centered design is directly linked to usability", a factor which makes us deduce a strong overlap between the two themes.

**Table 1.**  Principles of usability [13] and UD [14]

| Usability | Universal design | New denomination |
|---|---|---|
| Consistency, compatibility, evidence | Simple and intuitive use | Compatibility |
| Feedback, error prevention | Fault tolerance | Warning |
| Visual clarity, prioritization of functionality and information | Perceptible information | Communication |
| Transfer of technology | | Materialization |
| Capacity, user control | Flexibility of use and equitable use | Adaptability |
| | Minimum physical effort | Force |
| | Space dimensioning | Dimensioning |

The two first column of the following table (Table 1) come from the literature. They consist of the lists of the 7 principles of the Universal Design and the 10 principles of the

Usability. The last column is a proposition of reduction of variables with the perspective of using them inside the future proposed tool.

To be able to not only consider the human visual system in the research proposal, the use of the information acquisition given in [15] is chosen. Five sensorial systems are established: Basic Guidance System, Haptic System, Visual System, Auditory System and the Smell System/Taste. Such information can be obtained with any of the perceptual systems alone, or by any combination of them.

## 3.2   Requirements List

This requirements list was established from the two steps realized with groups of students in France and Brazil. Analysis of the answers of the questionnaires gave a certain number of functions that the future tool has to realize. Categories of requirements where proposed to better structure the result of the analysis (Table 2).

**Table 2.**  Requirements list established from the questionnaires

| |
| --- |
| *Functional requirements* |
| Enable the sharing of project activities performed by users of the tool in real time |
| Free choice of combinations of tool concepts and structuring principles to guide design actions |
| Be able to form database with developed projects |
| Allow insertion of written data (concepts) and images) |
| Possibility of printing the data entered in the tool (texts) and by means of boards (images) |
| Nonlinear use of the steps that make up the tool roadmap (allow advances, returns, etc.) |
| *Requirements on data* |
| Have the combination of a digital structure, paper and three-dimensional elements (physical) |
| Allow the possibility of printing papers (boards) of different sizes |
| Allow information about the concepts and principles of the structuring themes that form the theoretical content of the tool |
| *Environmental requirements and contexts* |
| Be able to use the tool on various platforms (digital and physical) |
| Easy access and storage (digital and physical) of data |
| Have feature of collaborative use |
| Possibility to securely store the collected and developed data |
| Easy and cost-effective organizational support (resources, training and management) |
| *Users requirements* |
| Appropriate information for users (architects, designers and engineers) |
| Simple and clear language |
| It can be used by a variety of users for projects with different characteristics |
| *Usability requirement* |
| Be efficient in its use |
| Be safe about the use and storage of data |
| Easy memorization and learning |
| Easy in use |

## 3.3   Digital Tool User Interface

The web oriented tool consists of a software that allow the architect (for example) to better manage its projects centered to the human use of the environment. After the creation of the project, the user friendly interface shown in Fig. 1 oblige the architect to select all combinations of principles before closing it. Thus, he has to think about solutions/proposals for each of them during the creativity phase of the project. Logically, he select the most adapted combinations of principles at first (depending of the project characteristics) … knowing that sometime, new solutions come from others points of view. He can add illustrations, comments, schemas etc. and then print the document to constitute the archives of the complete project.

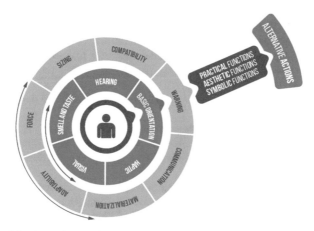

**Fig. 1.** Image of the dynamic user-friendly human interface the project manager has to manipulate to select a combination. For each combination (one blue principle with one green principle), he can propose one or more architectural solution to be used for the final proposal [16].

## 4 Conclusion and Perspectives

Design processes used by architects and industrial designers are complex and need frequent adaptation with the context of use in the future working environment. Products or working spaces have to be dedicated to all the characteristics of the final users. Theoretical approaches exist to follow and support these needs. The software proposed is based on 7 Universal Design and Usability principles and is complete by 5 human sensorial systems. The combination give 35 couples of principles that the architect or industrial design has to use to propose final solution.

## References

1. Herssens, J.: Designing architecture for more: a framework of haptic design parameters with the experience of people born blind. Hasselt: University Hasselt (2011)
2. De Moraes, D.: Metaprojeto: o design do design. Blucher, São Paulo (2010)
3. Di Russo, S.: Understanding the behaviour of design thinking in complex environments. Thesis (for the degree of Doctor of Philosophy) – Swinburne University of Technology, Melbourne (2016)

4.  Van Der Linden, J.C.d.S., Lacerda, A.P.: Metodologia projetual em tempos de complexidade. In: Martins, R.F.d., Van Der Linden, J.C.d.S. (org.). Pelos caminhos do design: metodologia de projeto. Londrina: Eduel, pp. 83–149 (2012)
5.  Cross, N.: A history of design methodology. In: Vries, M.J.d., Cross, N., Grant, D.P. (eds.) Design Methodology and Relationships with Science, pp. 15–27. Kluwer Academic Publishers, Dordrecht (1993)
6.  Fletcher, V., Crolius, W.A., Harada, F.J.B.: Improving senior ridership on public transit: an approach to working with user/experts to review the public transportation for the city of Cambridge/MA with emphasis on universal design. In: 2015 Universal Design Symposium. NC State College of Design (2015)
7.  Tschimmel, K.C.: Design as a perception-in-action process. In: Taura, T., Nagai, Y. (eds.) Design Creativity 2012 (ICDC). Springer, London (2012)
8.  Norman, D.A.: Human-centered design considered harmful. Interactions 12(4), 14–19 (2005)
9.  IIDA, Itiro; GUIMARÃES, Lia Buarque de Macedo. Ergonomia: projeto e produção. 3 ed. revista. Blucher, São Paulo (SP) (2016)
10. Lin, K.-C., Wu, C.-F.: Practicing universal design to actual hand tool design process. Appl. Ergon. 50, 8–18 (2015)
11. Boisadan, A. : Conception universelle pour une signalétique intuitive et accessible à tous. Thèse de doctorat en Ergonomie, École doctorale 261 «Cognition, Comportements, Conduites Humaines» Laboratoire Adaptations, Travail, Individu. Université Paris Descartes. Paris (2018)
12. Merino, G.S.A.D.: Metodologia para a prática projetual do designcom base no projeto centrado no usuário e com ênfase no design universal. Tese (Doutorado em Engenharia de Produção) – Universidade Federal de Santa Catarina, Florianópolis (2014)
13. Jordan, P.: An Introduction to Usability. Taylor & Francis, London (1998). 120 p.
14. CUD: The Center for Universal Design. Universal Design: product evaluation countdown. Raleigh, North Carolina State University (2002)
15. Gibson, J.J.: The Senses Considered as Perceptual Systems. Houghtan Miffin, Boston (1966)
16. Fonseca Livramento Da Silva, R., Dias Leão Costa, A., Thomann, G.: Design tool based on sensory perception, usability and universal design. In: 29th CIRP Design 2019, Póvoa de Varzim, Portugal (2019)

# Proof of Concept Experience in the SPES Experiment: First Solutions for Potentiometers Replacement in System Maintenance

Diego Paderno[1,2], Ileana Bodini[1,2(✉)], Aldo Zenoni[1,2], Antonietta Donzella[1,2], Lisa Centofante[1,3], and Valerio Villa[1,2]

[1] Department of Mechanical and Industrial Engineering, University of Brescia, via Branze 38, 25123 Brescia, Italy
ileana.bodini@unibs.it
[2] INFN, Pavia, Italy
[3] INFN, Legnaro, Italy

**Abstract.** SPES (Selective Production of Exotic Species) is a large facility, currently under advanced construction at the INFN-LNL (Istituto Nazionale di Fisica Nucleare, Laboratori Nazionali di Legnaro) for the production of Radioactive Ion Beams (RIBs). Coordinated efforts are being dedicated to the development and upgrading of both the accelerator complex and the up-to-date experimental set-ups. This paper describes a work of upgrading as far as the inspection and maintenance of the system is concerned, and it deals with human-centered design methods to reduce the time spent in the radioactive environment of the facility during ordinary maintenance operations and to simplify them, also considering stress conditions of the operator and the mandatory wearable radiation protection devices (such as tracksuit, gloves, oxygen tank mask) which make simple operations difficult.

**Keywords:** Proof of concept · Human-centered design · Functional design · Ergonomics · Design methods

## 1 Introduction

SPES (Selective Production of Exotic Species) is a large facility, currently under advanced construction at the INFN-LNL (Istituto Nazionale di Fisica Nucleare, Laboratori Nazionali di Legnaro) for the production of Radioactive Ion Beams (RIBs). Coordinated efforts are being dedicated to the development and upgrading of both the accelerator complex and the up-to-date experimental set-ups [1, 2].

Studies [3], such as the one presented in Fig. 1, have been performed to give indications about radiation exposure of operators who have to maintain the facility. In particular, Fig. 1 shows the results of a simulation to calculate the residual activation near the most radioactive components of the facility, in terms of ambient dose equivalent rate. The black circles correspond to locations where the operator is supposed to stand still in order to accomplish the ordinary and extraordinary maintenance operations.

L. Roucoules et al. (Eds.): JCM 2020, LNME, pp. 301–306, 2021.
https://doi.org/10.1007/978-3-030-70566-4_48

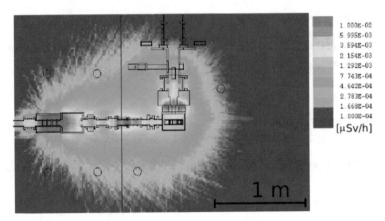

**Fig. 1.** Results of radiation exposure simulations.

The foreseen period of activity of the SPES facility, before the total or partial substitution of system structural components, is in the order of years. Component activation and consequent ambient equivalent rate progressively increase, due to the accumulation of medium-long lifetime isotopes.

Operator need to complete maintenance operations in stress conditions, due to the knowledge that the environment is potentially dangerous, that they are equipped with a tracksuit, two pairs of gloves, an oxygen tank mask, a helmet, and that the room is initially dark, and operators have to bring light with themselves. Moreover, operators know that they need to minimize the time they stay in the chamber, in order to minimize their exposure to dangerous radiations. All these conditions contribute to make simple operations difficult [4–6].

To minimize maintenance operation time, a sequence of simple operations is required, together with the reduction of tools.

The current inspection and maintenance plan presents some critical aspects, which need to be upgraded. In particular, this works analyzes the current mechanical solution for what concerns some diagnostics components and their substitution during ordinary maintenance operations. Critical aspects of the current solution are taken into account and possible human-centered designed solutions are proposed to simply operations, to reduce the stress conditions of the operators and to minimize the time for these operations. A human-centered designed proof of concept device of these possible solutions has been realized [7].

## 2   Case Study: Replacement of Potentiometers

To diagnostics purposes, on the SPES facility are assembled different components, such as the ones shown in Fig. 2. In these figures linear motors, potentiometers and limit switches are presented: these components are located in many parts of the facility and need ordinary maintenance, because they are useful and important to know and to make the right positioning of the crucial component of the facility. As is visible from the figures, some of these components are mounted on the system in positions which are critical to reach, or in which there is a small space to operate.

**Fig. 2.** Some ordinary components necessary as diagnostics: their maintenance is critical due to their current mounting and to their positions, near the most radioactive zones of the facility.

To diagnostics purposes, on the SPES facility are assembled different components, such as the ones shown in Fig. 2. In these figures linear motors, potentiometers and limit switches are presented: these components are located in many parts of the facility and need ordinary maintenance, because they are useful and important to know and to make the right positioning of the crucial component of the facility. As is visible from the figures, some of these components are mounted on the system in positions which are critical to reach, or in which there is a small space to operate.

In the study presented in this paper, the focus is on the maintenance of the potentiometers. These potentiometers are mounted in a lot of positions in the facility and they are supplied with all the measurement and alignment components of the system. Potentiometers have a passive role, but they are important for diagnostics.

The existing solution was analyzed to identify the parameters for a new project, in order to simplify the operation of potentiometers substitution.

The considered parameters were (i) the number and (ii) shape of the holding parts, (iii) the number and (iv) complexity of human operations for substitution, (v) the need for tools, (vi) the need of blocking degrees of freedom. In particular, the current holding system is based on a screw clamping ring which holds the body of the potentiometer and a hold on the sliding carriage for the threaded head of shaft of the potentiometer.

Figure 3 shows that, in order to assemble the potentiometer, the operator needs to align the body to the clamping ring, and then make it slide through the holding and keep it in position with one hand while the other tightens the screws using a key.

The next step is to align the shaft's head to the positioning hole, slide through it and fix it with an M3 nut tightened with a key.

The holding system is strong and fixes both axial movements and rotation of the body, but longitudinal alignment depends only on operator.

To mount the potentiometer in the considered system, a sequence of operations is required, two keys are needed, the nut is hard to handle with two gloves on each hand and aligning and sliding operations are not easy both for position of the holding ring and for

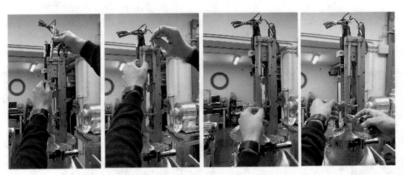

**Fig. 3.** Current potentiometer assembly.

the needed space to mount, which has to be double than the length of the potentiometer (not all the potentiometers in the facility are so easy to reach).

Operator's stress condition must also be considered.

The aim of the new solution was therefore identified as fixing both parts of the potentiometer (body and shaft) in one phase without using tools and without need to take manually care of alignment (Fig. 4).

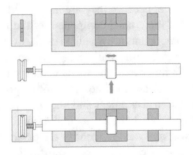

**Fig. 4.** Degrees of freedom of the potentiometers.

The method is based on the functional simplification of the system holding the potentiometers by splitting it in more parts, each of them both guide operator movements and block one degree of freedom shown in, and on the fact that screw and nut locks can be substituted by spring based locks, commercial latch clamps or gravity-based blocking parts.

Two proofs of concept have been developed and are shown in Fig. 5(a) and in Fig. 5(b):

– to correctly mount the sliding shaft, both the solutions are based on a couple of discs, held together by a spring, which is screwed on the head (before maintenance operation) and get mounted on an open slot on the sliding carriage as a clip. Choosing a spring stiffer than the force needed to make the shaft of the potentiometer slide is enough. Chamfering of all the corners makes easier the insertion.

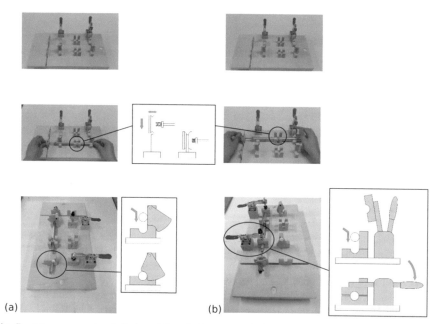

**Fig. 5.** A human-centered designed Proof of concept device has been realized for both the two proposed solutions assembly. (a) First proposed solution; (b) second proposed solution.

- To fix the body, both solutions are based on a screw clamping ring which is mounted on the potentiometer prior to maintenance operations. This part will provide exact aligning avoiding operator errors and block longitudinal degree of freedom.
- Holding parts are shaped in order to guide operator's movement and fix one degree of freedom at a time, being longitudinal alignment the fundamental, transversals secondary and rotation not so significant for potentiometer functioning.
- All corners were chamfered if needed to simplify positioning.
- Blocking the body is obtained by hanging mobile parts which are pushed away by potentiometer body during insertion and return to their position by gravity after the component reaches its final position blocking it, in the solution showed in Fig. 5(a), and by commercial latch clamps in the solution showed in Fig. 5(b).
- Both solutions are tools-free and allow operator to manage the potentiometer with two hands and mount it with one movement.

   A first discussion about the two solutions with SPES staff underlined that, despite of being based on one movement for both placing and blocking the potentiometer, the solution in Fig. 5(a) is perceived as less safe than the one based on the clamp, which need two movements (placing and blocking). This makes us add safety perception as an additional parameter to consider in the next phase of the project, which includes tests on users. A test desk for solutions comparison (existing one and both new solutions) is under construction to allow tests on a larger number of users.

## 3  Conclusions

Critical aspects of the design of the diagnostics system have been studied, and typical mechanical solutions, which are adequate in a common workshop, have been upgraded with other mechanical solutions which don't need tools and that allow for a simpler sequence of operations.

A human-centered designed proof of concept device has been realized for both the proposed solutions. Tests to objectively confirm the goodness of the proposed solutions and to compare these upgrades with the previous solutions need to be performed and have been already planned.

## References

1. Gramegna, F., et al.: The SPES exotic beam ISOL facility: status of the project, technical challenges, instrumentation, scientific program. Il nuovo cimento C **41**, 195–201 (2018)
2. Bisoffi, G., et al.: Progress in the realization and commissioning of the exotic beam facility SPES at INFN-LNL. J. Phys. Conf. Ser. **1067**, 052017 (2018)
3. Donzella, A., et al.: Shielding analysis of the SPES targets handling system and storage area using the Monte Carlo code FLUKA. Nucl. Inst. Methods Phys. Res. B **463**, 169–172 (2020)
4. Sonora, A., et al.: The impact of stress on surgical performance: a systematic review of the literature. Surgery **147**(3), 318–330 (2010)
5. Prabhu, A., et al.: Increased stress levels may explain the incomplete transfer of simulator-acquired skill to the operating room. Surgery **147**(5), 640–645 (2010)
6. Wetzel, C.M., et al.: The effects of stress on surgical performance. Am. J. Surg. **191**, 5–10 (2006)
7. Paderno, D., Bodini, I., Villa, V.: Proof of concept as a multidisciplinary design-based approach. In: Design Tools and Methods in Industrial Engineering, pp. 625–636 (2020)

# Human, Knowledge and Decision

# A New Smart Web Platform for Plastic Injection Molds in Industry 4.0 Environments

Cristina Martin-Doñate[1]([✉]), Sliman Shaikheleid[2], Abelardo Torres-Alba[1], and Jorge Manuel Mercado-Colmenero[1]

[1] Department of Engineering Graphics, Design and Projects, University of Jaén, Campus Las Lagunillas, s/n, 23071 Jaen, Spain
cdonate@ujaen.es

[2] Fakultät Mechanical and Medical Engineering, Hochschule Furtwangen University, Campus Villingen-Schwenningen, Schwenningen Jakob-Kienzle-Strasse 17, 78054 Villingen-Schwenningen, Germany

**Abstract.** This paper presents a new smart web platform for plastic injection molds for use in industry 4.0 environments. The new platform requires as its only input the CAD model of the plastic part in a discrete format, the accuracy of the analysis, the thermoplastic material of which the part will be manufactured and the number of parts to manufacture per year. Using this information and through a fully automated process based on hybrid algorithms developed by the authors the smart platform generates an extended CAD model of the mold with additional expert information useful for industry 4.0 environments. In this way, it is possible to design a mold with uniform heat transfer, balanced ejection and a uniform filling phase of the mold cavity. The presented platform differ from other applications for mold designing in that the resulting mold meets all the geometric, functional and technological requirements of mold designing without needing CAE simulation software for its validation. The presented platform is considered as the first smart platform that does not require the interaction of the designer in the process of dimensioning and designing the different subsystems that compound the mold, being a tool to reduce time and costs in the initial phases of plastic part design and with the ability to integrate into a flexible manufacturing environment 4.0.

**Keywords:** Smart manufacturing · Industry 4.0 · Injection molding · CAD

## 1 Introduction

Nowadays, plastic injection molding is the most widespread manufacturing method in the industry [1, 2]. With a conventional approach injection mold design is still performed empirically, based on the experience of the mold designer. The dimensioning tasks of the different mold components are highly dependent upon the knowledge of how the different geometrical and technological mold parameters interact with each other. In addition, mold designing requires highly qualified engineers from different disciplines. CAD designers deal with the dimensioning and modeling of each part, technicians in

© The Author(s) 2021
L. Roucoules et al. (Eds.): JCM 2020, LNME, pp. 309–315, 2021.
https://doi.org/10.1007/978-3-030-70566-4_49

rheological simulation validate mold design in terms of time cycle and part quality, and mechanical engineers ensure the structural integrity of the mold. On the other hand, in order to design and simulate the mold each designer requires specific software which the company must acquire. Due to diversified complications of mold design with mold manufacturing and the coupling effects of plastic with molding machine, it is quite difficult to consistently molding the products with higher quality and precision. Even with high-end commercial software and hardware, the seamless integration of mold design, mold manufacturing and molding process needs innovative thinking [3]. Cyberphysical systems in industrial processes are characterized by their ability to connect the physical devices that compose it with ICT systems capable of accessing and processing the set of technological data obtained from the process.

This paper presents a new smart application integrated into a web platform. The new application differs from other mold designing applications in that it only requires the CAD model of the plastic part in a discrete format, the precision of the analysis, the thermoplastic material of the plastic part and the number of parts to be manufactured per year as application inputs. With only this information, and through a fully automated process based on the use of several hybrid algorithms developed by the authors, the platform generates a complete CAD design of the injection mold associated with the plastic part under analysis. The set of developed algorithms are compound of a first phase of automated recognition of the plastic part surface and a second phase of expert optimization for mold dimensioning. As a result the mold design meets the geometric, technological and functional requirements being capable of guaranteeing an adequate and accurate manufacture of the plastic part. The application provides in a few minutes a standard CAD design in parametric format of the injection mold. The smart application is valid for any CAD format in which the plastic part has been modeled since it works in a discrete format. On the other hand, the algorithms for obtaining the mold designing can be applied in any CAD system, being independent of the model and the modeler. Thus, the presented application differs from other software linked to the CAD format of the model which requires access to the internal information of the part. The new platform is the first one capable of working completely autonomously and intelligently, performing the complete process of optimal dimensioning and mold design without requiring, at any time, mold designer interaction during the process. Furthermore, the design of the resulting mold does not require numerical simulations for functional validation, since the expert calculation algorithms incorporate technological requirements that guarantee the correct performance of the mold. The application is designed for future implementation in an Industry 4.0 system that allows analyzing the deviations in the manufacturing parameters of the product.

## 2    Methodology

### 2.1    Architecture of the Injection Molding Smart Platform

The injection molding smart platform is structured in two large blocks. The first block of the platform consists of the implementation of four geometric-expert algorithms [4–7] responsible for analyzing the geometric demoldability of the plastic part in an automated way as well as dimensioning and positioning the set of elements that constitute the feeding system, the system of cooling and the ejection system of the mold. The architecture of the algorithms is structured in two different stages: The first geometrical stage is in charge of recognizing in an automated way the topology of the plastic part, obtaining information of the geometric parameters necessary for the optimal dimensioning of the mold.

A second expert stage optimizes the different elements that constitute the three main systems of the mold (cooling, ejection and feeding), using genetic optimization algorithms. Finally, the platform connects the resulting expert variables with the algorithm responsible for optimally dimensioning the structural and mechanical elements that constitute the mold. This means that the mold obtained allows uniform heat transfer between the plastic part and the coolant fluid, balanced ejection of the plastic part from the mold and an equilibrated filling phase of the injection mold cavities. This fact differentiates the presented platform from other mold design applications since the methodology accomplishes geometrical, functional and technological requirements of the mold, independently of the additional use of CAE simulation software.

Figure 1 shows a block diagram of the different algorithms that compose the web platform. This uses as input data the discrete geometry of the plastic part, allowing a total independence of the CAD modeler in which the plastic part has been designed. In contrast to other applications, the new application does not require additional information or designer interaction during the analysis process, dimensioning and designing the mold in a fully automated process.

The presented platform incorporates four databases with information about technical characteristics of thermoplastic materials, standard components of injection molds, technical characteristics related to the injection manufacturing process such as the characteristics of the coolant fluid and a data base with information related to the injection machines. Figure 2 shows a logical operation scheme of the smart web platform, including the phase of data processing introduced by the user, the phase of calculation of the results from the input data, the phase of designing and executing the algorithms and the phase of management of the stored information in the databases.

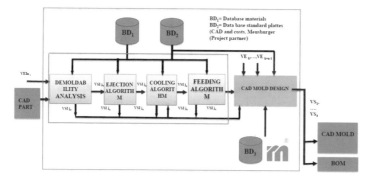

**Fig. 1.** Integration of the different geometric-expert algorithms of the smart web platform.

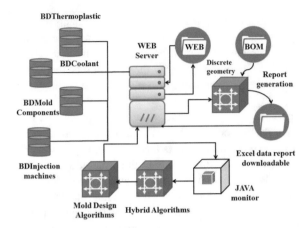

**Fig. 2.** Logical diagram of the functioning of the smart web platform.

# 3  Results

The injection molding smart platform generates a technological report that includes the results obtained for each geometric-expert hybrid algorithm [4–7, 8] and, in turn, the CAD modeling of the injection mold with the assembly of the elements which comprise it. The geometry of the study case and the result of the algorithm of the demoldability analysis is shown in Fig. 3a. As can be seen, the case study requires of two longitudinal slides along the X and Y coordinate axes for its manufacture. Figure 3b shows the location on the ejection map of the ejection points of the plastic part. Figure 3c shows the distribution of the cooling channels on the geometries of the plastic parts under study. The cooling system design obtained by the presented platform is dimensioned under three conditions: obtaining a uniform distribution of the exchange of heat flow between the melt plastic flow and the cooling system, guaranteeing that the difference between the heat flow provided by the melt plastic flow and the heat evacuated by the cooling system is zero and maintaining the cooling flow throughout the cooling system in turbulent conditions. In order to validate the final quality of the plastic parts of the case studies and the sizing and design of the cooling system. Figure 3d shows the results of the algorithm for the location of the injection point for each study case presented. Finally, Fig. 4a shows the temperature map after the cooling phase for each one of the plastic parts of the study. The temperature map of the plastic parts is uniform and homogeneous. It is thus demonstrated that the cooling quality, of the results obtained by this algorithm, is high. Figure 4b shows the CAD model of the injection molds of the plastic parts under study.

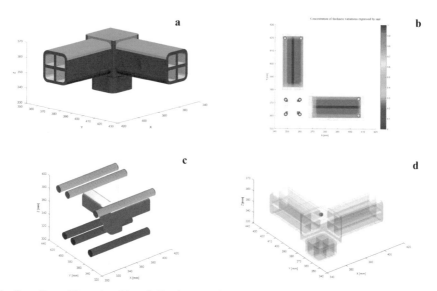

**Fig. 3.** **a** Demolding algorithm. **b** Design and dimensioning algorithm of the ejection system, **c** design and dimensioning algorithm of the cooling system, **d** location of the injection point

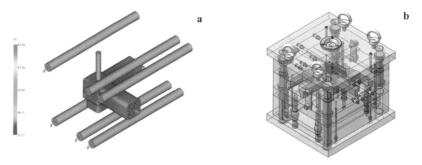

**Fig. 4.** **a** Results of the numerical simulation of the cooling, **b** CAD model of the injection mold

## 4  Conclusions

The work presented in this document shows a new smart web platform for the automated design of plastic injection molds based on geometrical-expert optimization algorithms. The presented smart platform has been tested through its application to a plastic part obtaining in a fully automated way a CAD model of the mold to manufacture the plastic part. The results obtained are presented as a basis for the implementation of a 4.0 platform in the molding industry.

The presented injection molding smart platform improves the mold design applications used up to now, since the resulting mold fulfills all the geometrical, functional and technological requirements of the mold independently of the use of CAE simulation software. Unlike other applications, it does not require additional information or the interaction of the designer during the process of analysis, sizing and design of the mold.

The new platform is valid for any CAD format in which the plastic part has been modeled, since it works in a discrete format. It functions independently of the CAD modeler used to model the plastic part and without requiring access to the internal geometric information of the part to be analyzed.

**Acknowledgments.** This research work was supported by the University of Jaen through the Plan de Apoyo a la Investigación 2019–2020-ACCION1 PAIUJA2019-20: TIC-159.-Research stay Cristina Martin-Doñate Fakultät Mechanical and Medical Engineering-Hochschule Furtwangen University. 2019.

# References

1. Martin-Doñate, C., Rubio Paramio, M.A.: New methodology for demoldability analysis based on volume discretization algorithms. Comput. Aided Design **45**(2), 229–240 (2013)
2. Torres-Alba, A., Mercado-Colmenero, J.M., Diaz-Perete, D., Martin-Doñate, C.: A new conformal cooling design procedure for injection molding based on temperature clusters and multidimensional discrete models. Polymers **12**(1), 154 (2020)
3. Jong, W.R., Chen, S.C., Wang, S.M., Liu, S.H., Liao, H.L., Ting, Y.H., Chen, H.T.: The implementation of cloud platform for injection molding process. Procedia CIRP **63**, 219–223 (2017)
4. Mercado-Colmenero, J.M., Paramio, M.A.R., Perez-Garcia, J.M., Martin-Doñate, C.: A new hybrid method for demoldability analysis of discrete geometries. Comput. Aided Design **80**, 43–60 (2016)
5. Mercado-Colmenero, J.M., Rubio-Paramio, M.A., Karlinger, P., Martin-Doñate, C.: A new procedure for calculating cycle time in injection molding based on plastic part geometry recognition. Int. J. Adv. Manuf. Technol. **98**, 441 (2018)
6. Mercado-Colmenero, J.M., Muriana, J.A.M., Paramio, M.A.R., Martín-Doñate, C.: An automated manufacturing analysis of plastic parts using faceted surfaces. In: Eynard, B., Nigrelli, V., Oliveri, S., Peris-Fajarnes, G., Rizzuti, S. (eds.) Advances on Mechanics, Design Engineering and Manufacturing. Lecture Notes in Mechanical Engineering. Springer, Cham (2017)
7. Mercado-Colmenero, J.M., Rubio-Paramio, M.A., de Juanes Marquez-Sevillano, J., Martin-Doñate, C.: A new method for the automated design of cooling systems in injection molds. Comput. Aided Design **104**, 60–86 (2018)

# Causality Learning Approach for Supervision in the Context of Industry 4.0

Kenza Amzil[1]([✉]), Esma Yahia[1], Nathalie Klement[2], and Lionel Roucoules[1]

[1] Arts et Metiers Institute of Technology, LISPEN, HESAM Université,
Aix-en-Provence 13617, France
kenza.amzil@ensam.eu
[2] Arts et Metiers Institute of Technology, LISPEN, HESAM Université, Lille 59046, France

**Abstract.** In order to have a full control on their processes, companies need to ensure real time monitoring and supervision using Key Performance Indicators (KPI). KPIs serve as a powerful tool to inform about the process flow status and objectives' achievement. Although, experts are consulted to analyze, interpret, and explain KPIs' values in order to extensively identify all influencing factors; this does not seem completely guaranteed if they only rely on their experience. In this paper, the authors propose a generic causality learning approach for monitoring and supervision. A causality analysis of KPIs' values is hence presented, in addition to a prioritization of their influencing factors in order to provide a decision support. A KPI prediction is also suggested so that actions can be anticipated.

**Keywords:** Industry 4.0 · Monitoring and supervision · Key performance indicators · Causality · Artificial intelligence · Machine learning · Decision support

## 1 Introduction

Over the last decades, the industrial world has known a wide emergence of Information Technologies that have led to a convergence towards a new industrial revolution commonly called Industry 4.0. This latter aims at exploiting the growing technologies as a backbone to integrate objects, humans, machines, and processes [1], in order to better overtop customers' requirements increasingly stringent about costs, quality and deadlines. All of this requires making right decisions at the right moment. Thus, the three following features should be considered [2]: (i) horizontal integration, which aims to optimize the value chain by connecting it beyond the company's perimeter; (ii) vertical integration of systems and subsystems, with production management tools through hierarchical levels, (iii) end-to-end integration of engineering across the value chain. The main enablers of vertical integration are real time monitoring and supervision [3]. Monitoring is achieved by collecting data to inform about the systems' current state, and does not have any direct action on decision making, while supervision must provide functions that may affect the con- duct of the monitored system, such as parameterization, re-planning, or optimization, based on the current state [4]. Supervision must also

© The Author(s) 2021
L. Roucoules et al. (Eds.): JCM 2020, LNME, pp. 316–322, 2021.
https://doi.org/10.1007/978-3-030-70566-4_50

be able to recognize and report abnormal situations, so that stakeholders can take well-founded decisions. Also, deviations should be detected preferably before they happen, so that actions can be taken on the factors that have caused it. In this context, KPIs serve as a strong monitoring mean [5] that quantifies processes and engaged action effectiveness [6]. The purpose of this paper is to provide an approach to conduct prediction and diagnosis in order to detect abnormal situations and identify their causes. For this, the focus will be on analyzing KPIs' values, by exhaustively identifying and prioritizing the root factors that affect them. Significant researches [7–10] confirm that the analysis of KPIs' values, and the identification of their influencing factors, often performed by experts in an empirical and descriptive way, do not allow exhaustive and exact identification of all direct and indirect causes of each KPI deviation. These empirical analyses conducted by following some specific problem solving approaches, may omit many factors that affect the investigated KPI, and mutual influence between the identified factors may not be noticed. Hence, this kind of analysis remains subjective and only represents the known part of reality, often leaving a hidden part we totally ignore [11]. Besides, KPIs may change over time and have new influencing factors. These reflections have led us to make our hypothesis, which is to take in consideration all available data, since experts may omit certain factors, mistakenly judging them as being uninvolved in the evolution of the KPIs' values. Thus, we propose to build a causal learning approach in order to identify and prioritize influencing factors. This approach is based on the strong assumption of collecting data from as many sources as possible, including physical world and information systems, by instrumenting as much as we can the systems to be monitored, as well as their environments, so that our analysis can exhaustively high- light all the leading causes among this data. The proposed approach can be applied in the production context as well as in other engineering contexts.

The rest of this article is organized as follows. In Sect. 2, we describe the global approach of the proposal, focusing on data analysis. Section 3 summarizes a use case and discuss the first results. Conclusion and future work are given in Sect. 4.

## 2 Methodology of the Proposal

The goal of the methodology is to be able to identify all the affecting factors of a given KPI, and to prioritize them in order to provide decision support by identifying actions which are more relevant to engage for improving the KPI value. These actions will have a more pronounced added value if they are taken before deviation occurs, hence the interest in predicting the KPI's values. The diagram shown in Fig. 1 describes the bricks that make up data analysis that will lead us to our goal.

(i) **Causal analysis** based on Bayesian Networks (BN), which aims at identifying the factors affecting the addressed KPI. The BN form a class of multivariate statistical models that has become popular as an analytical framework in causal studies, where causal relations are encoded by the structure of the network [12–14]. How- ever, the construction of the BN's structure is itself based on the experts' a priori knowledge. To cope with this, several algorithms exist to learn the structure [15] (constraint based algorithms, and score based algorithms), and to compute the structure's associated conditional probabilities.

(ii) **Prediction**: in order to anticipate actions to be taken on the factors identified by the causal analysis, the KPI value should be predicted and, if any deviation is detected, actions are engaged at the right time. In our case, this prediction is made possible by means of Artificial Neural Networks (ANN). ANN are used to solve complex problems and enable learning and modeling nonlinear and complex relationships between inputs and outputs. **(iii) ANN parameters' definition**: in order to predict one given KPI, the optimal structure of the ANN must be defined and must meet a reasonable computing time with good prediction results. For our case, a multilayer perceptron ANN is used, and for having a good compromise computing time/prediction accuracy, the authors followed, for one single KPI, an experimental approach to adjust the ANN's parameters (e.g. number of layers, learning rate, etc.). Since many KPIs need to be predicted, an optimal ANN structure needs to be defined for each KPI, thus, we need to provide greater genericity to our proposal in order to avoid following the same long experimental approach for each single KPI prediction ANN structure. For this, the authors suggest, in this brick, to adjust the ANN parameters using an optimization algorithm, so that optimal ANN for any KPI prediction can easily be generated.

**(iv) Prioritization of the impacting factors**: causal analysis provides us with the existing causality links between the factors and the addressed KPI, and mutual influences between the factors themselves so that we can go through the causality links until the root cause. In case of deviation, it would be wise to prioritize the impacting factors identified in (i). For this issue, weights of the used ANN are employed, since the weights represent the strength of connections between units of the ANN, and highlight the degrees of importance of the values of inputs [16]. **(v) Decision support:** given the outputs of precedent steps, a decision support can be pro- vided in case of deviation to adjust the implicated factors by priority order.

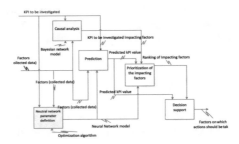

**Fig. 1.** SADT diagram representing the functions of the data analysis.

## 3   Use Case and Results

To implement the proposed methodology of data analysis, the authors have constructed a representative summary dataset to validate that the experiments are correct. This dataset respects, in a very flexible way, a certain amount of causality rules that have been previously defined to be compared with the resulting causality links.

The use case addresses one KPI: the production cycle time. The rest of the dataset is made of variables that may affect, or not, the addressed KPI: the day of the week, the time slot, the month, the indoor temperature, the operator's heart rate, his stress level, the defaults number, and the training level. The goal is to identify, among this dataset, variables that affect the production cycle time, and to prioritize them. To build a realistic and more relevant use case, the authors have made sure that the dataset does not follow these rules in an exclusive way. Basically, this KPI is calculated using two information: the remaining time for production, and the number of produced units. Given this, if deviation is detected, both information do not give answers neither to understand how did this happen nor to trace back to root causes. The above-presented approach was applied to this use case.

(i) **Causality analysis**: we used a constraint based algorithm (Peter and Clarck (PC) algorithm) to define the structure of the BN that will allow identifying the causal links. PC algorithm is a constrained based algorithm that begins with a complete undirected graph, and removes the edges between pairs which are not statistically significantly related by performing conditional independence (CI) tests, then it looks for the V-structures and directs the edges using two other rules (see [17] for more information). This algorithm was modified by adding one more constraint, in order to avoid having meaningless causality links (e.g. defaults that may cause day change is a meaningless link). The steps of this constraint verification are: (1) identify and define variables that can not be changed (e.g. the current hour or day), and (2), remove the edges which go to the nodes representing these variables. The final resulting graph corresponds to the starting assumptions (Fig. 2.a), but does not show a causality link between temperature and cycle time, even if the data was constructed assuming that (ii) **Prediction** and (iii) **ANN parameters' definition**: the KPI prediction was implemented using a multilayer perceptron ANN, with a K-fold cross validation. The predicted value is either 1 or 0 according to whether or not the cycle time will deviate or not, the prediction is 92% accurate (Fig. 2.b). The ANN that gave us these predictions is made up of 5 hidden layers. This ANN's parameters were found using an experimental approach, since the parameters' optimization algorithm is still being under development. Concerning (iv) **the impacting factors prioritization**, ANN's weights used for the prediction were employed to give a ranking of all available factors, and the ranking corresponds to the starting assumptions. Fig. 2.c shows that we obtain the same prioritization even if the prediction models' structures are different (conditioned on the fact that they have a predictive power upper than 85% in our case). The figure shows three lists, all with the same nodes ranking (4-0-3-1-2-5-6). To prove that this ranking is consistent, we have created a new dataset with the same rules and assumptions as the first one. First, we have used this dataset to test our prediction model; the resulting prediction accuracy was 89%. Then, we have replicated the same dataset five times, and each time, we changed the values of one influencing variable individually (day (1), temperature (3), stress (4), training (7), or defaults (6)), to evaluate the impact of each of these influencing variables. Each time, we replaced one variable (e.g. stress) by random values in the same range of variation as the initial values, letting the other variables as they were. Then, we predicted our KPI with the new dataset, using the same model that we have previously built, then we evaluated the prediction. We predicted five times, in addition to the first prediction, in order to see the impact of each of the five influencing variables. Finally, we compared the performances of each of the five predictions that have been run, by superimposing their receiver operating characteristic (ROC) curves, and we evaluated the performance of each prediction by calculating the area under the curve (Fig. 3.a). We can see that the curves reflect the ranking, and that the ranking of the areas under the curves corresponds to the ranking of the variables in the obtained list. Fig. 3.b shows the results obtained by repeating the same operation, but this time, instead of replacing the concerned variable values by random values in their range of variation, the concerned variable was treated as if its values were missing, and replaced the initial values with their mean, then we predicted the KPI. We have repeated the same operation for the five variables. ROC curves are obviously different from the ones in Fig. 3.a, but the ranking is the same. To better see the gaps between the curves, we only represented the four curves that do not intersect in Fig. 3.b, from where we can easily see the gaps between the prediction qualities even without calculating the areas under the curves.

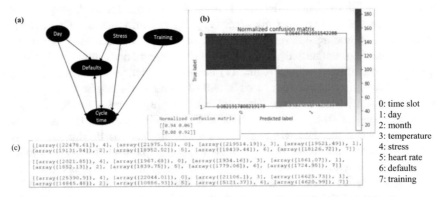

Fig. 2. Results: (a) causal graph, (b) prediction results, (c) factors prioritization.

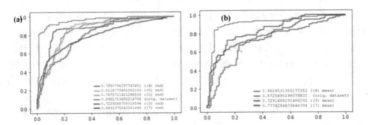

Fig. 3. Results: (a) ROC curves by replacing each time one variable by random values in its variation range, (b) ROC curves by replacing each time one variable by its values' mean.

## 4   Conclusion and Future Work

In this paper, we presented a causality learning approach for KPI supervision. The main idea was to collect as much data as possible in order to conduct an exhaustive data analysis. This analysis is based on BN structure learning, KPI prediction using ANN, and influencing factors prioritization. The proposal is actually in its early phases of development and many aspects need to be addressed, like the test and benchmarking of other algorithms and tools of learning BN structure. Moreover, we should enrich our causality learning with results of decisions taken by considering the impacting factors and their ranking proposed by the supervision system, to see if the KPI is evolving in the right sense, and hence to validate the analysis robustness. Also, a complete use case should be implemented using real industrial data to validate the proposed methodology performance beyond the constructed use case.

# References

1. Schumacher, A., Erol, S., Sihn, W.: A maturity model for assessing Industry 4.0 readiness and maturity of manufacturing enterprises. Procedia CIRP **52**, 161–166 (2016)
2. Wang, S., Wan, J., Li, D., Zhang, C.: Implementing smart factory of Industry 4.0: an outlook. Int. J. Distrib. Sensor Networks (2016)
3. Silva, P.R., Dionisio, R.A., Leitao, P., Barata, J.: IDARTS - Towards intelligent data analysis and real-time supervision for industry 4.0. Comput. Ind. **101**, 138–146 (2018)
4. Combacau, M., Berruet, P., Zamai, E., Charbonnaud, P., Khatab, A.: Supervision and monitoring of production systems. In: IFAC Proceedings Volumes, vol. 33, pp. 849–854 (2000)
5. Yin, S., Zhu, X., Kaynak, O.: Improved PLS focused on key-performance-indicator-related fault diagnosis. IEEE Trans. Industr. Electron. **62**, 1651–1658 (2015)
6. Neely, A., Gregory, M., Platts, K.: Performance measurement system design: a literature review and research agenda. Int. J. Oper. Prod. Manag. **25**, 1228–1263 (1995)
7. Moeuf, A., Pellerin, R., Lamouri, S., Tamayo-Giraldo, S., Barbaray, R.: The industrial management of SMEs in the era of Industry 4.0. Int. J. Prod. Res. **56**, 1118–1136 (2018)
8. Pérez-Álvarez, J.M., Maté, A., Gómez-López, M.T., Trujillo, J.: Tactical business-process-decision support based on KPIs monitoring and validation. Comput. Ind. **102**, 23–39 (2018)
9. Georgakopoulos, D., Jayaraman, P.P., Fazia, M., Villari, M., Ranjan, R.: Internet of Things and edge cloud computing roadmap for manufacturing. IEEE Cloud Comput. **3**, 66–73, July–August 2016
10. Nudurupati, S., Arshad, T., Turner, T.: Performance measurement in the construction industry: an action case investigating manufacturing methodologies. Comput. Ind. **58**, 667–676 (2007)
11. Belis, M.: Causalité, propension, probabilité - Intellectica, vol. 21, pp. 199–231 (1995)
12. Beinlich, I., Suermondt, G., Chavez, R., Cooper, G.F.: The ALARM monitoring system: a case study with two probabilistics inference techniques for belief networks. In: The Second European Conference on Artificial Intelligence in Medicine, pp. 247–256. Springer-Verlag, New York (1989)
13. Pearl, J.: Probabilistic Reasoning in Intelligent Systems: Networks of Plausible Inference. Morgan Kaufmann, San Mateo (1988)
14. Ellis, B., Wong, W.H.: Learning causal Bayesian network structures from experimental data. J. Am. Stat. Assoc. **103**(482), 778–789 (2008)
15. Niculescu, R.S., Mitchell, T.M., Rao, R.B.: Bayesian network learning with parameter constraints. J. Mach. Learn. Res. **7**, 1357–1383 (2006)
16. Han, S., Pool, J., Tran, J., Dally, W.: Learning both weights and connections for efficient neural networks. In: Advances in Neural Information Processing Systems, pp. 1135–1143 (2015)
17. Tsagris, M.: Bayesian network learning with the PC algorithm: an improved and correct variation. App. Artif. Intell. **33**, 101–1023 (2019)

# Digitization of Operational Processes: Use Case of Standardization in an Assembly Learning Factory

Florian Magnani[1,2(✉)]

[1] Aix Marseilla Université, Université de Toulon, CERGAM, Aix-en-Provence, France
florian.magnani@centrale-marseille.fr
[2] Ecole Centrale de Marseille, Marseille, France

**Abstract.** The objective of this research is to set the ground for new research to measure the impact of Industry 4.0, by combining two types of analysis, that of the performance, and that of the human behaviors, using experiments with digital instruments. A preliminary test, then a first experiment focusing on standardization of operational processes are presented showing promising results to understand how operational and human performance are impacted by the Industry 4.0. This study serves as a first step to a larger research project about productivity measurement of Industry 4.0 transformation.

**Keywords:** Industry 4.0 · Standardization · Behaviors · Productivity · Experiment

## 1 Introduction

Since the 1980s, the socio-technical and techno-economical environments have undergone profound transformations, leading enterprises to adopt numerous technological and organizational innovations. Research founds that these transformations performed successfully when they include the human dimension to the technical dimension associated with the chosen innovation. This human dimension is emphasized in all the studies dealing with the current industrial revolution named "Industry 4.0" [1] as a key element to positively influence the impact of these transformations on the enterprises' operational performance.

Study of technological innovations and its impact on performance have been a regular topic of interests. Thanks to fast improvements gathered by innovations, practices are constantly renewed, reshaped and put into perspective by the evolution of theoretical analysis. Today, given the gap between the expected performance and the real performance after the adoption of one of these innovations, it seems appropriate to study closely the characteristics of the so-called innovation with a focus on human behaviors. These behaviors studied throughout the utilization and appropriation of the innovation represent our units to explain this performance gap.

© The Author(s) 2021
L. Roucoules et al. (Eds.): JCM 2020, LNME, pp. 323–328, 2021.
https://doi.org/10.1007/978-3-030-70566-4_51

The recent trend in research guide us in conducting experiments in an observational platform, relatively similar to an industrial environment, gathering qualitative and quantitative data. The objective of this research is to set the ground for new research to measure the impact of the current industrial revolution, by combining two types of analysis, that of the performance, and that of the human behaviors, using digital instruments. The first experiment based on a multidisciplinary approach focused on standardized work transformation in operation processes.

## 2 State of the Art

Before embarking on experimentation, the literature review aims to define how Industry 4.0 influenced operations digitization and indirectly the human behaviors.

Hermann et al. [2] define Industry 4.0 as "a collective term for technologies and concepts of value chain organization." Industry 4.0 seems to develop more intelligent, flexible, and dynamic factories by equipping manufacturing processes with sensors, actors, and autonomous systems. In addition, manufacturing processes have the capacity of fulfilling more complex and qualified standards and requirements of products, making value-added integration occurs horizontally and vertically. Regarding the digitization of manufacturing process, Liao et al. [3] depicted research areas focusing on more human aspects: (1) Safety and Security: to ensure that production facilities and products themselves do not pose a danger either to people or to the environment; (2) Work Organization and Design: to implement of a socio-technical approach for work organization and design to offer workers the opportunity to enjoy greater responsibility and enhance their personal development. In line with these recommendations, conducting research on digitization of manufacturing processes require a socio-technical perspective.

Contrary to conventional belief, automation does not lead to less human interaction or workerless production facilities; but the workers' competencies requirements may evolve. The evolution of workers' role in their new industrial system is related to their ability to take quick and efficient decision on production, orders... through intelligently processed data. In this application, workers' role in Industry 4.0 is the one of a decision maker in production planning and control – with the help of a decision support assistance system – or of a smart and enhancers operator [4]. Moreover, Liao et al. [3] recommendations make it clear the relationship with the employee's responsibility and autonomy. This refers also to the work focusing on the relationship between Lean practices and digitization of manufacturing processes [5]. Nevertheless, it exists few articles that discuss the adaptation of Jidoka principles, the one related to employee's empowerment [5] or its further benefits. So, the tension between the standardization of work and the participation of employees in Industry 4.0 environment remains a crucial point in the design of work organization.

In the current organizations, execution of the operational tasks is generally documented in a very descriptive and detailed way, thus forming a standardized work, the first step to standardization. This standardized work enables processes to be consolidated in a consistent, accurate and repeatable way, in order to reduce their variability and simultaneously improve their performance. Maginnis [6] demonstrates that standardized work improves team productivity and team learning. Standardized work is not only a method of documentation, but it also allows the analysis of work situations. This denotes the relationship between standardized work and problem solving that jointly promotes incremental improvements. That is to say, looking at Industry 4.0, the interaction between standardization, continuous improvement and technology enabling team learning and coordination needs to be explored. Research provides empirical evidence that developing employee participation regarding their standardized work, fostering learning and knowledge sharing at an organization level can obtain greater benefits from technologies adoption. According to this, our study focuses on human behaviors at an operational level.

At the end of the literature review, the emergent questioning is formulated as follows: when a technological or organizational innovation occurs, what is its impact from a technical and behavioral point of view on the operational performance?

## 3   Methodology

The following methodology respects part of the research agenda proposed by Liao et al. [3]. Previous research looking at the impact of innovations focused solely on performance or solely on human behaviors. These same researches display either qualitative or quantitative methodological positioning but rarely mixed. This study, combining quantitative and qualitative data, is therefore built on an experiment including a series of trial / error phases. The experimental environment is that of a learning factory, a dynamometer lamp assembly line with a student audience to reduce bias related to the experiences of professionals [6].

The observational platform is built on a training platform named DynEO, which offers trainings in operations management. The experiment was built on the existing environment while integrating new technologies and digital instruments to measure industrial and human performance. Direct observations on operators' behaviors helped in the analysis of the influence of changing practices. The observational lab includes non-invasive tools to reduce unwanted influence on behavior (biased observations). The lab is also flexible, easily adaptable to various circumstances for testing a wide variety of hypotheses. During the preliminary test, the following measurement equipment were used: Tobii Eye Tracker, FaceReader, the Observer.

After the preliminary test, eye-tracking glasses were chosen because it helps studying how people interact with their environment, what catches their attention, and what influences decision making. This technology records gaze, saccades and fixation data of workers, and thus enables insights about the thinking process. For the case study, saccades represented an indicator for the visual search (stress level) and fixation represents an indicator of cognitive processing (concentration level).

During the preliminary test, the efficiency between operators (experienced vs not-experienced) in the assembly of a lamp was differentiated. Both workers wore the eye tracking glasses, and their faces were filmed for real-time analysis of facial expressions. A camera was placed in front of the face of the operator, whereas a first scene camera recorded the global area of the workplace. A second scene camera integrated in the portable eye tracker recorded in detail the direction of the gaze of the participant. It was decided to manually code in The Observer the video from the Tobii glasses. Behaviors coded were 'task' (start, stop) and 'Regions of Interest' (main body, other components, tools, elsewhere).

The preliminary test helped us design the first experiment (Fig. 1). The experimental protocol was defined in conjunction with the performance and behavior factors to be measured. In our study on the use of standardized work, the differentiation between the handwritten standards (paper standard) used until now and the introduction of a technological innovation is materialized by the use of an interactive tablet on which the standard is available (digital standard). Experimentation related to the introduction of this digital standard was initially carried out on reference groups and will be extended over several test groups to ensure the validity and potential generalization of the results.

**Fig. 1.** Preliminary test and raw data.

## 4   Results

The preliminary test showed us that the students are using standardized work (SW) during their training and are mostly looking at visual components: 96% of their time is spend on looking at the visual/images included in the standard (Table 1). However, the comparison between the measured value of emotions and the one verbalized by students showed that measured value itself is not relevant. The variability of measured emotions throughout the preliminary test can be related to the one verbalized. So, the emotion variability can represent an indicator of work's impact on the student. Emotions will not be measured in the following experiment.

The results of the experimentation can be summarized as follows: in general, students use the standard support (paper and digital) during the training phase and sporadically during the second phase (quality). During the third phase (performance), students perform the task entirely from memory without consulting the support, which can have an impact on the sequence followed and the resulting quality (Table 2). The observations showed that students following the paper support come back to it when necessary and produce better lamps in terms of quality requirements. Conversely, students following the digital support have difficulty in delivering flawless lamps. The paper support seems to help identify quickly quality issues.

**Table 1.** Comparative results from the preliminary test.

| Operator 1 | Operator 2 |
|---|---|
| Experimented | Beginner |
| Short execution | Long execution |
| Stable emotions | Variable Emotions |
| High concentration on the body | High concentration on components |
| Stable Heart Rate | Variable Heart Rate |

Student's perception was gathered throughout a survey: all consider that the tasks require a strong and permanent concentration. Students prefer the paper standard which gives them more flexibility. They showed less signs of incomprehension and seemed more focused. Depending on the standard used, a variation in the perception of the necessary time to achieve the task was noticeable (too short for those who use the digital standard). Those who used the digital standard questioned the given autonomy when they identified a problem to solve. If they identify a situation not foreseen by the standard, they find themselves stuck and call for outside help.

**Table 2.** Comparative data from paper standard and the digital standard utilization.

| Phase 1 (30') | | | | | Ph. 2 (30') | | Ph. 3 (30') | |
|---|---|---|---|---|---|---|---|---|
| | Time on support (t) (s) | Nb of saccades (s) | Nb of fixations (f) | Quality (Q) | t | Q | t | Q |
| Paper support | 31% $\sigma = 6,1$ | 37 $\sigma = 3,5$ | 15 $\sigma = 1,5$ | 50% $\sigma = 8,7$ | 4% $\sigma = 1,8$ | 92% $\sigma = 6,7$ | 3% $\sigma = 1,4$ | 89% $\sigma = 10,5$ |
| Digital support | 25% $\sigma = 3,5$ | 21 $\sigma = 3$ | 10 $\sigma = 1,7$ | 20% $\sigma = 18$ | 7% $\sigma = 2,7$ | 32% $\sigma = 17$ | 0% $\sigma = 0$ | 40% $\sigma = 21,9$ |

In the end, these three phases enabled us to observe a real difference in training, and therefore in performance according to the standards used. Because of its speed constraints, the digital standard does not attract students' attention during the key phases of learning. Conversely, the paper standard ensures a quality training, but slower. Finally, the paper standard is open to more criticisms or way to improve it which seems to foster student creativity while improving knowledge of the tasks performed. A next step would be to mix the utilization of digital and paper support according to the training and executing phases.

This experiment had some limitations: the sample was homogenous but small (about twenty students). Some of the data should be used with caution as some are based on subjective judgments (video coding, perception evaluated by a survey), on the capacity of each individual to follow a standard and the associated reactions (even if data were triangulated to limit the subjectivity). The results can vary greatly from one individual to another. It will be necessary to repeat the experiment on a larger sample. In addition, the measurement of emotions through image processing software can provide a better exhaustiveness of human behaviors interpretations.

## 5  Conclusion

This study serves as a first step to a larger research project about productivity measurement of Industry 4.0 transformation. The results showed promising data at the crossroad between performance and human behaviors. There are currently not a lot of studies about eye-tracking glasses technology applied in manufacturing environment, but the benefits of this technology are becoming a stimulating topic for researchers. A question remains unanswered: is the digital standardization about automate the management of work or automate the work itself?

## References

1. The Boston Consulting Group: "The Most Innovative Companies 2018 : Innovators go all in on digital" (2018)
2. Hermann, M., Pentek, T., Otto, B.: Design principles for Industry 4.0 scenarios. In: Proceedings of the 49th Hawaii International Conference on System Sciences, pp. 3928–3937 (2016)
3. Liao, Y., Deschamps, F., de Loures, E.F.R., Ramos, L.F.P.: Past, present and future of Industry 4.0 - a systematic literature review and research agenda proposal. Int. J. Prod. Res. **55**(12), 3609–3629, June 2017
4. Longo, F., Nicoletti, L., Padovano, A.: Smart operators in industry 4.0: a human-centered approach to enhance operators' capabilities and competencies within the new smart factory context. Comput. Ind. Eng. **113** (2017)
5. Rosin, F., Forget, P., Lamouri, S., Pellerin, R.: Impacts of Industry 4.0 technologies on Lean principles. Int. J. Prod. Res. **58**, 18 (2019)
6. Maginnis, M.A.: The impact of standardization and systematic problem solving on team member learning and its implications for developing sustainable continuous improvement capabilities. J. Enterprise Transform. **3**(3), 187–210 (2013)

# Enhanced Manufacturing Execution System "MES" Through a Smart Vision System

Chawki El Zant[1(✉)], Quentin Charrier[1], Khaled Benfriha[1], and Patrick Le Men[2]

[1] Arts et Metiers Institute of Technology, LCPI, HESAM Universite, 75013 Paris, France
chawki.el_zant@ensam.eu
[2] CFA Ingénieurs 2000, 77144 Montévrain, France

**Abstract.** The level of industrial performance is a vital issue for any company wishing to develop and acquire more market share. This article presents a novel approach to integrate intelligent visual inspection into "MES" control systems in order to gain performance. The idea is to adapt an intelligent image processing system via in-situ cameras to monitor the production system. The images are thus analyzed in real time via machine learning interpreting the visualized scene and interacting with some features of the MES system, such as maintenance, quality control, security, operations, etc. This novel technological brick, combined with the flexibility of production, contributes to optimizing the system in terms of autonomy and responsiveness to detect anomalies, already encountered, or even new ones. This smart visual inspection system is considered as a Cyber Physical System CPS brick integrated to the manufacturing system which will be considered an edge computing node in the final architecture of the platform. This smart CPS represents the 1st level of calculation and analysis in real time due to embedded intelligence. Cloud computing will be a perspective for us, which will represent the 2nd level of computation, in deferred time, in order to analyze the new anomalies encountered and identify potential solutions to integrate into MES. Ultimately, this approach strengthens the robustness of the control systems and increases the overall performance of industrial production.

**Keywords:** Artificial intelligence · Deep learning DL · Industry 4.0 · Machine learning ML · Smart Manufacturing · Automated Visual Inspection

## 1 Introduction

In the context of Industry 4.0, manufacturing systems are updated to an intelligent level. Intelligent or smart manufacturing aims to convert data acquired across the product lifecycle into manufacturing intelligence in order to yield positive impacts on all aspects of manufacturing [1]. It enables all physical processes and information flows to be available when and where they are needed across holistic manufacturing supply chains, multiple industries, Small and Medium-sized Enterprises SMEs, and large companies. Intelligent manufacturing requires certain underpinning technologies in order to enable devices or machines to vary their behaviors in response to different situations and requirements

L. Roucoules et al. (Eds.): JCM 2020, LNME, pp. 329–334, 2021.
https://doi.org/10.1007/978-3-030-70566-4_52

based on past experiences and learning capacities. These technologies enable direct communication with manufacturing systems, thereby allowing problems to be solved and adaptive decisions to be made in a timely fashion. Some technologies also have artificial intelligence AI, which allows manufacturing systems to learn from experiences in order to ultimately realize a connected, intelligent, and ubiquitous industrial practice [2].

More specifically, Industry 4.0 is a complex combination of innovative functionalities such as Intelligence, Prediction, Interoperability, and Modularity, assured and well implemented to the manufacturing systems through the integration of a bunch of different new technologies and techniques offered by the 4.0 as CPS, Advanced Robotics, IoT, and AI, in order to enhance the production performance, assure the security means of workers and machines, improve product quality, and develop smart products. Therefore, industrials must deal first with a big amount of data generated through the product lifecycle including visual inspection images and quality control processing [3]. Secondly, Manufacturing systems must be able to integrate such flexible and AI-Enabled modules in as less as possible intrusive way.

In this article, we propose a novel approach which improves the MES system with a visual inspection system having an intelligence property since it uses Artificial Intelligence AI/Machine Learning ML in order to optimize the production process. The new model proposed in this paper can be used as a reference for different applications to enhance several functionalities of manufacturing processes, such as maintenance, operations, Quality Control, and security.

## 2 State of the Art

It is widely accepted that Information and Communication Technologies ICT are essential to improve production systems. The successful integration of Industry 4.0 into the existing production systems necessitates the deployment of the Industrial Automation pyramid (including PLC, MES, ERP, and other key technologies) in production, or the Product Lifecycle Management supported by advanced CAD/CAM tools [4]. Otherwise, Industry 4.0 requires a new breed of MES that embraces the flexibility, agility, intelligence, and connectivity of the Smart Factory. To meet these needs, the way an MES operates needs to change to accept a comprehensive, decentralized way of operating throughout the plant and beyond. This requires new levels of connectivity and new ways of approaching how users interface with equipment [5]. The dynamics of shop-floor operations with smart materials and smart products are probably one of the biggest challenges for MES systems. Flexibility is enabled in a manufacturing system mainly through the integration of flexible machines and the implementation of flexible and alternative production routes [6]. In addition, this flexibility could not be achieved without having a modular and advanced shop floor control system SFCS. Having such an advanced and flexible manufacturing system, the opportunity to implement the intelligence aspects into the system is hence possible.

Moreover, Industry 4.0 is a complex combination of innovative functionalities such as Intelligence, Prediction, Interoperability and Modularity, assured and well implemented to the manufacturing systems through the integration of a combination of different new technologies offered by the 4.0 as CPS, Advanced Robotics, IoT, and AI, in order to enhance the production performance, assure the security means of workers and machines, improve the product quality, and develop smart products [2]. According to [7], AI is a cognitive science with rich research activities in the areas of image processing, natural language processing, robotics, and machine learning. Robotic systems are playing an interesting role in enabling Industry 4.0 implementation. Modern robots are characterized as systems offering autonomy, flexibility, cooperation, and reconfigurability [8]. A layer of flexibility and intelligence is offered through the integration of machine learning techniques to the image processing stage, and the interconnection between robots and machines which extend the decentralization aspect, and assure several benefits such as lower defect rate, higher quality and reliability [9, 10], and improved security.

## 3  Smart Visual Inspection System

The model given below in Fig. 1, presents the contribution of this work which figures in the successful implementation of a Machine Learning ML algorithm into an industrial production system, and the effectiveness of using ML to assure in real time visual inspection and rapid decision-making process. Thanks to the interoperability and modularity assured by the MES and the system architecture, the intelligence and the self-adoption of the manufacturing system is possible by adding smart CPSs and smart modules using machine learning and deep learning algorithms and techniques. The ML system runs in a specific machine and reports to the MES the processed data through the PLC using snap7 communication protocol. The MES takes into consideration the received signals and information and acts properly by modifying the initial program and upgrading it as it should. This smart system could play a role in being an edge computing module, which reports unexpected and untrained data to the cloud in order to train the model

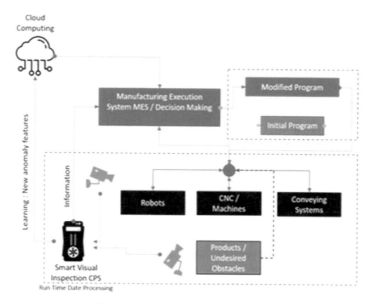

**Fig. 1.** Smart visual inspection model

for continuous data acquisition and analysis. This model can be used and developed in our ongoing project for several data processing tasks, providing valuable information regarding the predictive maintenance task, the quality control process, and the power consumption optimization.

## 4  Use Case

Our research is tested on an industrial production platform 4.0 developed and installed in the research laboratory Fig. 2. The aim of the platform 4.0 project is to create an intelligent and flexible manufacturing process and digital twins by integrating several technologies. Basically, this platform is built from different machines and robots: one fixed Kuka robot, one mobile Kuka robot, a Cognex camera system for raw material identification and classification, a lathe, a milling machine, and a laser cutting machine. At the very beginning, the platform can manufacture, assemble and transfer components with a closed and not-flexible automatic control, resulting in the production of a single product. This mode of operation corresponds to the conventional mass production mode and is subject to uncontrolled disturbances.

The development has focused on adding a flexible layer into this production line by enabling it to produce several product and following different manufacturing cycle in a rapid configuration. The platform is now operated through Wonderware System Platform WSP – Schneider product [11]. The production cycle is not anymore limited due to the high level of PLC intervention. The platform is controlled by the WSP, where alarms, trends, and all data from machines and sensors are collected. The production line, thanks to developing this flexible layer, is able to produce different personalized products (in the limit of machines capacity) and to follow different manufacturing plan, without the need of changing or modifying the manufacturing programs.

**Fig. 2.** Initial program running, no person detection

To test our model, we chose a security application, since we are using two robots in our line Fig. 2, we should not have any human presence inside the production line when it's functioning to avoid any accidents. So, to ensure security, the platform has a physical barrier and an access door that must be closed, preventing anyone from entering the interior when it is operating. However, if the door is closed and a person was present in the production facility or comes in through another unsecured access, no error is reported, and the person can be then in danger. To overcome this problem, we decided to conduct a smart visual inspection system which will detect the presence of the human within the production line using video streams. If a person is detected, the information will then be forwarded to the MES which will pause or reduce the speed of the robots and which will alert the operator who will be able to raise the alert after confirmation that no one is present in the line. The video streams are then fed back to this algorithm which will carry out the detection, and in case of detection of someone present will send the information back to the MES which will pass the line in safety mode Fig. 3 and will warn an operator who will have to raise this alert once the risk has passed.

**Fig. 3.** Program paused, person detection inside the platform

## 5   Conclusion

In this paper, we proposed a novel model for a smart visual system well integrated into the manufacturing execution system well implemented to improve the production system. This model is a concrete contribution that can be used for several applications and cases and assure the optimization of the production process and control. A visual inspection system application was presented as a successful implementation module. This model gives us wide opportunities to develop different parallel modules for data and image processing and integrate them into the manufacturing system in order to enhance and well optimize the whole production process.

## References

1. Tao, F., Qi, Q., Liu, A., Kusiak, A.: Data-driven smart manufacturing. J. Manuf. Syst. **48**, 157–169 (2018). https://doi.org/10.1016/j.jmsy.2018.01.006s
2. Zhong, R.Y., Xu, X., Klotz, E., Newman, S.T.: Intelligent manufacturing in the context of Industry 4.0: a review. Engineering **3**(5), 616–630 (2017). https://doi.org/10.1016/J.ENG. 2017.05.015
3. Bajic, B., Cosic, I., Lazarevic, M., Sremcev, N., Rikalovic, A.: Machine Learning Techniques for Smart Manufacturing: Applications and Challenges in Industry 4.0, p. 11
4. Posada, J., et al.: Visual computing as a key enabling technology for Industrie 4.0 and industrial internet. IEEE Comput. Grap. Appl. **35**(2), 26–40 (2015). https://doi.org/10.1109/MCG.201 5.45

5. Preparing Manufacturing Systems for Industry 4.0, 05 Mars 2019. https://www.connectivity 4ir.co.uk/article/167720/Preparing-Manufacturing-Systems-for-Industry-4-0.aspx (consulté le 31 janv. 2020)
6. Singh, A., Singh, D.J., Ali, D.M.: Some control strategies in a flexible manufacturing system-a simulation perspective. **13**(7), 8 (2018)
7. Lee, J., Davari, H., Singh, J., Pandhare, V.: Industrial Artificial Intelligence for industry 4.0-based manufacturing systems. Manuf. Lett. **18**, 20–23 (2018). https://doi.org/10.1016/j.mfg let.2018.09.002
8. Kamble, S.S., Gunasekaran, A., Gawankar, S.A.: Sustainable industry 4.0 framework: a systematic literature review identifying the current trends and future perspectives. Process Saf. Environ. Protect. **117**, 408–425 (2018). https://doi.org/10.1016/j.psep.2018.05.009
9. Ghobakhloo, M.: The future of manufacturing industry: a strategic roadmap toward Industry 4.0. J. Manuf. Tech. Manag. **29**(6), 910–936 (2018). https://doi.org/10.1108/JMTM-02-2018-0057
10. Gupta, S.: Artificial intelligence for enterprise networks. In: Artificial Intelligence for Autonomous Networks, 1 édn. Chapman and Hall/CRC, pp. 263–284 (2018)
11. Wonderware System Platform - Fondation de l'Industrie du Futur, Wonderware. https://www.wonderware.fr/produit/supervision-et-controle/ihm-supervision-et-controle/system-platform-2. consulté le 05 mai 2020

# Design Rules Application in Manufacturing Industries: A State of the Art Survey and Proposal of a Context-Aware Approach

Armand Huet[1]([✉]), Romain Pinquie[2], Philippe Veron[3], Frederic Segonds[4],
and Victor Fau[5]

[1] Arts et Metiers Institute of Technology, LCPI, HESAM Universite, 75013 Paris, France
armand.huet@ensam.eu
[2] Univ. Grenoble Alpes, CNRS, Grenoble INP, G-SCOP, Grenoble, France
[3] Arts et Metiers Institute of Technology, LISPEN, HESAM Université,
13617 Aix-en-Provence, France
[4] Arts et Metiers Institute of Technology, LISPEN, HESAM Université, 75013 Paris, France
[5] Capgemini DEMS, Toulouse, France

**Abstract.** [Context] In manufacturing industries, the design of a product needs to comply with many design rules. These rules are essentials as they help industrial designers to create high quality design in an efficient way. [Problem] However, the management of an ever-increasing number of design rules becomes a real problem, especially for new designers. Even if there exists some knowledge management tools for design rules, their capabilities are still limited and many companies continue to store their design rules in unstructured documents. Nowadays, design rule application is still a difficult task that needs a circular validation process between many expert services in a manufacturing company. [Proposition] In this paper, we will analyze the main existing approaches for design rules application from which we will demonstrate the need of a new approach to improve the current state-of-the-art practices. To minimize rule application impact on the design process, we propose to develop a Context-Aware Design Assistant that will perform design rule recommendation on the fly while designing using computer-aided technologies. Our Design Assistant relies on the modelling of the design rules and the design context in a single knowledge graph that can fuel a recommendation engine. [Future Work] In future work, we will describe the technical structure of the Context-Aware Design Assistant and develop it. The potential outcome of this research are: a better workflow integration of design rules application, a proactive verification of design solutions, a continuous learning of design rules, the detection and automation of design routines.

**Keywords:** Design rule · Product design · Knowledge graph · Context awareness · Cognitive assistant

## 1 Introduction

Designing a product is a knowledge-intensive activity. Thus, to prevent design errors, i.e., choices that make certain designs "not allowed" or inappropriate for their intended

© The Author(s) 2021
L. Roucoules et al. (Eds.): JCM 2020, LNME, pp. 335–340, 2021.
https://doi.org/10.1007/978-3-030-70566-4_53

use, design departments prescribe design rules. A design rule is a prescriptive statement – often an unstructured blend of text and graphical objects (equation, sketch, etc.) – aiming at assisting deployed designers for the achievement of a proof design, in compliance with best practices, applicable regulations, and DfX constraints. Many factors like the increasing complexity of design products or the number of norms per industrial domain tend to increase the number of designer rules an industrial designer has to comply with. The ever-increasing number of design rules leads to critical issues when industrial companies want to use and manage such "Big Data".

While a designer is working on a product, he or she often needs to check design rules if he or she wants to provide a design free of errors. An expert designer may have memorized all crucial rules of his domain. A non-expert designer does not have in mind all applicable design rules. He/she has no other alternative than to spend a large amount of time finding the subset of design rules that matches his design context by either searching in design manuals or asking expert-colleagues to guide him/her. This task is even more time consuming because design rules often come from different domains and are stored in documents or databases that may not be correctly organized. Unstructured documents are still common in the industry [1]. Many companies still store design rule knowledge in unstructured documents – mainly PDF format – which are over tens or hundreds of pages.

As we will see in the next sections, there already exists many industrial tools and researchers' proposals for managing design rules. For each approach, several researches have been performed on Google Scholar [2] as well as ScienceDirect [3] between the 12/17/2019 and the 12/20/2019. From the obtained results, we selected two types of contributions: Conceptual work to present a definition of the concept or approach, and application work in the domain of industrial design to illustrate pros and cons of the approach.

We will firstly present a design rule definition and classification (Sect. 2). Then, we will review the existing tools supporting the application of design rules (Sect. 3). Finally, we will focus on technologies that provide context awareness capabilities and briefly discuss how our Context-Aware Design Assistant would contribute to the current state-of-the-art solutions.

## 2   Design Rules

For this paper, design rule definition is based on the work of Fu et al. [4] who propose a formal definition of a Prescriptive Design Principle:

It is stated in the grammatical imperative form.
It includes a prescriptive action for a designer to take.
It increases the likelihood of reaching a desirable consequence.
It is situated within a particular context and point in time.

In this paper, we will reduce the scale of this definition and consider a more technical definition. Calkins et al. [5] in the field of Knowledge Based Engineering (KBE) consider that design rules synthesize the knowledge of a company and indicate how it should be used to create a proven design. In our case, we also want to put the emphasis on how this knowledge is stored in databases or in design books. After reviewing numerous design manuals, we assume that a design rule is a prescriptive statement in natural language, often associated with descriptive graphical objects (equation, table, chart, etc.) guiding the work of the designer in a specific design context to improve design quality.

Authors previously cited both emphasis on the heterogeneous aspect of design rules. Calkins et al. consider four different origins possible for design rules: Heuristic (rules from companies' best practices), Empirical (based on experimental data), Legislated constraints (established by official norms), Physic (based on analytical or numerical physical models). Design rules affect not only the design but also all the phase of a product lifecycle, such as product recyclability [6] or maintainability [7].

We conclude that design rules have various origins and objectives and that each one relates to a specific design context. They are essential for maintaining the quality level of industrial products and their number is significantly increasing.

## 3   State of the Art of Design Rule Application Tools

Most industrial design rule checkers [8–10] adopt procedural approach. They have built a set of rules for each possible rule origin. For example, a tool may have a set of a few dozen rules to check on a part machined using a milling process. Each rule is an algorithm that detects geometrical features in the CAD data that do not respect the rule. Bo Huang et al. [11] give a good example of the work realized in this field. Such a rule application strategy results in expert systems that have a pre-defined number of rules focusing on a very specific problem. They are efficient in detecting CAD quality errors and rules directly associated with a design issue like a manufacturing process. However, this approach has some flaws:

- Rule editing requires the work of experts in order to modify or create algorithms. As explained previously, a design rule is due to evolves with technology or even company habits. It is not trivial to maintain or personalize the design rule set with this technology.
- Design rule storage across multiple domain silos makes multi-domain rules application complicated. Design is a multi-domain process and many design rules do not fit in the pre-defined contexts fixed by those tools.
- Some rules represent unstructured knowledge that cannot be translated into algorithms. Traditional rule checking tools cannot process these rules but the designer still need to know and consider them for his design.

To improve design rule application, scientists explored declarative approach of design rule application. They use new technologies such as Semantic Networks (SN) to represent the knowledge associated with the product and the design rules. According to JF Sowa [12]: "A semantic network or net is a graph structure for representing knowledge in patterns of interconnected nodes and arcs." They are primarily used to represent conceptual knowledge as natural language expressions in a structured and unbiased way. This process is not straightforward for design rules but many scientists work to improve design rules representation in SNs like S. Kang et al. [13]. Using a SN could allow us to process all kind of design rules.

Various teams experimented SNs for design rule application in the manufacturing industry [14–16]. Their strategy is to build a data model to represent the type of product they focus on. Once this model is built, they implement a translation algorithm to automatically generate a semantic representation from the digital mock-up. Then, they can edit rules that will apply on the knowledge representation of the product. They are able to identify design errors as well as automatically correct some of them. In this method, design rules are not algorithms that will explore the geometry. They are descriptions of what the semantic representation of the product should be. This makes design rules easier to implement. Some teams even develop their own high-level language to facilitate new design rules editing.

However, this approach is not flawless. Each type of industrial product needs his own semantic representation to perform well with this approach. These representations are hard to develop and maintain [16] and are domain specific. Therefore, only specialized industrial domains can benefit from this approach. An industrial part often go through multiples manufacturing processes, needs to comply with assembly rules and has mechanical constraints. Creating the semantic representation of this part and the set of associated rules would be extremely difficult with the previous approach.

## 4   Introducing a Context-Aware Cognitive Design Assistant

In [17], A. K. Dey describes the context of a software user as: "Any information that can be used to characterize the situation of an entity. An entity is a person, place, or object that is considered relevant to the interaction between a user and an application, including the user and applications themselves." In our case, any information describing the design process, the digital mock-up or the team members in charge of the design, is part of the design context.

A. K. Dey gives a definition of a Context-Aware System (CAS): "A system is context-aware if it uses context to provide relevant information and/or services to the user, where relevancy depends on the user's task." Therefore, a context-aware application for rule application would have to provide rule information and design guidance when the user needs it to progress without mistakes in his design task. As context knowledge is multi-domain, such a tool would have a greater ability to process multi-domain design rules.

CAS are by definition user centered. Each user will have a different context and different needs in different times. The CAS will provide a suitable service for each case. This is crucial in industrial context where team members need to coordinate their work on a complex task. For example, M. A. Dhuieb et al. propose a context-aware architecture to present manufacturing knowledge to workers in factories [18]. Design context can be compared to this one as each designer has a specific task and level of expertise, but need to work with his/her team. Therefore, we propose to use context awareness in a rule application tool to improve multi-domain rules recommendation as well as the designer's productivity, which are the main flaws of state-of-the-art rule application tools previously presented.

On a practical aspect, context knowledge is represented in state-of-the-art CAS by a Semantic Network. The context graph is complex and evolves dynamically. A structural data model is essential to run context aware applications. J. Aguilar et al. propose a general data model for context modeling [19]. In our domain of application, R. Pinquie et al. [20] propose a data model to represent the user's design context in a context aware design assistant. This assistant would perform design rule recommendation on the fly to improve user's productivity. We propose to develop a proof of concept of such an assistant to demonstrate the efficiency of context aware technology for design rule application.

## 5   Conclusion and Future Work

The goal of this paper is to review the state-of-the-art of design rules and design rules application to demonstrate the need for a context-aware design assistant.

First, we discussed about the concept of design rules in the manufacturing industry. We demonstrated that the use of design rules is a crucial issue to improve design efficiency.

Then, we reviewed existing tools for managing rules. Tools based on a semantic network enable designers to better detect design flaws and ease the edition of rules. We raised the issue that all these product-centered tools failed to process efficiently multi-domain rules, which are very common in design activity. Moreover, semantic representations of industrial products are fastidious to produce and limited to a single domain.

Finally, we presented CAS and how they could be used to improve multi-domain rule application as well as user specific rule recommendation. This is why we propose in a future work to develop a Proof of Concept of this approach with an industrial set of data.

## References

1. Kassner, L., Gröger, C., Mitschang, B., Westkämper, E.: Product life cycle analytics – next generation data analytics on structured and unstructured data. Procedia CIRP **33**, 35–40 (2015). ISSN 2212-8271
2. Google Scholar. https://scholar.google.com/
3. ScienceDirect. https://www.sciencedirect.com/
4. Fu, K.K., Yang, M.C., Wood, K.L.: Design principles: literature review, analysis, and future directions. ASME. J. Mech. Des. **138**(10), 101103 (2016). https://doi.org/10.1115/1.4034105
5. Calkins, D.E.: Knowledge-based engineering (KBE) design methodology at the undergraduate and graduate levels. Int. J. Eng. Educ. **16** (2000)
6. Bralla, J.: Design for Manufacturability Handbook, p. 1368, McGraw-Hill (1998)
7. Pahl, G., Beitz, W.: Engineering Design: A Systematic Approach (1996)
8. DFMPRO. https://dfmpro.geometricglobal.com/
9. Siemens NX Checkmate. https://www.plm.automation.siemens.com/en_us/Images/2504_t cm1023-11882.pdf
10. Dewhurst, B.: DFMA. https://www.dfma.com

11. Huang, B., Xu, C., Huang, R., Zhang, S.: An automatic 3D CAD model errors detection method of aircraft structural part for NC machining. J. Comput. Des. Eng. **2**(4), 253–260 (2015)
12. Sowa, J.F.: Semantic Networks. www.jfsowa.com/pubs/semnet.html
13. Kang, S., Patil, F., Rangarajan, A., Moitra, A., Jia, T., Robinson, D., Dutta, D.: Extraction of manufacturing rules from unstructured text using a semantic framework. In: Proceedings of ASME 2015 International Design Engineering Technical Conferences and Computers and Information in Engineering Conference, pp. V01BT02A033 (2015)
14. Garcia, L.E.R., Garcia, A., Bateman, J.: An ontology-based feature recognition and design rule checker for engineering. In: Proceeedings of 10th International Semantic Web Conference, pp. 48–58 (2011)
15. Moitra, A., Palla, R., Rangarajan, A.: Automated capture and execution of manufacturability rules using inductive logic programming. In: Proceedings of the Twenty-Eighth AAAI Conference on Innovative Applications (IAAI 2016), pp. 4028–4034 (2016)
16. Fortineau, V., Fiorentini, X., Paviot, T., et al.: Expressing formal rules within ontology-based models using SWRL: an application to the nuclear industry. Int. J. Prod. Lifecycle Manag. **7**(1), 75–93 (2014)
17. Dey, A.: Understanding and using context. Pers. Ubiquitous Comput. **5**(1), 4–7 (2001)
18. Dhuieb, M.A., Laroche, F., Bernard, A.: Context-awareness: a key enabler for ubiquitous access to manufacturing knowledge. Procedia CIRP **41**, 484–489 (2016)
19. Aguilar, J., Jerez, M., Rodríguez, T.: CAMeOnto: context awareness meta ontology modeling. Appl. Comput. Inf. **14**(2), 202–213 (2018)
20. Pinquié, R., Véron, P., Segonds, F., Zynda, T.: A property graph data model for a context-aware design assistant. In: IFIP International Conference on Product Lifecycle Management, pp. 181–190. Springer, Cham (2019). https://doi.org/10.1007/978-3-030-42250-9_17

# Semantic Enrichment of 3D Models Based on Ontology Integration

Stéphane Nzetchou[1]([✉]), Alexandre Durupt[1], Benoit Eynard[1], and Sébastien Remy[2]

[1] Université de Technologie de Compiègne, Laboratoire Roberval FRE CNRS 2012, Rue du Dr Schweitzer, CS 60319, 60203 Compiègne Cedex, France
stephane.nzetchou@utc.fr
[2] Université de Technologie de Troyes, Institut Charles Delaunay - LASMIS FRE CNRS 2848, 12 rue Marie Curie, 10010 Troyes Cedex, France

**Abstract.** The rise of new technologies has led to a growth in the number of 3D models. They can come from various source, hence they are heterogeneous and complex. The level of 3D data access is often a function of the user's expertise since the 3D data are often registered to different file formats. Some file formats do not show the data tree, as IGES. For using information inside a 3D model, that does not show a data tree, each company adopts his own system that will allow him to access easily to 3D model in order to exploit the hidden knowledge within the models. In this article, we are going to speak about technologies that helps user to exploit and knowledge coming from different file formats. In addition, we are going to present a system named VAQUERO that uses ontology to access, store and share knowledge coming from 3D models.

**Keywords:** CAD · File format · Semantic representation · Ontology · PLM

## 1 Introduction

A digital continuity environment, as product data management (PDM) systems or product lifecycle management (PLM) systems have been introduced in recent years to help engineers manage the development of a product throughout its lifecycle, from the design phase to recycling [1]. This environment integrates all the company's departmental functions, structures the produced data and organizes design activities. It allows to manage 3D models information such as geometry, topology, metadata, specifications, analysis and simulation results, etc. [2].

Heterogeneity of 3D models: legacy formats (CATIA, CREO, NX, etc.) and neutral formats (STEP, IGES, JT, DXF, etc.), and their complexities during the development phases create an increasing need for knowledge and expertise derived from the 3D models and their simulation results. Links and dependencies, which appear between heterogeneous 3D models, become increasingly complex in the daily activities [3]. The legacy nature of data, specific to CAD vendors, limits the possibility of open-ended analysis of 3D model, sharing 3D data, modify or revise 3D model, sometimes the retrieval of the 3D model or any information related to the design intent.

© The Author(s) 2021
L. Roucoules et al. (Eds.): JCM 2020, LNME, pp. 341–346, 2021.
https://doi.org/10.1007/978-3-030-70566-4_54

One of the main problems faced by users of these systems is the lack of compatibility between systems. As a result, 3D models created using a particular software package may become inaccessible to its creators once that software is obsoleted. Incompatibilities of all kinds and at all levels impede the sharing of information and prevent the exchange of services between different systems. And developing interoperability means developing knowledge and solutions to eliminate incompatibilities [3]. Hence, we propose a conceptual system named VAQUERO capable of ensuring a better coverage of the product life cycle phases, by integrating additional information allowing improving and upgrading the low-level file formats such as STL, IGES, etc. in which the data were initially stored.

## 2   State of the Art: Management of 3D Data

Data and knowledge play an important role in the long-term sustainability and success of organization. The need for processes that facilitate the creation, sharing and leveraging of individual and collective knowledge has emerged for this reason. Knowledge Management has been introduced as one of the major activities for providing efficient and intelligent digital support as well as decision makers, to access any data required, across application borders [4].

### 2.1   Standards and Tools for Managing 3D Data

It is important for companies to provide one common standard for the integration and sharing of product data between various computer systems. Using a single standard as interoperability solution requires an effort by the users to understand the concepts and the methods of implementation. Therefore, the standards must be adapted to all users in a consensual manner, while, it is not impossible [5].

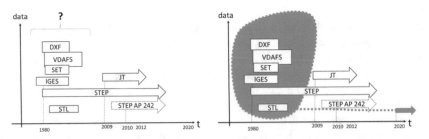

**Fig. 1.** a) Chronology of neutral file formats b) evolution and use of some formats.

In order to facilitate the flow of information in a heterogeneous digital continuity environment, implementing from several publishers' solutions, neutral file formats have been developed. Figure 1 shows a chronology of some neutral file formats that exist.

While it is true that 3D models created in a specific CAD system can always be accessed by the same system at different times, as far as data in neutral formats are concerned, this is still not the case. In a digital continuity environment, some user recommend file formats according to their needs and the CAD system they own. Thus, [6] has produced a work focusing on standards to be used in a PLM system. They highlight the choices to be made in order to determine which standard would be more appropriate or adequate for exchanging and sharing information in different digital continuity systems in order to reuse the data. Hence, [7] thought of using XML to convey information between different PLM systems. The standardization of conceptual and technical gateways between EXPRESS-STEP and UML-XML allowing the generation of efficient execution languages that specify the initial generic model and used in PLM systems. Following the STEP, JT, XML, etc. which are cited by some authors [8], for the extraction of geometric and topological information, attributes or properties, some platforms have been developed to manage the accumulation of data during the product design. We can find 3DEXPERIENCE from Dassault Systèmes, WINDCHILL from PTC and TEAM CENTER from Siemens that bring together some solutions from different companies and enable stakeholders involved in a project to work together without breaking the digital chain. These platforms offer to users intuitive applications to reuse existing parts instead of recreating parts, as they allow users to retrieve 3D models based on a reference part or keyword, make modifications, annotations and share them with other users of the same platform without losing time.

We have noticed based on Fig. 1 above, that the neutral file formats used from year 2009 are quite rich. That is why inside the digital continuity environment, apart from the legacy file format, most of the information needed were extracted from the rich neutral file formats such as STEP AP242, JT. The others file formats such as IGES, DXF, STL, etc. are not used (Fig. 1a). It seems that the data within the file formats produced before year 2009 are not used anymore. While these data can still been useful for starting a new project. It becomes necessary to find a way to use the data within these kind of low-level file formats (Fig. 1b).

## 2.2 Evolution of the Low-Level File Format

Based on the observation made in the previous section, where the so-called low-level file formats such as IGES or STL are no longer used in digital continuity environments, it appears necessary to identify the entire expert and business information extracted from rich files in order to upgrade and use the low-level files.

As a result, STEP AP 242 seems to be the most up-to-date and semantically rich file format today, allowing us to know what important information to extract and integrate. STEP AP 242 provides all the functionalities covered by the most commonly implemented and used Aps [8]. It additionally defines new structures for 3D parametric and geometric constraints design; geometric dimensioning and tolerance (GD&T); business object model; tessellation; kinematics; etc. The intent of STEP AP 242 is to support a manufacturing enterprise with a range of standardized information models that flow through a long and wide digital thread that makes the manufacturing systems in the enterprise smart.

Hence, the search for the granularity of the information to be integrated and the definition of all the features of the STEP AP242 and to characterize them with those of the low-level formats finally to carry out mappings and to make these files evolve.

Knowing the information to be integrated and the granularity of its information, a standardized ontology could be used to integrate all these information. We can quote for example OntoSTEP or ONTO-PDM [9], which would make it possible to integrate the information and to make evolve the low-level file formats into a STEP standard. For instance, by using ONTO-PDM that is more design oriented, not only ontology can evolve by increasing into other domains such as PMI (tolerance and annotation) etc., but also the semantics of the product by the features of the standard.

## 3   Semantic Enrichment of 3D Model: Proposal Vaquero

The needs of users for 3D models make it necessary to access and use the information contained within the file formats. However, as we have already said, the heterogeneity of the file formats make it almost impossible to access to all the information we want. The needs to have a rich file format that can contain the information related to geometry, topology, PMI and GD&T could help to process easily some tasks in company. That is why we propose a system named VAQUERO, which can help to integrate all the necessary information related to a 3D model in order to upgrade the file format.

For enriching the 3D model, Vaquero needs to access to the requested low-level file formats by using two types of queries. The queries are formulated based on a geometric descriptor or a semantic descriptor.

The user makes a query based on the geometric descriptor of the part that he would like to retrieve by using its shape. The geometrical descriptor could be one of those quoted in the literature as it can be found in the works of [10]. Once the result of the query is displayed like in Fig. 2a, the user selects the part to retrieve among the possible parts contained in the low-level file format.

After selecting, the enrichment of the file format is characterized by the treatment that the user will make. For instance, by identifying parts, surfaces and give names to any element composing his part. In addition, by characterizing information related to the component, being geometrical or topological information or simulation results. Manufacturing or functional information related to a specific component of the product can be expressed. The link between a PMI and the reference geometrical entity (axis, plane, surface, etc.) of the 3D model can be represented, and the graphical presentation of annotations and tolerance is returned to the user, thus facilitating comprehension (see Fig. 2b).

All the above-mentioned information will be saved in a tree structure in the form of an ontology that will be visible in another window of our system as we can see in Fig. 2b. The used ontology is a standardized one as OntoSTEP or ONTO-PDM that will allow the component to evolve. For example, if the 3D model is represented by IGES standard, by describing it in OntoSTEP, it will use the STEP terminology and will evolve in the STEP standard. The geometric and topological information as well as the graphical PMI will be defined in the ontology in order to provide a structural and semantic definition of each element. The semantic PMIs will allow access to all the information that characterizes it (reference elements, parameters, etc.), and possibly to modify them.

The second input is based on a semantic descriptor. Once some information are stored inside the ontology, he can directly query the system to retrieve a particular information based on the concepts already registered within the ontology.

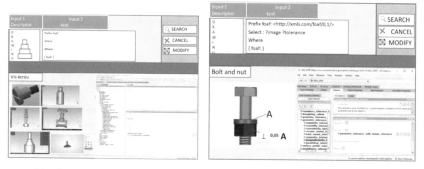

**Fig. 2.** a) Visualization result of a query; b) semantic annotation of the part.

Ontology, being considered itself as a database and knowledge base, the semantic definition and enrichment of the 3D model will allow an efficient and fast reutilization of the information contained within the enriched file format. The use of the ontology allows a quick annotation and modification of the information stored in the different classes. The fact that the ontology can be written in XML helps the stored information to be share.

## 4  Conclusion

The proposed system uses an ontology to store the information and semantically enrich 3D model coming from low-level file format such as IGES, STL, etc. with any kind of information. Ontology allows the definition of a domain and serves as bridge between different domains by identifying the different concepts that compose them and the links between them, their properties as well as axioms and rules concerning them. The information stored in a standardized ontology serves as a common and verified source of knowledge, that is used and which will be exploited in downstream processes, such as the realization of dimensional chains, dimensional control, etc. The granularity of the integrated information ensures the continuity through tools and solutions: Sharing information and services to maintain the semantic flow. It ensures the long-term sustainability of the information. We are thinking of interviewing some firms in the automotive industry to find out if there is still a need to look at this type of format in terms of its use.

## References

1. Penciuc, D., Durupt, A., Belkadi, F., Eynard, B., Rowson, H.: Towards a PLM interoperability for a collaborative design support system. Procedia CIRP **25**, 369–376 (2014)
2. Eynard, B., Gallet, T., Nowak, P., Roucoules, L.: UML based specifications of PDM product structure and workflow. Comput. Ind. **55**, 301–316 (2004)
3. Nzetchou, S., Durupt, A., Eynard, B., Remy, S.: Review of CAD visualization standards in PLM. In: IFIP 16th International Conference On Product Lifecycle Management. pp. 1–10 (2019)

4. Assouroko, I., Ducellier, G., Eynard, B., Boutinaud, P.: Knowledge management and reuse in collaborative product development – a semantic relationship management based approach. Int. J. Product Lifecycle Manag. 7(1), 54–74 (2014). https://doi.org/10.1504/IJPLM.2014.065460

5. Pirayesh, A., Etienne, A., Kleiner, M., Roucoules, L.: Performance evaluation of collaboration in the design process: using interoperability measurement. Comput. Ind. 72, 14–26 (2015). https://doi.org/10.1016/j.compind.2015.03.011

6. Rachuri, S., Subrahmanian, E., Bouras, A., Fenves, S.J., Foufou, S., Sriram, R.D.: Information sharing and exchange in the context of product lifecycle management: role of standards. Comput. Aided Des. 40, 789–800 (2008)

7. Taylor, P., Choi, S.S.: XML-based neutral file and PLM integrator for PPR information exchange between heterogeneous PLM systems. Int. J. Comput. Integr. Manuf. 23, 37–41 (2014)

8. Katzenbach, A., Handschuh, S., Vettermann, S.: JT format (ISO 14306) and AP 242 (ISO 10303): The step to the next generation collaborative product creation. In: IFIP Advances in Information and Communication Technology, vol. 411, no. ISO 14306, pp. 41–52 (2013)

9. Panetto, H., Dassisti, M., Tursi, A.: ONTO-PDM: product-driven ONTOlogy for Product Data Management interoperability within manufacturing process environment. Adv. Eng. Inf. 26(2), 334–348 (2012)

10. Tangelder, H., Veltkamp, R.C.: A survey of content based 3D shape retrieval methods. Multimedia Tools Appl. 39(3), 441–471 (2008)

# Essential Issues to Consider for a Manufacturing Data Query System Based on Graph

Lise Kim[1](✉), Esma Yahia[1], Frédéric Segonds[2], Philippe Veron[1], and Victor Fau[3]

[1] Arts et Metiers Institute of Technology, LISPEN, HESAM Université, Aix-en-Provence, France
lise.kim@ensam.eu

[2] Arts et Metiers Institute of Technology, LCPI, HESAM Université, Paris, France

[3] CapGemini DEMS, Toulouse, France

**Abstract.** Manufacturing industry data are distributed, heterogeneous and numerous, resulting in different challenges including the fast, exhaustive and relevant querying of data. In order to provide an innovative answer to this challenge, the authors consider an information retrieval system based on a graph database. In this paper, the authors focus on determining the essential functions to consider in this context. The authors define a three-step methodology using root causes analysis and resolution. This methodology is then applied to a data set and queries representative of an industrial use case. As a result, the authors list four major issues to consider and discuss their potential resolutions.

**Keywords:** Manufacturing data · Information retrieval · Graph database · Query system

## 1 Introduction

The volume of data generated by the manufacturing industry is large and increasing; it represents 3.6 EB in 2018 and will increase by 30% for 2025 [1]. The organization of companies in silos (justified by the need for specialization of the different business) generates data that is both distributed and heterogeneous. A part of data is managed by different information system (PDM, ERP, MES…) and generate structured data, while the other data are unstructured data (text, image, 3D…). In addition, the data can be explicitly linked to each other (like in the parent-child relations of a digital mock-up) or implicitly linked (like between the 3D of a component and its user manual).

To perform their work, employees have to query the data in order to retrieve the needed information. This task becomes complicated and time consuming due to the increasing volume of data, which are heterogeneous and are saved in distributed resources. To solve these issues, it is necessary to define a data querying system that deliver exhaustive and relevant data as fast as possible.

L. Roucoules et al. (Eds.): JCM 2020, LNME, pp. 347–353, 2021.
https://doi.org/10.1007/978-3-030-70566-4_55

To address this challenge, the authors worked to draw up the list of bare minimum issues to consider when defining the optimal framework. This paper is organized as follows: Sect. 2 defines the main orientations chosen based on a state-of-the-art analysis. Section 3 describes the methodology used to draw up the list of issues. Section 4 describes the experimental conditions and presents the results. Section 5 concludes with discussion.

## 2  Graph Database Consideration

Querying information can be achieved through Information Retrieval Systems that need to access to data in order to provide the most relevant one. This objective is reached by managing the data in NoSQL databases rather than traditional relational databases, as the former is faster, more efficient and flexible [2]. The main categories of NoSQL like column database, key-value store and document-oriented database includes indexing and quick access to the information but lack expressing of the relationships between data in their schema. The graph databases answer to this issue and consequently are the more suitable in our context.

To emphasize the benefit of the graph database, different researches have shown the importance of analyzing data with strong relational nature as in [3], applied in different manufacturing use cases as in [4]. Other works define a framework to allow data querying by transforming structured data [5] and unstructured data [6] into a graph, with enrichment by data linkage [6] with possible using ontologies for example in [7]. In this article, the authors aim to define the prerequisites for a manufacturing query-data system by answering this question: "What are the minimum issues to be taken into account for a querying manufacturing data system based on the graph database?"

## 3  Methodology

In order to define only the bare minimum issues to consider when defining the query system, an iterative method has been implemented. This method is detailed below:

(1)  **Integration of data** into a graph database. The data includes the minimum of information at initialisation (only metadata without text content). Metadata means all the properties carried by unstructured data and all metadata carried by structured data. Thus, each data is transformed into a node and each metadata integrates the properties of this node. On the other hand, the explicit relationships of relational databases are translated into relations between nodes (see Fig. 1).

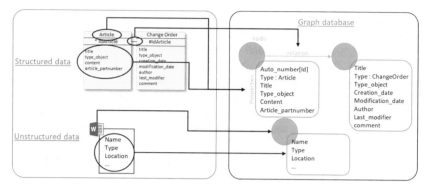

**Fig. 1.** Data set transformation into the graph at initialization

(2) **Application of queries** to the graph database, here refers to the translation of the user query adapted to the graph. Query transformation includes, in particular, the path of relations between data (e.g.: query = *employees related to 'additive manu-facturing'* become *finds the nodes mentioning 'additive manufacturing' and linked to nodes carrying employee information and return the employee information*) and the search for either a list of data (e.g.: query = *battery*) or a specific element within a data (e.g.: query = *price of battery*). The notion of an element is translated by the search for the associated property (value of the property 'price') and the sentences identifying the element ("the price of battery is […]"). Natural Language Processing (NLP) [8] tools will be used here to find the sentences.

(3) **The evaluation** of the proposal is conducted based on three requirements that are the response time (between the submission of the query and the result display), the completeness and the relevance of result (using precision[1] and recall[2] measures). The latter are calculated based on the expected results that are manually defined. When the results are below the accepted limits, the analysis of each error is then made (excess or missing data and too long execution data) in order to detect the root causes based on the Ishikawa diagram method[3] [9]). A score is then established by root cause according to its impact on the results (calculated with the number of errors associated with this root cause over the total number of errors). Once, this list of root causes is identified, it allows to define the main issues to be treated.

## 4   Experimentation Conditions and Results

The study was based on a dataset composed of 686 elements, representing data from a drone manufacturing company, and distributed as following: 47% unstructured data including spreadsheets, videos, photos and textual documents; 22% tree structure data;

---

[1] The precision is the number of relevant documents found compared to the total number of documents proposed in the result for a given request.

[2] The recall is defined by the number of relevant documents found with regard to the number of relevant documents in the database.

[3] Method of analysis used to research and to represent the different possible causes of a problem.

17% of data from relational databases and 15% of geometrical data. All these elements represent the data necessary for the development of a mechanical system (from design to prototyping through logistics, purchasing and project management). 19 queries have been written in response to innovative use cases characterized by Capgemini[4] (e.g.: a designer is looking to identify the product requirements or the justification for a product, a manager looking for identify an available team with the right skills, a salesperson looking for a customer's usage parameters, etc.). The expected performance thresholds are less than 1 s for time; this was fixed according to study conclusions on the impact of response latency in web search [10]. The tools used are *Neo4J*[5] for storage and querying in a graph database and *Standford CoreNLP*[6] for the exploitation of natural language. These tools are open source and relatively well documented.

After the first cycle of the methodology, the results are insufficient (see Table 1). Analysis of the results of this first cycle has shown that more than half of the anomalies are caused by the lack of textual content of the data in the graph database. For example, the search for the battery reference does not give any result because the information is carried by the content of an excel named "Bill of Materials". In order to treat this issue, a second cycle was therefore launched to integrate the text content of unstructured data. The text content is extracted using *Apache Tika*[7] as a parsing tool (to extract text from a document) and *Tesseract*[8] as an Optical Character Recognition Tool (to extract text from an image). Then the extracted text is integrated into the graph by adding a property named 'content' to each node.

The results of this second cycle are visible in Table 1 and the list of root causes is listed in Table 2. It is possible to remove the cause (6), only cause of which at least one element of the initial architecture is supposed to be resolved (an optimization of OCR is necessary). The remaining root causes therefore provide the list of the bare minimum issues to be resolved.

**Table 1.** First and second cycles results

|                      | Cycle 1 | Cycle 2 |
| -------------------- | ------- | ------- |
| Response time (ms)   | 5790    | 16978   |
| Precision            | 0.50    | 0.44    |
| Recall               | 0.01    | 0.31    |

[4] Company of digital services in the manufacturing industry.
[5] https://neo4j.com.
[6] https://standfordnlp.github.io/CoreNLP/.
[7] https://tika.apache.org/.
[8] https://opensource.google/projects/tesseract.

**Table 2.** Second cycle root causes

| Root causes | Anomalies distribution |
|---|---|
| (1) No highlighting performed on the most relevant result | 26% |
| (2) The information carried by the table format is not used | 24% |
| (3) Information is carried by a term close to the keyword | 19% |
| (4) The property carrying the information is close to the keyword | 13% |
| (5) Searching through the implicit relationships is impossible | 9% |
| (6) The OCR algorithm didn't extract the correct characters | 7% |
| (7) Extracting bulleted lists extraction is not correct | 2% |

## 5  Discussion

At the end of the second experimental cycle, 7 root causes remain present. The authors propose to classify them into 4 large families. Each one has a potential action plan in order to enhance the response time, recall and precision. The authors propose to prioritise, at first, the actions affecting both precision and recall:

(a) **Extracting text without format is not enough.** The cause (1) indicates that it is necessary to translate the information carried by the table format (rows and column) in order to use it in query. For example, to detect a reference contained in a specific cell of a bill of materials. Cause (7) indicates that bulleted lists processing is necessary for the performance of the chosen NLP tools. The table format and bulleted lists must be transformed to be used.

(b) **Searching for exact keyword or exact property is not enough.** Causes (2) and (4) indicate that reconciliation between different terms is necessary. For example, if the term 'reference' is used in the query, the term 'Part Number' must also to be searched. The use of a semantic network as an ontology could resolve part of the errors [11].

(c) **There is no order by relevance in the results.** Cause (3) indicates that unexpected results (but potentially relevant) are displayed in the same way as expected results. For example, searching for the battery reference provides many results with the terms "reference" and "battery" in the content, but these results are far from the information being searched for. Pre-labelling of data or additional filtering can be a solution.

(d) **The implicit links between data must be exploitable.** In some cases, related elements such as an element's functional reference and its supplier reference are disjoined in the different enterprise systems. In order to resolve the cause (5), the integration of the implicit relationships between data must be integrated into a graph.

The defined actions have a direct impact on increasing the precision and recall requirement and probably have a negative impact on the response time.

In conclusion, the above action list allows considering the essential functions for a query system construction, based on data graph and adapted to manufacturing data. This list was obtained according to the methodology described in Sect. 3, with a heterogeneous, distributed and relational data set and by applying queries in response to expected uses in the manufacturing industry.

## References

1. Reinsel, D., Gantz, J., Rydning, J.: The Digitization of the World From Edge to Core. IDC White Paper (2018)
2. Nayak, A., Poriya, A., Poojary, D.: Type of NOSQL databases and its comparison with relational databases. Int. J. Appl. Inf. Syst. (UAIS) **5**(14), 16–19 (2013)
3. Miller, J.: Graph database applications and concepts with Neo4j. In: Proceedings of the Southern Association for Information Systems Conference, Atlanta, GA, USA, March 2013.
4. Schalbus, S., Scholz, J.: Spatially-linked manufacturing data to support data analysis. J. Geogr. Inf. Sci. **1**(15), 126–140 (2017)
5. Paradies, M., Lehner, W., Bornhövd, C.: GRAPHITE: an extensible graph traversal framework for relational database management systems. In: Proceedings of the 27th International Conference on Scientific and Statistical Database Management, SSDBM 2015, La Jolla, June 2015, p. 12 (2015)
6. Groger, C., Schwarz, H., Mitschang, B.: The deep data warehouse: link-based integration and enrichment of warehouse data and unstructured content. In: Proceedings of the 18th IEEE International Enterprise Distributed Object Computing Conference, EDOC 2014, Ulm, September 2014, pp. 210–217 (2014)
7. Mordinyi, R., Schindler, P., Biffle, S.: Evaluation of NoSQL graph databases for querying and versioning of engineering data in multi-disciplinary engineering environments. In: 20th IEEE International Conference on Emerging Technologies and Factory Automation, ETFA 2015, pp. 1–8 (2015)
8. Chowdhury, G.G.: Natural language processing. Ann. Rev. Inf. Sci. Technol. **37**(1), 51–89 (2003)
9. Barsalou, M.A.: Root Cause Analysis: A Step-By-Step Guide to Using the Right Tool at the Right Time. CRC Press Taylor & Francis Group, Boca Raton (2014)
10. Arapakis, I., Bai, X., Cambazoglu, B.: Impact of response latency on user behavior in web search. In: Proceedings of the 37th International ACM SIGIR Conference on Research and Development in Information Retrieval, SIGIR 2014, New York, pp. 103–112 (2014)
11. Yan, L., Manoj, A.T., Kweku-Muata, O.-B.: Ontology-based data mining model management for self-service knowledge discovery. Inf. Syst. Front. **19**(4), 925–943 (2017)

# Design for Manufacturing and Assembly: A Method for Rules Classification

Claudio Favi[1]([✉]), Marco Mandolini[2], Federico Campi[2], Paolo Cicconi[3], Roberto Raffaeli[4], and Michele Germani[2]

[1] Università degli Studi di Parma, Parco Area delle Scienze 181/A, 43124 Parma, Italy
claudio.favi@unipr.it
[2] Università Politecnica delle Marche, Via Brecce Bianche, 60131 Ancona, Italy
[3] Università degli studi ROMA3, Via della Vasca Navale, 79, 00146 Roma, Italy
[4] Università degli studi di Modena e Reggio Emilia, Via Amendola 2, 42122 Reggio Emilia, Italy

**Abstract.** The paper provides a method to acquire, process, and represent DfMA rules to help designers and engineers in the development of mechanical products compliant with manufacturing and assembly technology. This research work wants to define a general method able to link DfMA design guidelines (knowledge engineering) with geometrical product features that are available by the investigation of the 3D model. Numerical parameters of design features are related to design guidelines for the identification of manufacturing and assembly issues within the analysis of the 3D model.

**Keywords:** DfMA · Design rules · Embodiment design · Knowledge formalization · Mechanical products

## 1 Introduction

Product Development Process (PDP) is a consolidated engineering activity that takes a product from conception to market. The steps in product development include drafting the concept, creating the overall design, developing detail design, and prototyping [1, 2]. One of the most recurring disciplines in the engineering design contexts relates the solid modelling and drawing (CAD - Computer-Aided Design). Nowadays, CAD tools couple the initial capability for which they were conceived, with the potentialities deriving from the integration of the multidisciplinary design methodologies (i.e., FEM, CAM) [3]. Design for X (DfX) is a target design methodology that gives designers a thought process and guidance for developing products oriented to a specific engineering challenge (i.e., manufacturability, assemblability, sustainability) [4]. For example, Design for Assembly (DfA) is a systematic procedure aiming at the reduction of assembly time through the reduction of the overall number of components in a given assembly [5]. At the same time, Design for Manufacturing (DfM) is an engineering practice aiming at the simplification of the manufacturing process for cost reduction of a given component [6]. DfMA (Design for Manufacturing and Assembly) is considered a comprehensive engineering practice

© The Author(s) 2021
L. Roucoules et al. (Eds.): JCM 2020, LNME, pp. 354–359, 2021.
https://doi.org/10.1007/978-3-030-70566-4_56

integrating manufacturing and assembly aspects. DfMA is a design methodology that suffers a real integration with 3D CAD systems. In particular, DfMA principles are currently applied downstream of the 3D modelling following available knowledge from the literature and the company's know-how (internal knowledge) [7]. Literature provides several attempts with regards to DfMA knowledge formalization using ontologies [8, 9]. Within the design context, the need to create a conceptual framework for data exchange is amplified by the nature of design information, which ranges from geometric descriptions of the part itself to manufacturing information such as material and cost. Ontologies overcome this limitation since their focus is not only on data or knowledge but also on the information context that lets specific access to detailed information parts to a latter phase [10]. However, the link with product analysis for geometrical features modification is still a grey area with possible potential for research activities. The goal of the paper is to provide a method that helps designers in the collection of DfMA design rules oriented to product manufacturing and assembly. In particular, this research work wants to define a general method able to link DfMA design guidelines (knowledge engineering) with geometrical product features that are available by investigating the 3D model. This method is leading to the development of a knowledge-based management tool that is accessible to designers for design updates. Numerical parameters of design features are related to design guidelines for the identification of manufacturing and assembly issues within the analysis of the 3D model. The novelty of this work is beyond the simple collection of DfMA rules by using an ontology method. The paper allows creating a list of design rules that can be integrated into a 3D system for the investigation of manufacturing and assembly issues early in the design step. The paper is structured as follows: after this introduction, Sect. 2 describes the knowledge-based system used for the classification of the DfMA rules, and Sect. 3 reports a case study of DfMA rule classification. Section 4 presents the concluding remarks and future perspectives on this subject.

## 2 Materials and Methods

A knowledge-based (KB) system is used for the classification of DfMA rules. The method for the classification of DfMA rules is grounded on three main pillars: (i) knowledge acquisition, (ii) knowledge processing, and (iii) knowledge representation. *Knowledge acquisition* refers to the literature analysis and industry best practices investigation for the collection of DfMA design rules. In particular, this phase is characterized by two main tasks: (i) collection of design rules for different manufacturing and assembly technologies (text), and (ii) identification of geometrical entities and numerical parameters involved in the design rules (numerical data). *Knowledge processing* refers to the link between the DfMA design rules previously collected within the *knowledge acquisition* phase and the geometrical features of a virtual 3D model. This phase is an essential task to transform a checklist (DfMA rules list) into a systematic design review of the product under development (3D model). *Knowledge representation* refers to the definition of a structured repository for the collection and the elicitation of DfMA knowledge. This phase encompasses the logical definition of DfMA design guideline (syntax) and all the necessary information to store within the repository, including example about possible changes to guarantee product manufacturability and assemblability.

## 2.1 Knowledge Acquisition Phase

Knowledge acquisition phase starts with the analysis of the literature (e.g., book, research papers, technical reports, master and Ph.D. thesis) related to the DfMA topic. In particular, in this research work, the following books were analyzed and reported in the references section: (i) Boothroyd et al., [11], (ii) Bralla [12], (iii) Caimbrone [13], (iv) Poli [14], and (v) El Wakil [15]. In some references, DfMA rules are available as a list of actions. Designers should consider such rules during the design phase of a mechanical component subjected to a given technology. On the other hand, for some other books, the DfMA rules are not explicitly stated, and a deeper analysis is necessary to extract design rules. Another important source for the acquisition of DfMA rules concerns the use and access to the available documentation of commercial tools developed for DfMA analysis. For example, DFMA® tool, and DFMPro® are two software tools developed to help designers and engineers in developing assembly-compliant products. Besides, several meetings in design departments of mechanical industries permitted the collection of the best practices and rules dedicated to given manufacturing technologies.

## 2.2 Knowledge Processing Phase

The knowledge processing phase starts with the definition and classification of manufacturing technologies associated with a given DfMA rule. Classification of manufacturing technologies requires the characterization of different clusters: (i) Manufacturing technology class (i.e., machining, sheet metal stamping, metal forming, metal casting, plastic forming, welding, assembly), (ii) Manufacturing technology type – level I, and (iii) Manufacturing technology type – level II. These three groups are necessary to classify DfMA rules that are generic for a technology class or specific for a manufacturing type (operation). Indeed, a DfMA rule may be valid for a manufacturing technology class (i.e., machining) regardless of the specific operation (i.e., turning, milling, drilling). Conversely, a DfMA rule may be valid only for a specific operation (i.e., drilling) and cannot be generalized for the manufacturing technology class that contains the operation (i.e., machining). The two levels of manufacturing technology type allow classifying DfMA rules base on a list of operations (i.e., turning) or for a single operation (i.e., drilling, external cylindrical turning, internal cylindrical turning). After the definition of manufacturing technologies, the authors established a new classification method based on the material involved within the DfMA rule. Again, two different clusters have been defined: (i) material class, and (ii) material type. These two clusters permit allocating a given DfMA rule to a generic class of materials (i.e., stainless steel) or to a specific type (i.e., AISI 304). The last definition deals with 3D CAD features to recognize according to a given DfMA rule. Three clusters are necessary: (i) 3D CAD features to recognize, (ii) PMI – Product manufacturing Information to read, and (iii) Dimension/geometry. The CAD feature can be identified through different methods such as the 3D Convolutional Neural Networks proposed by Zhang et al. [16] allowing to learn machining features from CAD models of mechanical parts (Fig. 1).

| 1 | Through hole |
| 2 | Rectangular pocket |
| 3 | Rectangular passage |
| 4 | Rectangular through slot |
| 5 | Rectangular blind slot |
| 6 | Blind hole |

**Fig. 1.** Example of a 3D CAD feature recognition [16].

### 2.3   Knowledge Representation Phase

The knowledge representation phase starts with the definition of a structured form for a DfMA rule. Indeed, a guideline syntax is necessary to keep consistency among different guidelines and to provide the same level of details and information to a mechanical designer. DfMA guideline syntax requires necessary information and optional information. For example, necessary information provides the minimum set of information to perform a design improvement. Necessary information consists of: (i) the design action to do (*verb*), and (ii) the subject which requires modification (*name*). Optional information provides additional data that allow clarifying the context of the required design action. Optional information consists of: (i) the manufacturing process, (ii) the type of feature involved, (iii) the type/family part, and (iv) the type of material.

## 3   Case Study: DfMA Rules Classification

This section provides an example of the application of the proposed method in the development and classification of DfMA rules in machining technology. Figure 2 highlights the knowledge processing phase, while Fig. 3 reports the knowledge representation phase in a case of deep holes for the drilling process.

| RULE # | MANUFACTURING TECHNOLOGY | | | MATERIAL | | CAD FEATURES RECOGNITION | | |
| | Manufacturing technology class | Manufacturing technology type - level 1 | Manufacturing technology type - level 2 | Material class | Material type | CAD features to recognize | PMI to recognize | Dimensions and rules to verify |
|---|---|---|---|---|---|---|---|---|
| 1 | Machining | Turning | Drilling | Steel | Stainless Steel - AISI 304 | - Hole | - Roughness | D - Hole diameter<br><br>L - Hole length<br><br>Ra - Hole roughness<br><br>Ra < 0.8 micron<br><br>L/D > 5 |

**Fig. 2.** Example of DfMA rule collected for machining class (knowledge processing phase).

Avoid deep holes with internal roughness (Ra) of 0.8 micron or less. Deep holes are defined when the ratio between hole length (L) and diameter (D) is equal or bigger than 5.

→ L/D ≥ 5

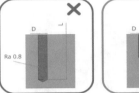

**Fig. 3.** Example of DfMA rule syntax and picture (knowledge representation phase).

It is worth noting that the presented approach allows checking the possibility to perform a given process for a specific technology. Thus, the validity of a rule is ensured by checking the manufacturing technology and the material classes, as well as the CAD feature recognition class. The rule model does not allow designers to verify possible investment vs. Available machines and equipment.

## 4   Conclusions and Future Outlook

This paper investigates the possibility to gather and to classify DfMA rules to create a list of design rules that can be integrated into a 3D system for the investigation of manufacturing and assembly issues early in the design step. The method is applicable in each context of manufacturing technology (i.e., machining, sheet metal stamping, metal forming, metal casting, plastic forming, welding, assembly). Furthermore, it provides a systematic approach to link product features (3D CAD model features) with manufacturing and assembly concerns early in the embodiment design. In conclusion, the method is an efficient way to collect manufacturing knowledge and to re-inject this knowledge in the development of mechanical components. In this first step of the research activity, a set of DfMA rules have been retrieved by analysing literature as well as by interviewing different manufacturing companies (internal knowledge). More rules can be defined with an efficient interview campaign and the involvement of company suppliers. The setting of further rules is the first outlook of this research work. Another significant outlook relates to the possibility of developing a software system able to integrate the retrieved DfMA design rules with a CAD tool for an efficient application of DfMA in the embodiment design phase (CAD-integrated DfMA tool).

## References

1. Pahl, G., Beitz, W., Feldhusen, J., Grote, K.H.: Engineering Design: A Systematic Approach, 3rd edn. Springer, Cham (2007)
2. Ulrich, K.T., Eppinger, S.D.: Product Design and Development, 5th edn. McGraw-Hill Education, USA (2011)
3. Robertson, B.F., Radcliffe, D.F.: Impact of CAD tools on creative problem solving in engineering design. Comput. Aided Des. **41**(3), 136–146 (2009)
4. Benabdellah, A.C., Bouhaddou, I., Benghabrit, A., Benghabrit, O.: A systematic review of design for X techniques from 1980 to 2018: Concepts, applications, and perspectives. Int. J. Adv. Manufact. Tech. **102**(9–12), 3473–3502 (2019)

5. De Fazio, T.L., Rhee, S.J., Whitney, D.E.: Design specific approach to design for assembly (DFA) for complex mechanical assemblies. IEEE Robot. Autom. **15**(5), 869–881 (1999)
6. O'Driscoll, M.: Design for manufacture. J. Mater. Process. Technol. **122**(2–3), 318–321 (2002)
7. Hoque, A.S.M., Halder, P.K., Parvez, M.S., Szecsi, T.: Integrated manufacturing features and Design-for-manufacture guidelines for reducing product cost under CAD/CAM environment. Comput. Ind. Eng. **66**, 988–1003 (2013)
8. Chang, X., Rai, R., Terpenny, J.: Development and utilization of ontologies in design for manufacturing. J. Mech. Des. **132** (2010)
9. Debord, S., Segonds, F., Pinquié, R., Veron, P., Croué, N.: Proposition of a design rules framework. In: Proceedings of 25ème colloque des Sciences de la conception et de l'innovation, CONFERE 2018, Budapest, Hungary (2018)
10. Molcho, G., Zipori, Y., Schneor, R., Rosen, O., Goldstein, D., Shpitalni, M.: Computer aided manufacturability analysis: Closing the knowledge gap between the designer and the manufacturer. CIRP Ann. Manufact. Tech. **57**, 153–158 (2008)
11. Boothroyd, G., Dewhurst, P., Knight, W.A.: Product Design for Manufacture and Assembly, 3rd edn. CRC Press, Boca Raton (2010)
12. Bralla, J.G.: Design for Manufacturability Handbook. McGraw Hill Professional, New York (1999)
13. Ciambrone, D.F.: Effective Transition from Design to Production, 1st edn. Auerbach Publications, Boca Raton (2007)
14. Poli, C.: Design for Manufacturing: A Structured Approach. Elsevier Science, Amsterdam (2001)
15. El Wakil, S.D.: Processes and Design for Manufacturing, 3rd edn. CRC Press, Boca Raton (2019)
16. Zhang, Z., Jaiswal, P., Rai, R.: FeatureNet: Machining feature recognition based on 3D Convolution Neural Network. Comput. Aided Des. **101**, 12–22 (2018)

# Education in Circular Economy: Focusing on Life Cycle Thinking at the University of the Basque Country

Rikardo Minguez[✉], Erlantz Lizundia, Maider Iturrondobeitia, Ortzi Akizu-Gardoki, and Estibaliz Saez-de-Camara

University of the Basque Country UPV/EHU, Bilbao, Spain
rikardo.minguez@ehu.eus

**Abstract.** Since 2002, the University of the Basque Country (UPV/EHU) has supported several teaching experiences related to the so-called Life Cycle Thinking or Ecodesign in collaboration with local Institutions. The kick off was the Ecodesign Learning Center aiming to promote environmental education at the Faculty of Engineering in Bilbao. In this framework, the last effort has been the implementation of a Master's Degree entitled *Circular Economy: Business Application*. This course has been successfully implemented in the 2019–2020 academic year and has been specifically designed to provide training in circular economy for people with backgrounds as varied as product manufacturing engineering, environmental engineering or economics. These studies are aimed to become a European reference in its goal of promoting circular economy, life cycle thinking, ecodesign and sustainable development. This paper analyzes the learning issues and characteristics of this Master's degree placing a special emphasis on its novel competencies and learning outcomes for our society. It can be concluded that the Master's degree is a pioneering teaching experience, being the forefront of Circular Economy Education in Southwestern Europe.

**Keywords:** Environmental education · Life cycle thinking teaching experience · Circular economy · Life cycle assessment

## 1 Introduction

Besides the world health pandemic crisis we are enduring in spring 2020, our planet lives in an environmental crisis, with alarming social and economic inequalities, increasing rates of loss of natural and cultural biodiversity, and a senseless growing pressure on natural resources and systems.

In this context, since the end of the decade of the 80s, aware of the necessity for a deep change, the concept of sustainability has been strengthened. At the same time, it seems to have become a multipurpose and valid term for any context, objective, argumentation or ideological-political current [1].

Currently the United Nations's Global Action Programme on Education for Sustainable Development (ESD) [2] aims to contribute substantially to the 2030 development agenda, through two objectives:

L. Roucoules et al. (Eds.): JCM 2020, LNME, pp. 360–365, 2021.
https://doi.org/10.1007/978-3-030-70566-4_57

– Reorienting education and learning so that everyone has the opportunity to acquire the knowledge, skills, values and attitudes that empower them to contribute to a sustainable future.
– Strengthening education and learning in all agendas, programmes and activities that promote sustainable development.

In parallel with this, the European Commission has proclaimed the European Green Deal [3] which constitutes a new growth strategy aiming to transform the European Union (EU) into a fair and prosperous society, with a modern, resource-efficient and competitive economy with no net emissions of greenhouse gases by 2050 and where the economic growth remains decoupled from resource use.

At regional level, the government of the Basque Autonomous Community strongly supports EDS in higher education by means of the Agenda Euskadi - Basque Country 2030 [4]. The two main objectives of the program are to encourage the market to act in the interest of the environment, and to promote ESD in higher education in coordination with the United Nations's Global Action Programme. In January 2020, the Basque Government has made public the keys of the circular economy strategy 2030 [5], with many strategic objectives and challenges, among others promoting degree studies and research at university level.

Aligned with all these policies, the University of the Basque Country (UPV/EHU) has supported different teaching experiences related to the so-called Life Cycle Thinking or Ecodesign in collaboration with some local Institutions.

The kick off for these experiences was established in 2002 with the foundation of the Ecodesign Learning Center at the Faculty of Engineering in Bilbao [6]. More recently, these efforts have been focused on the implementation of a new Master's Degree entitled *Circular Economy: Business Application*, which is going to be described in the following chapters, with especial emphasis on the aspects related to ecodesign and life cycle thinking [7].

## 2   Origin of the Master Degree in Circular Economy

The Basque Ecodesign Center [8] is a partnership framework between firms in the private sector and the Basque Government. It aims to foster the design and execution of innovative ecodesign projects. Ihobe [9] is a publicly-owned company under the auspices of the Basque Government's Ministry for the Environment, Territorial Planning and Housing. Ihobe supports the Basque Government in the implementation of its environmental policy and in the spreading of the environmental sustainability culture in the Basque Autonomous Community.

Boosted by Ihobe, the Basque Ecodesign Center and the Sustainability directorate of the University of the Basque Country, in October 2019 the university launched the first edition of the postgraduate course entitled *Circular Economy: Business Application* (consisting of 36 ECTS) and for the next academic year a new master degree with 60 ECTS is on its way.

The European studies that have inspired and served as a reference for the Master's Degree proposal have been the CIRCLE programme [10], which is an Erasmus Mundus

International Master's Programme on Circular Economy, with several labeled master programmes certified by the EIT (European Institute of Innovation and Technology) [11].

## 3   Facts of the Master Degree in Circular Economy

The Master's Degree in Circular Economy is projected to begin teaching during September 2020 and to finish these lectures on June 2021 (Table 1). The Master's Thesis to complete the 60 ECTS necessary to obtain the degree, is expected to be defended on September 2021.

The maximum number of students has been fixed to 25 and more than 35 lecturers and professionals from the industrial sector will take part in it. Student candidates will be sought among engineers and economists working in local industry, service companies or public administration. Subjects will be taught at the Faculty of Engineering and the Faculty of Economics and Business in Bilbao.

**Table 1.**  Subjects of the Master Degree in Circular Economy for the 2020–2021 academic year.

|    | Subject | ECTS |
|----|---------|------|
| 1  | Circular Economy: General Context | 6 |
| 2  | Eco-innovation as a Business Opportunity. Regulatory Framework and Concept | 3 |
| 3  | Environmental Impacts Derived from the Production-consumption of Products and Services | 3 |
| 4  | Environmental Management Tools in the Company | 3 |
| 5  | Ecodesign and Circular Economy | 3 |
| 6  | Life Cycle Thinking 1: Tools for Calculation and Communication | 3 |
| 7  | Life Cycle Thinking 2: Quantification of the Product Environmental Footprint | 3 |
| 8  | Circular Economy in the Company: From Ecodesign to Product Commercialization | 6 |
| 9  | Circular Economy in the Company: Sustainable Entrepreneurship, Business Models in Circular Economy | 3 |
| 10 | Circular Economy Initiatives. Business Cases | 9 |
| 11 | Practical Workshops on Circular Economy | 6 |
| 12 | Master's Thesis | 12 |

### 3.1  Life Cycle Thinking: Ecodesign and Life Cycle Assessment

The subjects closely related to Life Cycle Thinking make up 9 ECTS divided into 3 subjects consisting of 3 ECTS:

*Ecodesign and Circular Economy.*

1. Introduction to the concept of ecodesign: basic principles and implications for the design of products and services.

2. Standards in ecodesign.

3. Ecodesign methodology.

4. Ecodesign and companies: Integration of the Ecodesign in the enterprise management systems. Product-service systems.

*Life Cycle Thinking 1: Tools for Calculation and Communication.*

1. Product life cycle: background and concepts. Life cycle and extended producer responsibility. Implications for the product designer.

2. Methodological principles of Life Cycle Analysis based on international standards. Life Cycle Assessment Software: Open LCA and SIMAPRO.

3. Communication and marketing with a Life Cycle approach.

*Life Cycle Thinking 2: Quantification of the Product Environmental Footprint.*

1. Social and environmental impacts.

2. Calculation of the environmental footprint of products and organizations. Combination of Life Cycle Assessment tools and Input-Output sector tables. Global Multi Regional Input-Output methodology (GMRIO).

3. Sustainable Product Design Tools. Product design cycle for a circular economy.

4. The R's principles in Circular Economy. Reuse, Repair or Recycle + Restructure, Revalue, Relocate and Reconceptualize.

Overall, emphasis is made on how ecodesign serves to reduce the environmental impacts associated with a product along its whole life cycle. The relevance of servitization as a more sustainable new business model is also shown. Within an ecodesign framework, the methodology for LCA is thoroughly shown using different software. To get further knowledge, the calculation of different footprints is also shown. All this would enable the students to get a wide and deep overview on the Environmentally Conscious Design concept.

## 4  Specific Learning Outcomes or Competencies

When drawing up the Master's programme, the aim was to ensure that the students become experts in the challenges that the new circular economy poses for local companies. Consequently, the most relevant specific competencies (knowledge, skills, and attitudes) that students must have acquired at the end of their Master's studies are the following:

- To define the concepts of Circular Economy, Sustainable Economy, Low Carbon Economy and Eco-Innovation.
- To understand and explain the changes involved in the transition from a linear to a circular type of economy, as well as to recognize the opportunity they represent.
- To identify the current applicable environmental regulations (air, water, waste and soil) in order to carry out proper environmental management in the company.
- To define the concept of eco-design and to become aware of the environmental, economic and social implications of product design.

- To apply the ecodesign methodology and manage the tools available for ecodesign in the industrial field
- To define the concept of life cycle and identify the phases of the life cycle of a product, as well as list the regulations of the Life Cycle Analysis.
- To know and apply the evaluation methodologies and software tools for product life cycle analysis.
- To formulate guidelines for communication and marketing with a life cycle approach.
- To define and understand corporate and competitive strategies to integrate the environmental variable in the company and select the most appropriate strategy for each particular case.
- To identify business opportunities in the field of the Circular Economy for different sectors, stakeholders and design proposals.

These competencies are fully aligned with those proposed by some European universities that are considered to be at the leading edge in this area of knowledge and education [12–14]. The Master's degree covers the proposed seven circular economy competencies for design to a greater or lesser extent: Circular Impact Assessment, Design for Recovery, Design for Multiple Use Cycles, Circular Business Models, Circular User Engagement, Circular Economy Collaboration, and Circular Economy Communication.

## 5    Conclusions

It can be concluded that postgraduate studies in Circular Economy are an ongoing reality at the University of the Basque Country.

Furthermore, these studies are ambitious and highly developed in the fields of Ecodesign and Life Cycle Thinking. It is thus expected that the graduated students will help advancing towards a sustainable society as they are incorporated into academia or industry. This would enable not only a more efficient integration of the circular economy concepts into our community but also may allow enterprises to get ahead of the many legislative changes that are expected to occur in the near future.

The initiative is pioneering in Southwestern Europe and in the near future alliances will be sought with other European universities to offer a joint degree or at least to be able to offer mobility to students so that they can experience other academic realities.

**Acknowledgments.** The authors would like to thank Ihobe and the Basque Ecodesign Center for promoting economically and technically the Master Degree in Circular Economy: Business Application in the Faculty of Engineering in Bilbao.

## References

1. McGregor, S.L.T.: Alternative Communications about Sustainability Education. Sustainability, vol. 5, p. 3562 (2013)

2. UNESCO (United Nations Educational, Scientific and Cultural Organization). Education for Sustainable Development Goals: Learning Objectives; UNESCO: Paris, France (2017)
3. https://ec.europa.eu/info/files/communication-european-green-deal_en. Accessed 4 May 2020
4. https://www.euskadi.eus/pdf/agenda-euskadi-basque-country-2030.pdf. Accessed 4 May 2020. (in Spanish)
5. https://www.ihobe.eus/publications/the-keys-of-the-circular-economy-strategy-of-the-bas que-country-2030-draft. Accessed 4 May 2020
6. Prado-Trigo, A., et al.: Eco-design classrooms project: life cycle assessment and eco-design in Basque industry. Dyn. Spain **86**(1), 74–79 (2011)
7. https://www.ehu.eus/es/web/graduondokoak/master-economia-circular-aplicacion-empresa. Accessed 4 May 2020. (in Spanish)
8. https://www.basqueecodesigncenter.net. Accessed 4 May 2020
9. https://www.ihobe.eus/home. Accessed 4 May 2020
10. https://www.jointdegree.eu/de/circle-erasmus-mundus-international-masters-programme-on-circular-economy/. Accessed 4 May 2020
11. https://eitrawmaterials.eu/eit-rm-academy/labelled-masters/. Accessed 4 May 2020
12. Lambrechts, W., et al.: The role of individual sustainability competences in eco-design building projects. J. Clean. Prod. **208**, 1631–1641 (2019)
13. Sumter, D., et al.: Circular economy competencies for design. Sustainability **12**, 1561 (2020)
14. Korhonen, J., et al.: Circular economy: the concept and its limitations. Ecol. Econ. **143**, 37–46 (2018)

# Generation of Entrepreneurship Environments for New Product Development

Cristina Martin-Doñate[1]([✉]), Fermín Lucena-Muñoz[2],
Lina Guadalupe García-Cabrera[3], and Jorge Manuel Mercado-Colmenero[1]

[1] Department of Engineering Graphics, Design and Projects, University of Jaén,
Campus Las Lagunillas, s/n, 23071 Jaen, Spain
cdonate@ujaen.es
[2] Department of Management, Marketing and Sociology, University of Jaen,
Campus Las Lagunillas, s/n, 23071 Jaen, Spain
[3] Department of Computer Sciences, University of Jaen, Campus Las Lagunillas, s/n, 23071
Jaen, Spain

**Abstract.** The economic environment demands companies to be able of innovating and presenting new products and technologies. However, current industrial environments are composed of big established companies, small or medium family businesses and regional clusters. This business map does not comply with the dynamism required for detecting needs and proposals in form of new products that meet the current customer requirements. The creation of start-ups in the field of engineering is considered as a possibility to cover a growing market in designing high technological products. However, despite the importance and economic impact of the start-ups for less industrialized environments the courses in the engineering field do not include activities that contextualize theoretical knowledge and entrepreneurship. The paper presents the results of creating an entrepreneurial environment focused on the student as entrepreneurial unit. In this context, a series of new activities based on new design proposals have been developed on the basis of new technical solutions for industrial companies. The new designs have been presented as minimum viable products. The technical knowledge required for making the new designs has been complemented with entrepreneurial training. The results show an increase in students' willingness to create small companies based in new products as an employment option at the end of their studies.

**Keywords:** Industrial design · Entrepreneurship · CAD · Product development

## 1 Introduction

An entrepreneur in new product design must have the skills of both an industrial designer and a business manager [1]. Currently the field of entrepreneurship has taken importance as a mean of self-employment in the field of new product development. Unfortunately, academic education mainly focuses on the learning of CAD modelers and technical subjects. Design learning has focused on preparing professionals according to this technical profile. The four main components of NPD (New Product Development) are classified

L. Roucoules et al. (Eds.): JCM 2020, LNME, pp. 366–371, 2021.
https://doi.org/10.1007/978-3-030-70566-4_58

into: Research, Design (including engineering and industrial design) [2, 3], Marketing [4] and Manufacturing [5, 6]. These tasks are usually developed by specialists, especially in big companies with large-scale projects. However, the demand for small-scale projects increases every day, there being a great need for engineers with an adaptable profile and flexible skills. The business professional in new product design demands a changing mod-el of learning. This trend presents very different boundary conditions compared to the past industrial landscape that created the field of product design. Currently there are many more options available for engineers, in addition to working for big companies or design consultancies. In business environments engineers must have skills related to creativity, sustainability [7] and management of human resources. Industry requires professionals with technical experience, and the ability to pro-vide new ideas generating demand through innovation. At the moment students can produce and manufacture their own designs or work with colleagues to create a start-up company.

The development of skills in entrepreneurship for engineering environments requires a change of the educational process. However, and although university institutions are the most important part of this process, actions for the development of entrepreneurship are usually complemented by initiatives by other institutions outside the education system. The entrepreneur engineer must be able to evaluate possible solutions, use software tools adequately, learn to work in team, and use efficiently communication skills to work with other people. In addition, the engineer must have the ability to conceptualize and turn the concept into reality. The development of the entrepreneurship in engineering requires a big change in the learning processes linking educational and research areas to enhance creative processes and practices [8].

This paper presents the results of the activities carried out with the objective of devel-oping business capacity in mechatronic engineering students through the introduction of new activities related to entrepreneurship in the field of new product development. The designs presented technical solutions to problems performed for industrial companies. The design results have been presented in the form of prototypes complementing the projects with product business models.

## 2  Methodology

A great quantity of engineering students wants to develop their work in big companies. However, it is not always possible to find a qualified job in this area especially in less industrialized environments. Additionally, engineering students usually do not have the knowledge to create a new small company, mainly be-cause they have a technical profile. In this line, a series of activities have been proposed, consisting of create and design innovative products obtaining its business plan.

The objective of these activities has been to motivate and prepare engineering stu-dents in using industrial design as a tool for self-employment. Several industrial compa-nies have collaborated in the entrepreneurship activities. These companies presented to the engineering student's technical problems related to needs to design to solve Fig. 1.

As a complement to the student's training in entrepreneurship issues several work-shops related to business model methodologies were presented. These work-shops were taught by lecturers of the area of business and industrial engineers experts in

**Fig. 1.** Presentation of the technical problems

entrepreneurship. The objective was to help the students to understand that the product concept is only one part of the process of creating a startup. The tool to propose the business model based on product design was the Canvas methodology. The Canvas method is a practical and simple tool that allows obtaining a global analysis of all the important aspects that make up the business model, working on problems as a team. This method is based on the analysis of the boundary conditions in which the company is going to work: identifying the target client; how to reach customers and finally the means of payment. The set of entrepreneurial activities were complemented with workshops taught by staff experienced in entrepreneurship (Fig. 2), and several visits to entrepreneurs where students were able to discuss their doubts with entrepreneurs with experience in creating a startup.

**Fig. 2.** Workshops taught by staff experienced in entrepreneurship

Team students created different product proposals by using CAD tools and new products methodology. Finally, the designs in form of a minimum viable product together with the business plan were presented to the companies which assessed the students, commenting on possible improvements in both product and business area. A total of 20 different proposals were presented by teams formed by five students. In Fig. 3 two of the products presented to the companies are shown.

**Fig. 3.** Presentation of the developed products by the students

## 3 Results

In order to evaluate the influence of the performed activities in the entrepreneurial skills and the intention for creating a new start up, a survey was conducted on the ILIAS platform for the students that had participated in these activities. ILIAS is an open-source web-based learning management system (LMS) that supports learning content management (including SCORM 2004compliance) and tools for collaboration, communication, evaluation and assessment [9]. The inter-views were completed voluntarily by all the students that have participated in the activities, being a total number of 15 participant students (13 males and 2 female) from different working design teams.

**Fig. 4.** Survey results

Students were asked about considering creating a start up at the end of their studies. 66,6% of the students that participate in the activities wanted to create a new start up after they finish their studies, with a 66,6% totally agreeing and 33,3% agreeing. The results of the survey have been indicated in Fig. 4.

## 4  Conclusions

Develop entrepreneurial skills in engineering needs of a big change in training activities in order to integrate educational and industries to generate new creative processes and practices. The paper presents the results of creating an entrepreneurial environment focused on the student as entrepreneurial unit. In this context, a series of new activities based on new design proposals have been developed on the basis of new technical solutions for industrial companies. The new designs have been presented as minimum viable products. The technical knowledge required for making the new designs has been complemented with entrepreneurial training. The results show an increase in students' willingness to create small companies based in new products as an employment option at the end of their studies. Students were asked about considering creating a start up at the end of their studies. 66,6% of the students that participate in the activities would like to create a new start up after they finish their studies.

**Acknowledgments.** This research work was supported by the University of Jaen with the Plan de Apoyo a la Transferencia del Conocimiento, el Emprendimiento y la Empleabilidad (2019–2020) AC- 10 through the project "Creation of an entrepreneurial ecosystem of microstart-ups based on innovative products".

## References

1. Huang-Saad, A.Y., Morton, C.S., Libarkin, J.C.: Entrepreneurship assessment in higher education: a research review for engineering education researchers. J. Eng. Educ. **107**(2), 263–290 (2018)
2. Mercado-Colmenero, J.M., Rubio-Paramio, M.A., la Rubia-Garcia, M.D., Lozano-Arjona, D., Martin-Doñate, C.: A numerical and experimental study of the compression uniaxial properties of PLA manufactured with FDM technology based on product specifications. Int. J. Adv. Manufact. Technol. **103**(5–8), 1893–1909 (2019)
3. Fayolle, A., Lamine, W., Mian, S., Phan, P.: Effective models of science, technology and engineering entrepreneurship education: current and future research. J. Technol. Transfer, 1–11 (2020)
4. Martin-Doñate, C., Lucena-Muñoz, F., Gallego-Alvarez, J.: Integration of marketing activities in the mechanical design process. In: Advances on Mechanics, Design Engineering and Manufacturing, pp. 961–969. Springer, Cham (2017)
5. Mercado-Colmenero, J.M., Martin-Doñate, C., Rodriguez-Santiago, M., Moral-Pulido, F., Rubio-Paramio, M.A.: A new conformal cooling lattice design procedure for injection molding applications based on expert algorithms. Int. J. Adv. Manufact. Technol. **102**(5–8), 1719–1746 (2019)

6. Doñate, C.M., Paramio, M.R.: New methodology for demoldability analysis based on volume discretization algorithms. Comput. Aided Des. **45**(2), 229–240 (2013)
7. Kuckertz, A., Wagner, M.: The influence of sustainability orientation on entrepreneurial intentions—investigating the role of business experience. J. Bus. Ventur. **25**(5), 524–539 (2010)
8. Poggesi, S., Mari, M., De Vita, L., Foss, L.: Women entrepreneurship in STEM fields: literature review and future research avenues. Int. Entrepreneurship Manag. J. **16**(1), 17–41 (2020)
9. https://www.ilias.de/en/

# Engaging in Product Development as Means to Understand the Basics of Design

Sergio Rizzuti and Luigi De Napoli[✉]

DIMEG, University of Calabria, Ponte Pietro Bucci 46/C, 87036 Rende, CS, Italy
luigi.denapoli@unical.it

**Abstract.** Teaching product design is not a trivisal task. Considering the experience done along 20 years of teaching at master level class in mechanical engineering it is possible to take stock. The model followed is Project-Based Learning and this method can be licensed as the model that gives greater satisfaction to all attendees. Students give high score to the survey organized by university at the end of the course to assess didactic validity. Also, teachers have many stimuli when discussing with students the activities proposed. The course is based on the development of an industrial product that solves a problem, eventually posed by industry or emerged by customers. Based on the course schedule, the different phases of product development put in evidence the steps that require *divergent thinking* and those where it is necessary to employ *convergent thinking*. A case study allows explaining all the phases of product design.

**Keywords:** Project-Based Learning · Project-Based Pedagogy · Product design and development · Integration of design methodologies

## 1 Introduction

Design education is an experiential path on which students should be engaged in having the development of a product, as main task. After 20 years of teaching on such a matter, present authors said explicitly that there is no other method for students to understand the basis of design. As underlined by Dym, Agogino et al. [1] Project-Based Learning is the most-favored pedagogic model for teaching design. Discussion on Project-Based Learning (PBL) started on the basis of the Aalborg experiment, when the University was established in 1974. Still now Aalborg University is organized on such a method [2] and PBL is the pedagogical paradigm used in all courses organization. PBL's main assumption is the workgroup, as a means by which a set of people are involved in a discussion on a problem and searching for solutions to that problem. To find solution to problems is here focused on the engineering context, since the knowledge that students learned along with their course of study converges towards engineering solutions. The course in which students are asked to find a solution to a problem is placed in the second year of the Master's Degree in Mechanical Engineering and the name is Product Design and Development. As already reported in the past [3], the course is focused on the development of a product, considered as an engineering solution to a problem. After the

L. Roucoules et al. (Eds.): JCM 2020, LNME, pp. 372–378, 2021.
https://doi.org/10.1007/978-3-030-70566-4_59

definition of the problem to be solved, proposed by industry or emerged as a relevant theme, perceived inside the class, a set of activities are scheduled along the semester on which all the attendees are involved, working typically in a team. The paper describes the general organization of the course and the underlying methodologies employed along with some pedagogical considerations. Furthermore, a case study is briefly reported to give an insight to the commitment requested to students, that in all course editions have always positively rated it, by an anonymous survey at the end of semester.

## 2 Course Organization for Master Level

The course is proposed in the last semester of the Master's Degree in Mechanical Engineering. This allows students to employ all the subjects they have studied and learned along with their course of study. The course of Product Design and Development is one of the few offered at UNICAL (University of Calabria), that employs the teamwork as condition to attend it. Only marginally this is due to avoid the terribly amount of work, in term of deliverables, to be done by one person alone. The real motivation to promote teamwork is emphasized by problem solving and decision making tasks, which originates in all phases of the design work. Further it is important to push students towards improving quality of communication and consciousness of their own potentialities. The articulation of the course provides a set of activities that are classified into three learning-induced causes: instructor-directed learning; team-directed learning and individual self-directed learning [4]. During the 14 weeks of the semester, every week the activities are scheduled with two hours of lecture and two hours of teamwork in classroom on different days and at least eight hours of homeworking (in team and/or alone).

### 2.1 Questioning

At the basis of product design, or design in general, there is the need to understand the actual questions that a design solution would try to solve. This phase cannot be performed as individual activity. Students are not able to explore all the aspects visible and/or not visible behind a problem. Surely, well-experienced designers are able to investigate alone all the aspects often hidden behind a problem. A lot of heuristics is required to identify the true factors, discarding all the things not required or confusing. The problem clarification is generally performed by investigating on customer needs and functional decomposition.

### 2.2 Divergent Thinking and Convergent Thinking

In the last decades, the design practice has been analyzed and codified to make it an applicable approach in various areas. An important variation is *design thinking* [5, 6]. This may be summarized as a series of phases, which may seem chaotic at certain moments, that is resolved naturally passing from *divergent thinking* to the *convergent thinking*. *Divergent thinking* involves the typical creative phases in which alternatives are created, sometimes in the presence of more or less planned stimuli. *Convergent thinking* covers the phases in which analyzes are carried out and choices are synthesized. In *Divergent* steps, the previously assimilated knowledge represents the support layer. Therefore, students move from problem framing and diagnosis [7] to a process of exploration of needs and possible solutions starting from extensive questioning. *Convergent* steps, on the other hand, allow making the right choices with the aid of engineering knowledge. According to this scheme and working in team, sometimes supported by the teacher, interesting results may be obtained in a product design process. In the timetable shown in Table 1 it can be seen how in each phase of the design process, *divergent* and *convergent* approaches are identified. In the Table these are marked there with the symbols (<) or (>). The decades of experience in teaching product design has led the authors to identify and continuously update the most suitable practices and tools, also referring to the literature [8–13]. Lectures are delivered in order to illustrate methods and tools for stimulating and fostering the *divergent* phases. Furthermore, teachers introduce the most suitable methods to manage the *convergent thinking* in order to orient students towards an acceptable design solution [14]. For this purpose, they encourage students to resume those basics that they have acquired during their studies, sometimes integrating them.

## 2.3 Pedagogical Aspects of Teaching Design by PBL

At the early editions of the course, it was almost a gamble to organize a design activity on something that was not already designed. In the beginning, the starting point was to decide which kind of device one people would have been liked to design and a design process would be started with few variations, i.e., in terms of dimension or operation condition or productivity. Design activity consisted of an ordered sequence of steps, with few feedbacks, to converge towards a defined solution. The change of paradigm emerged when a problem was assigned instead.

A lot of discussion about the investigation around the problem would have required the need of more people involved in the process because one person alone would have been fruitless. During the course evolution over the years, according to the phases summarized in Table 1, students' aggregation in team came as a spontaneous requirement. And the optimal number of the team, for this class, has been settled around three elements per group (minimum two and maximum four). It was recognized that, generally, the workload during homework activity seems to be distributed equally among the team members. At the end of the product development process, to attend the exam students must write a mandatory essay in which all the steps followed are reported and discussed. This part can be written collegially or as a collage of many parts each one assigned to a team component. The exam is passed after a meeting in which all teams show their own product. A certain time (typically 1/2 h) is assigned to each project and every student talks about some topics. This part is necessary to verify either the level of agreement among team members and the ability of each one. A second part is dedicated to an individual exam in which each student must answer on the complementary topics he/she did not treat previously. The project assessment is a sort of combination of the ranking of the project discussion (2/3) and the individual skill (1/3). Basically the way in which the problems have been faced and solved are assessed, instead of the product itself. Also when the theme is proposed by firms, the pedagogical aspects are considered as more relevant.

**Table 1.** Timetable of the course lectures.

| Topic | Week | Lecture | Convergent (>) VS Divergent (<) |
|---|---|---|---|
| 1. Clarify the task | 1 | The nature of design and the need for systematic approaches. | - |
| | 2 | The steps of product design and development process. The product life cycle. | - |
| | 3 | Identification of market opportunity. Identification of customer needs; product planning; the requirement list. | < |
| 2. Conceptual design | 4 | The functional approach to design; the functional net: clustered graph and function tree. | < |
| | 5 | Concept generation: search internally, individually and in-group; externally activities. The design creativity. | < |
| | 6 | Concept selection. Matrix of relative and absolute importance: screening and scoring. | > |
| 3. Embodiment | 7 | Product architecture and product structure. | < > |
| | 8 | Introduction to DSM (Design Structure Matrix) | > |
| | 9 | Axiomatic design and design matrix. | > < |
| | 10 | DfM (Design for Manufacturing) and DfA (Design for Assembly). | < > |
| | 11 | DfE (Design for Environment). | < > |
| 4. Detail design | 12 | Robust Design (Taguchi method) and ANOVA. | > |
| | 13 | House of Quality. Customer satisfaction (Kano model). | > < |
| | 14 | Detail design and the geometric quality of the industrial products | > |

# 3   Case Study

The 2019 edition of the course proposed the theme "Transporting liquids for medium-term excursions maintaining the temperature". The final product proposed has been a canteen. One of such products can be seen in Fig. 1, that has been also proposed by a rapid prototyping process (FDM). The theme was decided by students after some time on which many alternatives were arranged. The theme has been particularly relevant in 2019 considering that the movement "Friday for future" demanded the governments on planet survival. "NOplanetB" [15] has been, and is still now, the slogan that explains the crude reality for the battle against climatic changes. After a research on such matter, Brainwriting [16] allowed students to generate an average of 10 ideas per team. The employment of *divergent thinking* was applied by the student for the first time along with their curriculum and the emotional trouble reported in the literature [17, 18] was confirmed. A lot of works followed this first phase and a set of methods, employed for selection, comparison, trade-off, optimization, and finally, the drafts, complete with the checks on Geometric Dimensioning and Tolerancing, allows to give birth to the device.

## 4 Discussion

Over the years, the experience matured by teachers convinced them of the opportunity to emphasize the importance of teamwork. The team is considered the nucleus on which all elaborations must be referred. The people involved in a course of product design are many. Teacher, Students, Customer/s. The way in which they are related is at the basis of the goal of such kind of course: *Teach to Design* and above all *Learn to Design*. Indubitably a set of information must be transferred to students and a certain number of lectures are required. Many are the approaches proposed worldwide and it can be said that no one may solve completely all questions. It remains to teachers the selection of the methodologies more appropriate for the solution of a defined problem. Concerning Table 1, some topics could be treated only as information, considering that a particular problem probably does not require those methods. Immediately after each lecture, teamwork is required to verify if the theoretic introduced have been matured and well applied.

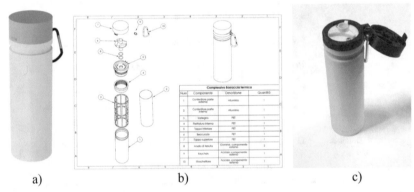

a)                              b)                                              c)

**Fig. 1.** First prototype of the canteen: a) 3D model; b) exploded draft; c) 3D printing prototype.

Discussions inside the design team allow students to clarify if their understanding is well established. Teamwork is essential to search for alternative products already present on the market, for consulting of patents and literature. It is fundamental for organizing interviews with potential customers of the proposed device. Conceptual design is mandatory for the team, where brainwriting methodology is applied. In brief, many phases of design must be done inside the team, because the amount of work must be shared among pairs and performed with the same degree of consciousness. Individual work is also natural along the semester because each team component continues to think all aspects of the problem also when he/she remains alone. Each team member is involved in the creative process and it is not clear when the bulb start lighting for him/her, as Wallas described in the incubation phase of his model of the creative process [19]. Typically, not only during workshops. A designer looks like a lover that thinks to his/her love over and over again. Product design is a fascinating activity. Or one person is involved in it, or NOT. A designer cannot be considered a cold lover. Project-based design is a stimulating atmosphere that involves all attendees and produces a circulation of ideas.

Teachers must also answer to subjects or solutions that they never thought and that can reveal new ways of solution.

## 5    Conclusion

The program of the course of Product Design and Development held at the University of Calabria can now be considered consistent and well arranged. Along the time new experiments have been tried and constantly students answered positively to the organization. The course completes student skillfulness at the end of their curriculum because the course is put in the second semester of the Master's Degree. Anyway, the strength's point is the organization of the work in teams. The discussion, the questioning, the sharing of tasks introduces a different way of solving problems, first of all, because all work must be generated, conceived, organized, shared, assessed. Not only during *Divergent* phases in which several points of view must be considered but also in *Convergent* phases where the right criteria must be chosen in order to validate the promising ideas. The approach Project-Based Learning is ideal for Product design teaching. Moreover, it has been noticed over the years how this type of teaching engages students to such an extent that, in general, all of them pass the exam on the first attempt and with a very high mark.

## References

1. Dym, C.L., Agogino, A.M., Eris, O., Frey, D.D., Leifer, L.J.: Engineering design thinking, teaching and learning. J. Eng. Educ. **94**(1), 103–119 (2005)
2. https://www.en.aau.dk/education/problem-based-learning. Accessed 28 Jan 2020
3. Rizzuti, S.: A coherent teaching program for a course of product design at master degree level. In: XXIII International Conference on Graphic Engineering, Madrid, 19–21 June 2013, pp. 1–5 (2013). ISBN 978-84-695-7930-5
4. Larsen, S.B., Kjaersgaard, N.C., Bigum, P.V., Jacobsen, P.: Understanding how students learn in project-based courses: a review of literature. Paper presented at the 47th SEFI Annual Conference, Budapest, Hungary. DTU Library (2019)
5. Brown, T.: Design thinking. Harvard Bus. Rev. **86**(6), 84–92 (2008)
6. Razzouk, R., Shute, V.: What is design thinking and why is it important? Rev. Educ. Res. **82**(3), 330–348 (2012)
7. Jackson, N.J., Buining, F.: Enriching problem-based learning through design thinking. In: Barrett, T., Moore, S. (eds.) New Approaches to Problem-Based Learning: Revitalising Your Practice in Higher Education, pp. 269–293. Routledge, London (2010)
8. Otto, K.N., Wood, K.L.: Product Design: Techniques in Reverse Engineering and New Product Development. Prentice-Hall, Upper Saddle River (2001)
9. Taguchi, G., Chowdhury, S., Wu, Y.: Taguchi's Quality Engineering Handbook. Wiley, Hoboken (2005)
10. Suh, N.P.: Axiomatic Design: Advances and applications. Oxford University Press, Oxford (2001)
11. Ulrich, K.T., Eppinger, S.D.: Product design and development. McGraw-Hill, New York (2012)
12. Cross, N.: Engineering Design Methods, Strategies for Product Design. Wiley, Hoboken (2000)

13. Pahl, G., Beitz, W., Feldhusen, J., Grote, K.H.: Engineering Design: A Systematic Approach. Springer, Heidelberg (2007)
14. Rizzuti, S., De Napoli, L.: ANOVA applied to the taguchi method: a new interpretation. In: Rizzi, C., Andrisano, A., Leali, F., Gherardini, F., Pini, F., Vergnano, A. (eds.) Lecture Notes in Mechanical Engineering. Springer, Cham (2020)
15. https://www.noplanetb.net/. Accessed January 2020
16. Rizzuti, S., De Napoli, L.: Proposal of a framework based on continuous brainwriting to expand mindfulness in concept generation. In: Rizzi, C., Andrisano, A., Leali, F., Gherardini, F., Pini, F., Vergnano, A. (eds.) Design Tools and Methods in Industrial Engineering, ADM 2019. Lecture Notes in Mechanical Engineering. Springer, Cham (2020)
17. Baruah, J., Paulus, P.B.: Collaborative creativity and innovation in education. In: Mullen C. (ed.) Creativity Under Duress in Education? Creativity Theory and Action in Education, vol. 3. Springer, Cham (2019)
18. Henriksen, D., Richardson, C., Mehta, R.: Design thinking: a creative approach to educational problems of practice. Thinking Skills Creativity **26**, 140–153 (2017)
19. Wallas, G.: The art of thought. J. Cape, London (1926)

# Investigating Students' Perception on the Outcomes of Technical Drawing University Courses

Barbara Motyl[1]([⊠]), Stefano Filippi[1], Gabriele Baronio[2], and Domenico Speranza[3]

[1] DPIA, University of Udine, Udine, Italy
`barbara.motyl@uniud.it`
[2] DIMI, University of Brescia, Brescia, Italy
[3] DICEM, University of Cassino and Southern Lazio, Cassino, Italy

**Abstract.** This paper presents the results of a survey carried out with students enrolled in the first two years of the BS in Engineering at three Italian university locations. The study is part of a wider range of methods, tools and aids for the improvement of teaching and learning of technical drawing at university level developed by the University of Brescia, Udine, and Cassino and Southern Lazio. In particular, this work analyses the results of questionnaires related to the basic technical drawing outcomes, taking inspiration from previous research work in this field. What emerges is a positive picture that shows students' interest in 3D CAD modeling topics such as part or assembly construction, but also their interest in more traditional subjects like sketching and dimensioning.

**Keywords:** Technical drawing · CAD · Engineering education · Students' surveys

## 1 Introduction

This work is part of a research aimed at improving teaching and learning of technical drawing (TD) for BS courses in Engineering. As mentioned in previous works [1–5], the Universities of Brescia, Udine and Cassino and Southern Lazio have been collaborating on these themes since 2014. In this paper, the authors analyze the results of questionnaires proposed to first years students in Industrial or Mechanical Engineering during the academic year 2016–2017. One of the goals of the survey is to clarify what students think and what are their opinions on some outcomes related to TD.

After the background section summarizing the reasons for this survey and the sources chosen as starting points, section two describes the survey contents and the working methodology adopted. In section three the results are reported while in section four they are analyzed and discussed. Finally, conclusions and possible future developments are pointed out.

L. Roucoules et al. (Eds.): JCM 2020, LNME, pp. 379–385, 2021.
https://doi.org/10.1007/978-3-030-70566-4_60

## 2  Background

The starting point for structuring the questionnaire and for selecting the questions to use was the work of Barr [6, 7], where he discusses the current state of engineering graphical communication in Engineering Education in the US. In particular, Barr defined some educational outcomes for technical drawing or engineering graphics (as this discipline is called in USA). He chose fourteen potential outcomes and conducted a survey by interviewing a few colleagues to assess their importance and to understand which one could be introduced as part of an engineering graphics course. In particular, these fourteen outcomes are related to traditional technical drawing, 2D and 3D CAD topics.

**Table 1.** Barr's fourteen outcomes and the results of the 2004 and 2012 surveys [6, 7].

| Label | Description | Average score 2004 | Average score 2012 |
|-------|-------------|--------------------|--------------------|
| O1 | Ability to sketch engineering objects in the free hand mode | 4,67 | 4,54 |
| O2 | Ability to create geometric construction with hand tools | 2,13 | 2,71 |
| O3 | Ability to create 2D computer geometry | 4,21 | 4,08 |
| O4 | Ability to create 3D solid computer models | 4,75 | 4,75 |
| O5 | Ability to visualize 3D solid computer models | 4,46 | 4,54 |
| O6 | Ability to create 3D assemblies of computer models | 4,29 | 4,54 |
| O7 | Ability to analyze 3D computer models | 3,71 | 4,13 |
| O8 | Ability to ability to generate engineering drawings from computer models | 4,33 | 4,29 |
| O9 | Ability to create section views | 4,13 | 4,33 |
| O10 | Ability to create dimensions | 4,38 | 4,38 |
| O11 | Knowledge of manufacturing and Rapid Prototyping methods | 3,42 | 3,63 |
| O12 | Ability to solve traditional descriptive geometry problems | 2,29 | 2,75 |
| O13 | Ability to create presentations | 3,42 | 3,46 |
| O14 | Ability to perform design projects | 3,96 | 4,08 |

In his surveys, Barr proposed to evaluate their importance to twenty-four engineering graphics faculty and members of the Engineering Design Graphics Division of the ASEE, (https://edgd.asee.org/), asking to rank each outcome using a numerical scale from 1 to 5, where 1 corresponds to "Not Important at all" and 5 to "Very Important". The survey was conducted twice, first in 2004 and again in 2012. The results of these surveys can be summarized as follows. In addition to the predominance of CAD-related outcomes in the results of the surveys, some traditional outcomes like sketching, dimensioning and sectioning, are also perceived as important by the interviewed people. Therefore, as reported by Barr, there is still a perception of the need to teach graphic fundamentals in Engineering Education [6]. Table 1 reports the fourteen outcomes and the average scores obtained in the 2004 and in the 2012 surveys [6, 7].

## 3   The Survey

Thus, taking inspiration from the Barr's outcomes, from the literature on latest technological and educational trends [8, 9] and from the experience of the authors as instructors and researchers [10], a questionnaire has been developed to propose online to students of the first years of Engineering courses in the three Universities involved. The questionnaire consisted of twenty-six questions. The first fourteen questions were related to the Barr's outcomes and the students were asked to evaluate them using the same one-to-five scale. Ten questions were related to Industry 4.0 topics, to the use of educational support tools such as smartphones, tablets, personal computers, learning management systems and specific CAD software. In addition, there were two open questions to collect the students' opinion about their possible participation at design competitions and about the TD course in general.

**Table 2.** Data relating to the sample of students interviewed.

| | Students | Male (#; %) | | Female (#; %) | |
|---|---|---|---|---|---|
| Brescia | 97 | 88 | 90,7 | 9 | 9,3 |
| Cassino | 128 | 105 | 82,1 | 23 | 17,9 |
| Udine | 70 | 55 | 78,6 | 15 | 21,4 |
| TOTAL | 295 | 248 | 84,1 | 47 | 15,9 |

This paper focuses the analysis on the results obtained thanks to the fourteen questions concerning the Barr's outcomes to put our recent results (academic year 2016–2017) into relationship with the Barr's recent ones (2012), although the former refers to students and the latter to teachers. The questionnaire has been proposed to 295 students during the first lessons of the TD courses at the three university locations during the academic year 2016–2017. Table 2 shows the data relating to the sample of students interviewed.

## 4   Results

Table 3 shows the results of the survey; they appear both divided by university and as totals. The average scores for each single question have been calculated to get an easier reasoning about these results and in order to make the comparison to the values obtained by Barr feasible.

**Table 3.** Results obtained for the 14 outcomes during the survey.

| Outcome | Average scores | | | |
|---------|---------|---------|-------|-------|
|         | Brescia | Cassino | Udine | Total |
| O1  | 3,68 | 4,16 | 3,70 | 3,89 |
| O2  | 3,87 | 3,96 | 3,70 | 3,87 |
| O3  | 4,36 | 4,55 | 4,36 | 4,44 |
| O4  | 4,52 | 4,54 | 4,29 | 4,47 |
| O5  | 4,56 | 4,45 | 4,31 | 4,45 |
| O6  | 4,19 | 4,06 | 3,84 | 4,05 |
| O7  | 4,34 | 4,35 | 4,24 | 4,32 |
| O8  | 4,03 | 4,05 | 3,81 | 3,99 |
| O9  | 4,11 | 4,29 | 3,90 | 4,14 |
| O10 | 4,42 | 4,55 | 4,20 | 4,42 |
| O11 | 3,72 | 3,85 | 3,66 | 3,76 |
| O12 | 3,70 | 4,06 | 3,56 | 3,82 |
| O13 | 4,33 | 4,02 | 4,13 | 4,15 |
| O14 | 3,76 | 4,38 | 4,10 | 4,11 |

## 5  Discussion

The results are analyzed both individually, for each of the three sites involved, and in an aggregate way, reasoning on the total average scores and comparing them with the values of the 2012 survey conducted by Barr [7]. Considering each university, what emerges from the analysis of the average scores obtained for the various outcomes is as follows.

The students at the University of Brescia considered O5, O4, and O10 as the most important outcomes, while O1 was the least important. In Cassino, the outcomes considered as the most important were O3, O10, and O4, while O11 was the least important. In Udine, the O3, O5, and O4 outcomes were considered the most interesting, while O12 was considered the least attractive. All of this highlights quite clearly the students' interest in 3D CAD modeling (O4), even if the dimensioning (O10) also represents one of the most interesting topics in two out of the three locations. On the other hand, considering the data in the aggregated way, the totals confirm O4, O5 and O3 as the most interesting outcomes, followed by O10. Conversely, the least attractive are O11, O12 and O2. If these data report the opinion of students at the time of their entry into university courses, please consider now what can be said by comparing these data with those obtained by Barr in 2012, referred to teachers' opinions rather than students' ones. Figure 1 reports the histogram comparing the average scores of the three Universities, the total average scores and the Barr's values, all of this for each of the fourteen outcomes.

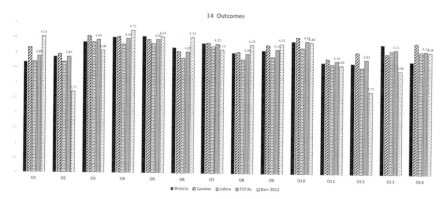

**Fig. 1.** Average scores in the three Universities, as totals and from Barr's 2012 survey [7].

Basically, there is a correspondence between the students and teachers' opinions, despite the difference in time. In fact, both show greater interest in 3D CAD topics in general (modeling of parts, visualization, creation of assemblies), even if some more traditional topics such as hand drawing (O1) and dimensioning (O10) are reputed as interesting as well. Only O1, O2, O12 and O13 show discrepancies. O1 seems to be more interesting for teachers while the other three appear as more interesting for students. This could be due to several factors; for example, the time distance between our investigation and the one carried out by Barr, but above all the different level of experience with TD related topics held by the students and the teaching staff.

## 6 Conclusion and Future Developments

The survey investigating students' perception on TD topics proposed during the BS courses had interesting results. It allowed getting an updated and wider picture of what students think about TD, showing that they are aware of the different outcomes related to it and that even if there is a clear interest in the most modern topics like 3D and 2D CAD, more traditional and fundamental topics such as sketching and dimensioning must not be neglected.

These results pushed the authors to plan to propose the questionnaire every year from now on and to make it available also to other interested Universities. The data collected throughout will allow verifying the goodness of the questionnaire in terms of completeness and robustness. Moreover, all of this will allow highlighting and dealing with possible bias due to the design and selection of the questions.

**Acknowledgments.** The authors would like to thank all the engineering students who participated at the survey during the Academic Year 2016–17.

# References

1. Speranza, D., Baronio, G., Motyl, B., Filippi, S., Villa, V.: Best practices in teaching technical drawing: experiences of collaboration in three Italian Universities. In: Eynard, B., Nigrelli, V., Oliveri, S.M., Peris-Fajarnes, G., Rizzuti, S. (eds.) Advances on Mechanics, Design Engineering and Manufacturing. LNCS, pp. 903–913. Springer, Cham (2017). https://doi.org/10.1007/978-3-319-45781-9_90

2. Baronio, G., Motyl, B., Paderno, D.: Technical Drawing Learning Tool-Level 2: an interactive self-learning tool for teaching manufacturing dimensioning. Comput. Appl. Eng. Educ. 24(4), 519–528 (2016). https://doi.org/10.1002/cae.21728

3. Villa, V., Motyl, B., Paderno, D., Baronio, G.: TDEG based framework and tools for innovation in teaching technical drawing: the example of LaMoo project. Comput. Appl. Eng. Educ. 26(5), 1293–1305 (2018). https://doi.org/10.1002/cae.22022

4. Baronio, G., Bodini, I., Copeta, A., Dassa, L., Grassi, B., Metraglia, R., Motyl, B., Paderno, D., Uberti, S., Villa, V.: Integrated approach to the innovation of technical drawing teaching methods. In: Cavas-Martínez, F., Eynard, B., Fernández, C.F., Fernández-Pacheco, D., Morer, P., Nigrelli, V. (eds.) Advances on Mechanics, Design Engineering and Manufacturing II. LNCS, pp. 705–713. Springer, Heidelberg (2019). https://doi.org/10.1007/978-3-030-12346-8_68

5. Motyl, B., Baronio, G., Speranza, D., Filippi, S. TDT-L0 a test-based method for assessing students' prior knowledge. In F. Cavas-Martínez et al. (Eds.): INGEGRAF 2019, LNME, p. 10 (2020). https://doi.org/10.1007/978-3-030-41200-5_50

6. Barr R.E.: The current status of graphical communication in engineering education. In: Proceedings of Proceedings - Frontiers in Education Conference, FIE, 2004, vol. 3, pp. S1D-8-S1D. https://doi.org/10.1109/FIE.2004.1408688

7. Barr R.E.: Engineering graphics educational outcomes for the global engineer: an update. Eng. Design Graph. J. 76(3) (2012)

8. Motyl, B., Baronio, G., Uberti, S., Speranza, D., Filippi, S.: How will change the future engineers' skills in the Industry 4.0 framework? A questionnaire survey. Procedia Manuf. 11, 1501–1509 (2017). https://doi.org/10.1016/j.promfg.2017.07.282

9. Ciolacu, M., Tehrani, A.F., Binder, L., Svasta, P.M. Education 4.0 - artificial intelligence assisted higher education: early recognition system with machine learning to support students' success. In: 2018 IEEE 24th International Symposium for Design and Technology in Electronic Packaging, SIITME 2018 - Proceedings, art. no. 8599203, pp. 23–30. https://doi.org/10.1109/SIITME.2018.8599203

10. Metraglia, R., Baronio, G., Villa, V.: Issues in learning engineering graphics fundamentals: shall we blame cad? In: Proceedings of the International Conference on Engineering Design, ICED 2015 (DS 80-10), vol. 10, pp. 31–40 (2015)

# Parametric Cost Modelling for Investment Casting

Marco Mandolini[1](✉), Federico Campi[1], Claudio Favi[2], Paolo Cicconi[3], Michele Germani[1], and Roberto Raffaeli[4]

[1] Department of Industrial Engineering and Mathematical Sciences, Università Politecnica delle Marche, Via Brecce Bianche 12, 60131 Ancona, Italy
m.mandolini@univpm.it
[2] Department of Engineering and Architecture, Università degli Studi di Parma, Parco Area delle Scienze 181/A, 43124 Parma, Italy
[3] Dipartimento di Ingegneria, Università degli Studi Roma Tre, Via della Vasca Navale 79, Roma, Italy
[4] Department of Sciences and Methods for Engineering, Università degli Studi di Modena e Reggio Emilia, Via Amendola 2, 42122 Reggio Emilia, Italy

**Abstract.** This paper presents a parametric cost model for estimating the raw material cost of components realized employing the investment casting process. The model is built using sensitivity analysis and regression methods on data generated by an analytic cost model previously developed and validated by the same authors. This is the first attempt of developing a parametric cost model for investment casting based on activity-based costing. The proposed cost model accounts component volume, material density and material price. The error in estimating the raw material cost for components whose volume is within the common range of investment casting is around 11%.

**Keywords:** Parametric cost modelling · Sensitivity analysis · Investment casting · Cost estimating relationship · Cost estimation

## 1 Introduction and Literature Review

The lost wax casting technique is one of the oldest and most advanced of the metallurgical arts [1]. The precision casting process is capable of producing complex castings with tight tolerances and good surface finish, and can meet the highest performance standards, such as those typical of turbojet engines. Other advantages of this process include the ability to melt materials that are impossible to forge and difficult to process employing other technologies. Wax casted components have weight ranging from few grams to a couple of dozens of kg and dimensions ranging from 5 mm to 300 mm. The tolerance ranges normally achievable in investment cast casting is ±1% of the nominal size, with a minimum of ±0,10 mm for dimensions lower than 10 mm [2], with a minimum roughness of 3.2 μm [3].

The manufacturing cost is one of the most important design requirements, for many kinds of products, which drives most of the technical and technological solutions [4].

L. Roucoules et al. (Eds.): JCM 2020, LNME, pp. 386–392, 2021.
https://doi.org/10.1007/978-3-030-70566-4_61

For casting processes, different analytic cost models have been developed based on the process's peculiarities. Several researchers [5–7] have identified the major cost items of a casting process, such as material, tooling, labour, energy and overheads. The DFM Concurrent Costing® software developed by Boothroyd Dewhurst Inc. [8] has an analytic and detailed investment casting cost estimation module aimed at product designers. It considers most phases of the process (pattern and core manufacturing, pattern and cluster assembly, cleaning and etching, investment operation, melting, sintering, break out, blast cleaning and cut off).

However, analytical cost estimation methods and related cost models can be used only during the embodiment and detailed design phases, when the product is almost completely defined. For estimating manufacturing cost at the conceptual design phase, parametric cost models seem more suitable for this aim. The parametric cost estimation method involves formulating relations between product characteristics and its cost using available data. The issues of using analytic cost models for estimating the investment casting process during the design phase is even stronger for investment casting. This manufacturing process consists of several manufacturing phases and product/process cost drivers, which are too many for being managed by a design engineer.

Nowadays, industry 4.0, internet of things and data analytic are paradigms that foster the popularity of parametric cost modelling and cost estimating relationship methods [9]. Nonetheless, in literature, only few works presenting parametric cost models for investment casting. Among them, the most significant is that one presented by Creese [10], which considers the following product drivers: number of surface patches, number of patterns, volume of the part, floor area, material density and price. The parametric cost model is built according to historical data and this activity is also called activity-based costing [11, 12]. However, it is not recognizable any scientific paper presenting a parametric cost model for investment casting process, based on data got from an analytic model. This is the novelty of the paper.

Based on a well-established cost estimating relationship approach [9], this paper presents a parametric cost model for investment casting. The model is built using sensitivity analysis and data regression methods on data generated by an analytic cost model previously developed and validated by the same authors.

## 2  Cost Estimation Relationship Building Process

The estimation of the manufacturing cost for wax-casted products generally requires complex cost estimation relationship, which need various inputs by the designer. Often, the designer knows this data only at an advanced stage of part design. The purpose of this section and in general of the study is to obtain a parametric relationship which allows a cost estimation of the part using data already knows in the early stages of component design. The cost estimating relationship (CER) is the distinguishing feature of parametric cost estimation. A CER is a mathematical expression that describes how the values of, or changes in, a "dependent" variable are partially determined, or "driven," by the values of, or changes in, one or more "independent" variables [9]. The CER building process described later is divided in three parts and it uses the method developed by the International Society of Parametric Analysts [9].

## 2.1 Data Collection

All parametric estimating techniques, including CERs and complex models, need credible data before they can be used effectively. Parametric techniques require the collection of historical cost data and the associated non-cost information and factors that describe and strongly influence those costs (technical non-cost data). Technical non-cost data describes the physical, performance, and engineering characteristics of a system, subsystem or individual item. Cost data are obtained using an analytical cost model for investment casting, developed in a previous work, which is an improvement of that one proposed by Boothroyd and Dewhurst [8] and taken as reference. The approach for this improvement consists of four steps, hereunder summarized.

The *first step* consisted in collecting equations and data, available in Boothroyd and Dewhurst [8], within an electronic spreadsheet in which investment casting process cost has been divided in 13 phases: (i) core manufacturing, (ii) pattern manufacturing, (iii) pattern assembly, (iv) cluster assembly, (v) dissolving core, (vi) cleaning and etching, (vii) investing pattern cluster, (viii) melt out, (ix) sintering, (x) melting, (xi) break out, (xii) blast cleaning and (xiii) cut off. The costs of each phase are composed by *process costs* (manufacturing, setup and consumables costs) and *raw material costs* (the costs related to the casted metal).

The *second step* consisted in organizing two workshops with foundries to improve the cost model.

During the *third step* all the improvements to the investment casting cost model proposed in DFM have been integrated in the electronic spreadsheet developed at step 1. In this manner, it was possible to draft the cost breakdown for investment casting and detect the most cost expensive phases: (ii) *pattern manufacturing* ($12 \div 20\%$), (iv) *cluster assembly* ($11 \div 12\%$), (vii) *investing pattern cluster* ($14 \div 15\%$), (ix) *sintering* ($\sim11\%$) and (x) *melting* ($32 \div 39\%$), which globally represents the $87 \div 90\%$ of the product cost.

In the *fourth step* the cost model has been evaluated by comparing the costs obtained with the proposed model and the actual values of around twenty components provided by the foundries. For all the components it can be noted a deviation lower than 10%.

Starting from the cost breakdown and taking into consideration the most cost expensive phases, cost drivers are the following: *batch size; part volume; part thickness; raw material price; raw material density; labour cost*. For each cost driver a reference value was fixed. These reference values refer to a component of a food packaging machine (weight: 2 kg, dimensions: $160 \times 172 \times 15$ mm) realized in stainless steel AISI316 and manufactured in Italy.

For having enough data for the parametric relationship development, each cost driver indicated above was varied between a maximum and a minimum value and raw material and process costs derived from the analytical model were collected. This sensitivity analysis was carried out in two phases: in the first phase (a) only one cost driver at a time was changed, keeping the others at reference value; instead in the second phase (b) two cost drivers together were changed. The maximum, minimum and reference values of each cost driver are indicated in Table 1.

**Table 1.** Reference, maximum and minimum values of cost drivers

| Cost driver [U.O.M.] | Reference value | Max value | Min value |
|---|---|---|---|
| Batch size [ad] | 1000 | 2000 | 10 |
| Part volume [m$^3$] | 0.00026300 | 0.00063120 | 0.00001325 |
| Part thickness [mm] | 25 | 50 | 2 |
| Raw material price [€/kg] | 3.4 | 5.4 | 0.8 |
| Raw material density [kg/m$^3$] | 7850 | 15000 | 2000 |
| Labour cost [€/hour] | 30 | 70 | 3 |

## 2.2 CER Development

The cost estimating relationship development follows the two-phases analysis indicated above. With the data obtained in the first phase (a) it is possible to evaluate the relationships among the process/raw material costs and the cost drivers. From this first phase it can be concluded that volume and density are the only cost drivers that significantly influence both the cost of the process and the cost of the raw material. The batch size, part thickness, and labour cost affect only the process cost, while a change in the material price only affects the raw material cost. This first step lets to understand the cost drivers to use in the second step to obtain parametric cost relationships. In this study, the focus is about the cost of raw material, which relationship with raw material price, part volume, and density is shown in Fig. 1. For this figure it is possible to observe that raw material cost linearly depends on its unitary price (the higher the unitary price the higher the cost). Raw material cost is related to part volume and density via a step function because, while increasing the weight, pouring and handling systems may change discontinuously.

To build a parametric relationship function of these three cost drivers, it was firstly calculated 26 linear least square regressions (one for each discrete levels of raw material price) between raw material cost, the dependent variable, and part volume, the independent variables. For each regression it was obtained 2 coefficients: the slope and the intercept. Two other linear least square regressions were subsequently made between the obtained coefficients, in this case the dependent variable, and the discrete raw material price levels, the independent variable. By combining the data of the two regressions, a parametric parabolic equation was obtained (1). To develop the previous equation, material density was fixed to the reference value ($\rho = 7850$ kg/m$^3$), allowing the raw material cost calculation only for part which have the same material density of reference value.

$$\begin{aligned} raw\ material\ cost\ (\rho = 7850) = 23760.18 \cdot (part\ volume) \cdot (raw\ material\ price) \\ -2056.62 \cdot (part\ volume) + 0.96 \cdot (raw\ material\ price) - 0.12 \end{aligned} \quad (1)$$

Therefore, in order to account the effects of using different materials (e.g. melting temperature, gate volume, etc.) on the raw material cost, a density factor coefficient ($f_\rho$) it was introduced. Its value varies in function of material density and is equal to 1 for the reference material (AISI 316), less than 1 for materials with a density lower than the reference (0.31 for Aluminium 1100; 0.98 for G 1800 Gy cast iron; 0.99 for 42CrMo4), greater than 1 for material whit a density higher than the reference (1,01 for 39NiCrMo4). Then the cost of raw material can be calculated by Eq. (2).

$$raw\ material\ cost = f_\rho \cdot raw\ material\ cost\ (\rho = 7850) \tag{2}$$

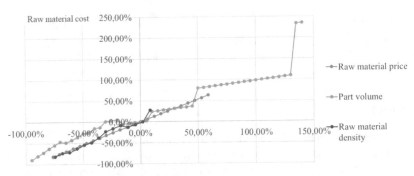

**Fig. 1.** Sensitivity analysis of raw material density, price and part volume on raw material cost.

## 2.3  Validation

The cost estimating relationship must produce, to a given level of confidence, results within an acceptable range of accuracy. The validation of the previous relationship was carried out comparing its results with those deriving from analytical cost model of investment casting. Table 2 presents the absolute cost deviation between estimated values and the data obtained using the analytical model.

**Table 2.** Comparison between analytical model (reference) and parametric model (calculated)

| Component | Material | Reference cost | Calculated cost | Absolute deviation |
|---|---|---|---|---|
| 1 | Aluminium 1100 | € 3.77 | € 4.40 | 17% |
| 2 | G 1800 grey cast iron | € 2.53 | € 2.44 | 4% |
| 3 | 42CrMo4 | € 5.85 | € 6.48 | 11% |
| 4 | 39NiCrMo4 | € 8.29 | € 9.44 | 14% |
| 5 | AISI 316 | € 20.03 | € 23.77 | 19% |
| Average absolute deviation | | | | 11% |

Results show a low deviation between the raw material cost estimated using the parametric model and the data obtained using the analytical model (the average absolute deviation is approx. 11%). This result demonstrates the goodness of the proposed parametric model.

## 3   Conclusions

This paper presented a parametric cost model for investment casting, developed according to the method presented in the parametric estimating handbook. The elementary data used for developing the cost model was obtained from an analytic cost model developed in a previous work and validated with the cooperation of two foundries. By employing sensitivity analysis and regressions methods, the paper presented a parametric cost model for estimating the raw material cost of wax casted parts. The error in estimating the raw material cost for components whose volume is within the common range of investment casting is around 11%.

Future research, following the method used in this work, should aim to develop parametric cost models for the estimating variable (i.e.: consumable, process and setup) and fixed costs (i.e. tooling cost).

## References

1. Pattnaik, S., Karunakar, B.D., Jha, P.K.: Developments in investment casting process—a review. J. Mater. Process. Technol. **212**(11), 2332–2348 (2012)
2. ASM Handbook, Volume 15: Casting, ASM International (2008)
3. Ashby, M.: Materials Selection in Mechanical Design, 3rd edn. (2004)
4. Sohlenius, G.: Concurrent engineering. CIRP Ann. Manuf. Technol. **41**(2), 645–655 (1992)
5. Chronister, T.G.: The Cost of making castings. Casting Eng. **7**(4), 8–11 (1975)
6. Jain, P.L.: Analysis of cost effectiveness for selection of pattern material. Indian Foundry J. **33**(1), 27–30 (1987)
7. Kulkarni, A.: Cost control in foundries through innovation. Indian Foundry J. **34**(1), 21–25 (1988)
8. Boothroyd, G., Dewhurst, P., Knight, W.A.: Product Design for Manufacture and Assembly, 3rd edn. CRC Press (2011)
9. Parametric Estimating Handbook, 4th edn. International Society of Parametric Analysts (2008)
10. Creese, R., Adithan, M.: Estimating and Costing for the Metal Manufacturing Industries, 1st edn. CRC Press (2019)
11. Langmaak, S., Wiseall, S., Bru, C., Adkins, R., Scanlan, J., Sóbester, A.: An activity-based-parametric hybrid cost model to estimate the unit cost of a novel gas turbine component. Int. J. Prod. Econ. **142**(1), 74–88 (2013)
12. Ardiansyah, R., Sutopo, W., Nizam, M.: A parametric cost estimation model to develop prototype of electric vehicle based on activity-based costing. In: 2013 IEEE International Conference on Industrial Engineering and Engineering Management, pp. 385–389 (2013)

# Initial Proposal for a General Systems Engineering Methodology to Early Design Phase Cost/Value Estimation

John Wilson[1], Thècle Alix[2], Elise Gruhier[3], Nicolas Maranzana[4], Nicolas Perry[3(✉)], Gérard Magnin[5], Nicolas Parrod[5], Rémy Servonnat[5], and Julien Tuery[5]

[1] Arts et Métiers Institute of Technology, I2M Bordeaux, 33400 Talence, France
[2] University of Bordeaux, CNRS, IMS UMR 5218, 33400 Talence, France
[3] Arts et Métiers Institute of Technology, University of Bordeaux, CNRS, Bordeaux, INP, INRAE, I2M Bordeaux 33400 Talence, France
nicolas.perry@u-bordeaux.fr
[4] Arts et Métiers Institute of Technology, LCPI, HESAM Université, 75013 Paris, France
[5] CIMPA SAS, 31700 Blagnac, France

**Abstract.** We propose that a systems engineering methodology may be applied in an effective interactive design environment for lifecycle cost estimation and value optimization in the context of a manufacturing enterprise. In order to optimize a product design for value, engineering and manufacturing businesses need to be able to estimate accurately product lifecycle costs during the early design phases of its development, because this is when the majority of these costs are determined. Systems engineering defines realizing value as meeting stakeholder requirements and emphasizes formalizing these in order to link coherently the individual estimated costs of a design to the needs it fulfils. Furthermore, formalized requirement and design parameters are suitable for modelling and simulation, and we envision a systems model implemented within existing knowledge-based engineering tools embedded in a design environment. The results of this model may support design decisions, as well as reinforce systems engineering analyses in evaluating processes for value chain simulations.

**Keywords:** Design-to-cost · Design-to-value · Systems engineering · Knowledge based engineering · Design decision-making · Early design phase

## 1 Introduction

Businesses whose revenues derive from sales of engineered products of their own design and manufacture often suffer from a lack of means to accurately estimate and allocate the costs of product development during the initial design phases of its lifecycle. It is during these early conceptual design phases that decisions are made that dictate the majority (at least 70% [1]) of overall costs of the product. These costs are only fully revealed several phases later however, when they are incurred to put the product into production. This is summarised in a graph shown in Fig. 1, which illustrates how the cost incursion of a

L. Roucoules et al. (Eds.): JCM 2020, LNME, pp. 393–399, 2021.
https://doi.org/10.1007/978-3-030-70566-4_62

product lags the cost commitment across its lifecycle phases [2]. During later lifecycle phases, a business has fewer opportunities to revise the product design and optimise for cost. The result is that the final products may generate sub-optimal values to the business, its customers and other stakeholders.

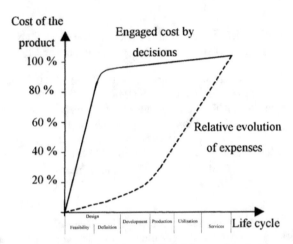

**Fig. 1.** Evolution of committed (engaged) and incurred costs (expenses) [2].

Despite the apparent importance of the design phase for controlling value, it is most commonly only at later lifecycle phases, once the product has entered production, where product value is closely scrutinised and metered by means of statistical process control paradigms for quality management such as Taguchi methods, Six Sigma, and Lean Manufacturing [3]. These focus on increasing value and reducing costs by reducing manufacturing defects and non-value-added expenditures, rather than optimizing the product design itself.

Industry standards on Value Engineering [4, 5], provide methods to formally take account of cost to evaluate designs during design phases, however they do not provide guidance on cost estimation. Value Engineering considers a technical definition of value derived from the general relationship value = function/cost. It insists that value is relative and *"viewed differently by different parties in differing situations"* [4]. The task of optimizing value is therefore achieved by balancing the extent to which needs are satisfied against resource expenditure [4]. Systems engineering, the theory and practice of realising and modelling systems—be they products, services, or organisations—to fulfil a particular purpose, can apply to this task because it is centred on a multidisciplinary interpretation of value as meeting needs, namely 'stakeholder requirements' [3], and that cost may be understood as a member or subcategory of such needs. We therefore propose to investigate how to apply systems engineering to a methodology of concept evaluation and cost estimation. This proposal accordingly identifies three challenges to meet:

1. Implementing a means to estimate product cost and value from design concepts.
2. Integrating a means to display estimate results within the concept design workflow in an interactive environment that responds to evolving product concept data, and which can simulate alternative scenarios.
3. Integrating cost estimate information into design decision and concept selection processes.

The following sections are dedicated to explaining the challenges as well as the approaches, methods and tools that can be used in the first instance to address these issues. Finally, the paper ends with a conclusion and some perspectives to explain how this work will continue.

## 2   Cost Estimation—A Systems Engineering Approach

While a study from Saravi et al. [6] indicates that Taguchi methods can apply to cost estimation in the concept design phases, the literature discusses several dedicated cost estimation techniques for product and service designs. Elmakkaoui [7], and Datta and Roy [8] review and classify these techniques in detail, while Farineau et al. [2] outline a more concise classification summarised by Table 1 with Fig. 2, comparing the approaches of the techniques and indicating to which lifecycle phases they are suited.

**Table 1.** Fundamental cost estimation techniques.

| Name | Description | Analysis methods |
|---|---|---|
| Analogical | Estimation by comparison with similar previously completed projects and existing systems | Case based reasoning |
| Analytic | Estimation by the classification of a system's lifecycle into evaluable constituent processes | Business process modelling Value stream mapping of similar systems |
| Parametric | Estimation by the classification of a system into evaluable components, features or functions which are used to build a cost model supported by statistically derived scaling formulae | Bill of Materials (BOM) analysis Functional analysis Regression |

Two characteristics of cost estimation imply that it is well suited to a systems engineering approach to value optimization:

1. Cost estimation techniques take account of all product lifecycle phases so the domain of design engineering cannot solve the problem of cost estimation alone. This is because manufacturing enterprises integrate multiple operations centres of various disciplines, not just engineering but also project management, production, planning,

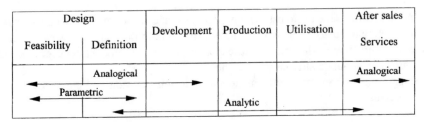

**Fig. 2.** Recommended lifecycle phases for performing cost estimation techniques [2].

supply chain, marketing, finance, customer services etc., which each have different degrees of responsibility over different lifecycle phases. They have complex interrelationships and the decisions and actions of one are liable to affect any of the others. Systems engineering is the domain of engineering theory and practice that analyses these relationships. Fundamentally it does not consider an engineered product and its lifecycle to be the central object of value, but rather the system of interactions between the product, its components, and the various organisational apparatus constructed for its development (thus acknowledging that the latter is itself 'engineered' concurrently with the former) [3]. Precisely, systems engineering is defined by the International Council of Systems Engineering as *"A transdisciplinary and integrative approach to enable the successful realization, use, and retirement of engineered systems, using systems principles and concepts, and scientific, technological, and management methods."* [3].

2.  In order to optimize for value, estimated costs must be representative of the costs of meeting needs. Systems Engineering is advantageous because it focuses on realizing value as opposed to "value proxies" such as technical performance or efficiency, which it achieves when stakeholder requirements are met. Systems engineering therefore emphasizes formally integrating these requirements into the specification of a product system design, in order to link them coherently to its estimated costs [3, 9].

Applied studies of systems-focused cost estimation are described in the literature [10–12], of these, COSYSMO [12] is a parametric model deriving from cost modelling theory applied to both product development and software projects. This model evaluates a system by calculating the person-months required to develop it from 'size drivers' and 'cost drivers', coefficients that estimate the complexity of the system. This model also features in the commercial systems engineering cost estimation tools SEER-H and SEER-SYS from Galorath Inc. [13], assimilating with aforementioned product and service cost models. A blind validation study comparing the former with a competing tool found it had an average error of 23% estimating the cost of twelve NASA science missions, comprising the project management, systems engineering, safety and mission assurance, payload, spacecraft bus, and systems integration and test components [14].

## 3 Cost Estimation in the Interactive Design Environment Workflow—Knowledge-Based Engineering

Systems engineering methods are deployable in most engineering and manufacturing organisations because of the availability of substantial volumes of enterprise records and data. These may be processed and classified to model the product and enterprise systems and simulate alternative scenarios. This is possible thanks to the ubiquity of IT systems in engineering enterprises, which can augment their computing resources as needed by scalable cloud services.

Knowledge-based engineering (KBE) software tools, reviewed by La Rocca et al. [15] and Verhagen et al. [17] are a proven technology to organise, process, and generate systems data for aiding the rapid generation of new designs. In some implementations, these are embedded tools in computer-aided design (CAD) software environments and thus directly integrate into a design workflow. Such tools may be used to structure and classify costs data (among others) for an arbitrary set of system entities and processes, as well as costs calculation rules for use in making estimations for new designs. Verhagen et al. [18] and Zhao et al. [19] both present KBE implementations for cost estimations of aerospace designs. These implementations demonstrate how a product cost model may be constructed and executed. They may be extended and adapted to a systems model, accommodating information flows from multiple functional units (operation centres within the manufacturing business), and generating results for multiple measures of value. Such a KBE implementation may have multiple interfaces embedded within the workflows of the functional units as, for example, modular extensions to existing CAD and enterprise resources planning (ERP) tools.

## 4 Cost Considerations in Design Decisions

Estimating cost during design phases allows it to be a factor in design decisions. The literature describes several concept selection and design decision tools from the mature Pugh concept selection [20] to more recent advances on multi-criteria decision aiding methods [21], and joint probability distribution techniques [22]. These methods compare alternative product concepts over a number of selection phases by scoring them on arbitrary sets of criteria relating to customer/stakeholder requirements. The methods also provide guidance on the selection of criteria, how their relative values are weighted and interact in scoring, and the rules of competition. Such models may consider estimated cost as a single criterion. For the systems engineering view, Bosch-Mauchand et al. [23] present value-based decision support tools for manufacturing processes.

## 5 Conclusion

We propose that an effective methodology for cost and value estimation in the early design of phases of engineered products should follow a systems engineering approach. Systems engineering emphasizes formally defining stakeholder requirements before providing a value analysis of product components and functionality, while considering the costs of its supporting development, manufacturing and servicing apparatus to determine overall generated value. Individual dedicated models exist for product, service and system cost/value estimation techniques. Integrating these may provide an extensive, holistic, and more accurate, methodology or model of cost/value estimation.

A product design workflow may implement such a model via established knowledge-based engineering tools. These tools may query product and process cost and value data to compute evaluations for new designs, and generate results to support design decisions in concept selection by including costs and values as determining criteria. These results may feedback and reinforce systems engineering analyses in evaluating processes for value chain simulations.

## References

1. Asiedu, Y., Gu, P.: Product life cycle cost analysis: State of the art review. Int. J. Prod. Res. **36**, 883–908 (1998)
2. Farineau, T., Rabenasolo, B., Castelain, J.M., Meyer, Y., Duverlie, P.: Use of parametric models in an economic evaluation step during the design phase. Int. J. Adv. Manuf. Tech. **17**, 79–86 (2001)
3. SEBoK Editorial Board: The Guide to the Systems Engineering Body of Knowledge (SEBoK), vol. 2(1). In: Cloutier, R.J. (Editor in Chief). The Trustees of the Stevens Institute of Technology, Hoboken (2019). www.sebokwiki.org. Accessed 26 Dec 2019
4. Association Française de Normalisation: NF EN 12973 Value Management. AFNOR (2000)
5. SAVE International: Value Standard and Body of Knowledge (2009). https://web. archive.org/web/20090319204415, http://www.value-eng.org/pdf_docs/monographs/vmstd. pdf. Accessed 23 Jan 2020
6. Saravi, M.E., Newnes, L.B., Mileham, A.R., Goh, Y.: Estimating cost at the conceptual design phase to optimize design in terms of performance and cost. Loughborough University (2012). https://hdl.handle.net/2134/10730
7. Elmakkaoui, M.: Cost estimation in early design phases of projects: Development of a R&D and investments costing tool for vehicle projects. Arts et Métiers ParisTech (2019)
8. Datta, P.P., Roy, R.: Cost modelling techniques for availability type service support contracts: a literature review and empirical study. CIRP J. Manuf. Sci. Tech. Ind. Prod. Serv. Syst. **3**, 142–157 (2010)
9. Boehm, B., Jain, A.: A value-based theory of systems engineering. In: INCOSE International Symposium, vol. 16, pp. 840–854 (2006)
10. SEBoK Editorial Board: Economic Value of Systems Engineering - SEBoK [WWW Document] (2019). https://sebokwiki.org/wiki/Economic_Value_of_Systems_Engineering. Accessed 05 May 2020
11. Sydor, P., Shehab, E., Mackley, T., John, P., Harrison, A.: Improvement of system design process: towards whole life cost reduction. In: Procedia CIRP, Proceedings of the 3rd International Conference in Through-life Engineering Services, vol. 22, pp. 293–297 (2014)
12. Valerdi, R.: The Constructive Systems Engineering Cost Model (COSYSMO). University of Southern California (2005)
13. McRitchie, K., Kha, K.: Addressing the Challenges of Systems Engineering Estimation. Galorath Incorporated (2016)
14. Friz, P.D., Klovstad, J.J., Leser, B.B., Towle, B.C., Hosder, S.: Blind study validating parametric costing tools PRICE true planning and SEER-H for NASA science missions. In: 2018 AIAA SPACE and Astronautics Forum and Exposition. Presented at the 2018 AIAA SPACE and Astronautics Forum and Exposition, American Institute of Aeronautics and Astronautics, Orlando, FL (2018)

15. La Rocca, G.: Knowledge based engineering: between AI and CAD. Review of a language based technology to support engineering design. Adv. Eng. Inform. **26**, 159–179 (2012)
16. Verhagen, W.J.C., Bermell-Garcia, P., van Dijk, R.E.C., Curran, R.: A critical review of knowledge-based engineering: an identification of research challenges. Adv. Eng. Inform. **26**, 5–15 (2012)
17. Verhagen, W.J.C., Garcia, P.B., Mariot, P., Cotton, J.-P., Ruiz, D., Redon, R., Curran, R.: Knowledge-based cost modelling of composite wing structures. Int. J. Comput. Integr. Manuf. **25**, 368–383 (2012)
18. Zhao, X., Verhagen, W.J.C., Curran, R.: Estimation of aircraft component production cost using knowledge based engineering techniques. Adv. Eng. Inform. **29**, 616–632 (2015)
19. Design Institute: Concept Selection. Xerox Corporation (1987). http://edge.rit.edu/edge/P10 505/public/Pugh%20Concept%20Selection.pdf. Accessed 07 Jan 2020
20. El Amine, M., Pailhes, J., Perry, N.: Integration of concept maturity in decision-making for engineering design: an application to a solar collector development. Res. Eng. Design **28**, 235–250 (2017)
21. El Amine, M., Pailhès, J., Perry, N.: Selection and use of a multi-criteria decision aiding method in the context of conceptual design with imprecise information: application to a solar collector development. Concurr. Eng. **24**, 35–47 (2016)
22. Bosch-Mauchand, M., Siadat, A., Perry, N., Bernard, A.: VCS: value chains simulator, a tool for value analysis of manufacturing enterprise processes (a value-based decision support tool). J. Intell. Manuf. **23**, 1389–1402 (2012)

# Industry 4.0 and Decision Making

Frédéric Rosin[1](✉), Pascal Forget[2], Samir Lamouri[3], and Robert Pellerin[4]

[1] LAMIH UMR CNRS 8201, Arts et Métiers, Aix-en-Provence, France
frederic.rosin@ensam.eu
[2] Department of Industrial Engineering,
Université du Québec à Trois-Rivières, Trois-Rivières, Canada
[3] LAMIH UMR CNRS 8201, Arts et Métiers, Paris, France
[4] Department of Mathematics and Industrial Engineering, École Polytechnique de Montréal,
Montreal, Canada

**Abstract.** Industry 4.0 is an ubiquitous term that suggests significant impacts on the productivity and flexibility of production systems. But to what extent do the various technologies associated with Industry 4.0 contribute to enhance autonomy of operational teams by helping them make better and faster decisions, particularly in the context of Lean production system? This paper proposes a model of different types of autonomy in the decision-making process, depending on whether or not the steps in the decision-making process are enhanced by technologies. This model will be tested afterwards in a use case implemented in a learning factory offering Lean management training before being tested in a real production unit.

**Keywords:** Industry 4.0 · Decision-making · Decision types · Autonomous production system

## 1 Introduction

Recent years have been marked by a growing interest in Industry 4.0 and related technologies by manufacturing companies of all sizes. This major technological shift has been proposed to manufacturing companies under various names: L'Industrie du futur in France, Smart Industry in the United States, Industrie 4.0 in Germany, Made in China 2025 in China, etc. In all cases, greater agility and profitability could be achieved through the inter-connectivity of products, machines, supply chains and customers, and the growing decision-making capabilities of systems [1–3].

As there are multiple definitions of the concept of Industry 4.0, the associated technologies have not been definitively identified. The Boston Consulting Group [4] proposes grouping them around nine pillars of Industry 4.0, i.e., autonomous robots, simulation, horizontal and vertical integration of systems, the Internet of Things, cyber security, cloud, additive manufacturing, augmented reality and big data analysis. Other authors have presented different lists of technologies, including Moeuf et al. [3], CEFRIO [5], Dombrowski et al. [6], Mayr et al. [7], Sanders et al. [8], and Wagner [9].

© The Author(s) 2021
L. Roucoules et al. (Eds.): JCM 2020, LNME, pp. 400–405, 2021.
https://doi.org/10.1007/978-3-030-70566-4_63

However, companies are struggling to define the new forms of work organization that will enable them to make the most of this technological shift. In particular, how can they take advantage of these technologies to improve decision-making and to increase the degree of autonomy of operators and machines at the operational level?

Highlighting the difficulty of industrial companies in establishing a strategic roadmap in the jungle of different terminologies, ideas and concepts, Osterrieder et al. [10] proposed a research model about the smart factory, structured around eight distinct thematic perspectives, including decision-making. They underlined the crosscutting nature of the decision-making problems that also appear in the other thematics and affect an extremely wide range of activities in manufacturing. Indeed, the use of Industry 4.0 technologies such as the Internet of Things, cobotics, augmented reality, massive data analysis, machine learning and Artificial Intelligence, can increase the autonomy of production systems, including operators and production equipment, by supporting problem-solving and other decision-making processes. This plays a central role in the concept of autonomous intelligent factories and it is not surprising to note a number of research studies focuse on data-based decision-making in production over a very broad scope covering design, scheduling, planning and process control [10]. But the proposed models do not put into perspective the different types of autonomy in the decision-making process made possible by the use of all the technologies associated with Industry 4.0.

Fortunately, decision-making models have been proposed in other contexts. Simon [11] was one of the first to propose a decision-making model, which is the most concise, yet comprehensive, characterization of a rational decision-making approach [12]. Using the ideas from Simon's model and analysing 25 decisions from different companies, Mintzberg [13] proposed a model for strategic decision-making in companies. This model is non-sequential and allows for the possibility of bypassing certain routines and interrupting the process or providing feedback.

In a similar approach, the Naturalistic Decision Making (NDM) stream [14] has worked on describing the actual decision-making process in an operational context. In this context, some authors have focused on the biases and limitations in human decision-making in situations with strong temporal constraints [15] or crises [14].

In the literature related to the development of artificial intelligence and intelligent agents, other ideal decision-making models have been proposed and are inspired by human decision-making models. This is notably the case of the BDI (Beliefs-Desire-Intention) models [16], which aim to design artificial decision-making systems. However, no model has been proposed yet to analyse or evaluate the impact of Industry 4.0 technologies on decision-making processes.

Recognizing the potential benefits of Industry 4.0 technologies on operational decision-making, this paper proposes a generic model covering several types of autonomies that can be adopted by management, depending on the decision-making process steps that they want to reinforce in the use of technologies. The proposed decision-making process and the different types of autonomies that can be achieved are presented in Sect. 2. Section 3 discusses future research prospects and concludes this article.

## 2  Decision Making Model

Based on the Mintzberg model [13], an ideal decision-making process is proposed (Fig. 1). It is adapted to a decision-making process that is required by the detection of a problem in the broad sense (problem or opportunity) in an operational-level manufacturing production context.

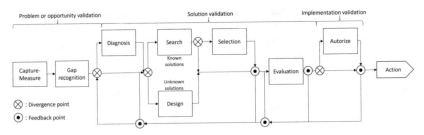

**Fig. 1.** Decision making process in an operational context.

This process includes three phases, similar to Mintzberg [13]: Problem or opportunity validation, Solution validation and Implementation validation. The Problem or opportunity validation phase includes the *Capture-Measure* and *Gap recognition* steps. The *Capture-Measure* step consists of collecting information in real time in the production system. The recognition of an abnormal situation (a problem), i.e. a discrepancy between the current situation and the desired situation, takes place in the *Gap recognition* stage. The Solution Validation phase includes the *Diagnosis, Search, Selection* or *Design* steps. The *Diagnosis* step corresponds to the analysis of the cause and effect relations in the studied situation. Depending on whether or not solutions are known to address the identified problem, the next step is to choose between the *Search* or *Design* steps. If solutions are known, the *Search* step allows one to look for a set of possible solutions by drawing upon the organization's memory. The *Selection* step is then used if the number of known solutions is too high, in order to eliminate a certain number of them and thereby lighten the work to be carried out in the *Evaluation* step. If no solution is known, the *Design* step aims to design a new solution for the problem or to modify a known solution. The *Evaluation* step ensures that the solution chosen will solve the situation, while respecting the constraints of the environment. The third phase includes a single step, *Authorize*. Here, if necessary, an authorization is issued by a higher hierarchical entity, such as a team leader, a manager or a centralized IT system. This step can be bypassed if the work center itself (an operator or a machine) has the authority to validate the implementation.

This decision-making process model can then be used to define different types of autonomy and uses of Industry 4.0 technologies. Such technologies can help a work centre, which include operators and/or machines, to enhance one or more steps in the decision-making process.

Inspired by the four capability areas of Porter and Heppelmann's smart products [2], we propose six types of autonomy in the decision-making process based on Industry 4.0 technologies for manufacturing systems: 1) Cyber Monitoring, 2) Cyber Search, 3) Cyber Control, 4) Cyber Design, 5) Decision Support and 6) Cyber Autonomy (Fig. 2).

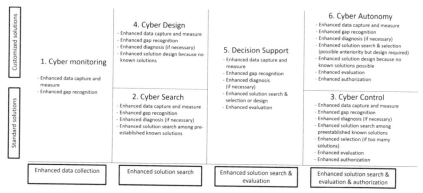

**Fig. 2.** Model of types of autonomy: an Industry 4.0 decision-making support model.

These six types of autonomy are not incremental, in the sense that they are not mutually inclusive and do not present a gradation in terms of intelligence or autonomy. Rather, they respond to different needs for decision support, depending on whether or not one is aware of a more or less limited set of possible solutions and the specific steps reinforced or supported by the types of Technology 4.0 involved. The Cyber Monitoring type corresponds to the reinforcement of the *Capture and Measure* and *Gap recognition* steps facilitating the collection of production data and the analysis of this data to enable the detection of an abnormal situation or an opportunity for improvement. The other stages of the decision-making process are then carried out by men according to usual practices.

The Cyber Search type corresponds to the reinforcement of the *Capture and Measure* and *Gap recognition* steps by using 4.0 technologies but adding optional support for the *Diagnostic* step to facilitate the identification and analysis of the reasons that underly the observed gap. The *Search* step is also strengthened to find possible solutions that may be applied from a set of known solutions.

The Cyber Control type is an extension of the Cyber Search type, where the final stages of the decision-making process are also assisted, including the final stage of authorizing the action. The *Selection* stage is reinforced if the number of possible solutions is too large. The *Evaluation* and *Authorize* steps are also reinforced; a human can then be assisted in all of the steps mobilized in this case.

The Cyber Design type is similar to the Cyber Search type but corresponds to a case in which no possible solution is known for the situation. It is the *Design* stage that is reinforced. In this case, the preferred 4.0 technologies must be able to assist in the design of a new solution to reduce the observed gap.

The Decision Support type is similar to the Cyber Design type, with the addition of the reinforcement of the *Evaluation* step. Only the *Authorization* step remains entirely performed by a human without any assistance.

Finally, the Cyber Autonomy type is based on the Decision Support type, with a reinforcement in the *Authorize* step. As for the Cyber Control type, the human is likely to be assisted in all of the stages of the decision process mobilized in this case.

## 3 Conclusion

In this article, we have proposed an operational decision-making process model describing several types of autonomy that can be achieved with the adoption of Industry 4.0 technologies. This model is a part of a more global study aiming at studying the integration of Industry 4.0 technologies in Lean production systems.

In this regard, a previous study investigating the links between Industry 4.0 and Lean approaches showed that some Lean principles currently appear to be little or not at all improved by Industry 4.0 technologies. These include principles related to employees and teamwork, continuous improvement, stable and standardized processes, and the Toyota philosophy [17].

In order to test the proposed model of different types of autonomy in the decision-making process, a use case is being formalized. It is based on a learning factory and existing Lean management training modules designed in partnership with several manufacturers. A panel of Industry 4.0 technologies such as IoT, cloud, big data analysis, machine learning, simulation, augmented reality and data vizualisation will be progressively deployed. Within this framework, the different types of enhancement of the autonomy of operational teams in decision-making will be tested in order to manage in real time the production problems encountered. This will constitute a first step in validating our model before attempting its implementation within a real production unit.

## References

1. Buer, S.V., Strandhagen, J.O., Chan, F.T.: The link between Industry 4.0 and lean manufacturing: mapping current research and establishing a research agenda. Int. J. Prod. Res. **56**(8), 2924–2940 (2018)
2. Porter, M.E., Heppelmann, J.E.: How smart, connected products are transforming companies. Harv. Bus. Rev. **93**(10), 96–114 (2015)
3. Moeuf, A., Pellerin, R., Lamouri, S., Tamayo-Giraldo, S., Barbaray, R.: The industrial management of SMEs in the era of Industry 4.0. Int. J. Prod. Res. **56**(3), 1118–1136 (2018)
4. Rüßmann, M., Lorenz, M., Gerbert, P., Waldner, M., Justus, J., Engel, P., Harnisch, M.: Industry 4.0: the future of productivity and growth in manufacturing industries **9**(1), 54–89. Boston Consulting Group (2015)
5. CEFRIO: Prendre part à la révolution manufacturière? Du rattrapage technologique à l'Industrie 4.0 chez les PME. Centre francophone d'informatisation des organisations (2016)
6. Dombrowski, U., Richter, T., Krenkel, P.: Inter-dependencies of Industrie 4.0 & lean production systems - a use cases analysis. In: 27th International Conference on Flexible Automation and Intelligent Manufacturing, FAIM 2017 (2017). Procedia Manuf. **11**, 1061–1068 (2017)

7. Mayr, A., Weigelt, M., Kühl, A., Grimm, S., Erll, A., Potzel, M., Franke, J.: Lean 4.0-a conceptual conjunction of lean management and Industry 4.0. Procedia CIRP **72**, 622–628 (2018)
8. Sanders, A., Elangeswaran, C., Wulfsberg, J.P.: Industry 4.0 implies lean manufacturing: research activities in Industry 4.0 function as enablers for lean manufacturing. J. Ind. Eng. Manag. **9**(3), 811–833 (2016)
9. Wagner, T., Herrmann, C., Thiede, S.: Industry 4.0 impacts on lean production systems. Procedia CIRP **63**, 125–131 (2017)
10. Osterrieder, P., Budde, L., Friedli, T.: The smart factory as a key construct of Industry 4.0: a systematic literature review. Int. J. Prod. Econ. **221**, 107476 (2019)
11. Simon, H.A.: The new science of management decision (1960)
12. Lin, H.W., Nagalingam, S.V., Kuik, S.S., Murata, T.: Design of a global decision support system for a manufacturing SME: towards participating in collaborative manufacturing. Int. J. Prod. Econ. **136**(1), 1–12 (2012)
13. Mintzberg, H., Raisinghani, D., Theoret, A.: The structure of 'unstructured' decision processes. Adm. Sci. Q. **21**(2) (1976)
14. Klein, G.: Naturalistic decision-making. Hum. Factors **50**(3), 456–460 (2008)
15. Power, D.J., Cyphert, D., Roth, R.M.: Analytics, bias, and evidence: the quest for rational decision-making. J. Decis. Syst. **28**(2), 120–137 (2019). Special issue on cognitive bias
16. Rao, A.S., Georgeff, M.P.: BDI agents: from theory to practice. In: Proceedings of the 1st International Conference on Multi-Agent Systems, ICMAS-1995, SanFrancisco, USA, June 1995, vol. 95, pp. 312–319 (1995)
17. Rosin, F., Forget, P., Lamouri, S., Pellerin, R.: Impacts of Industry 4.0 technologies on Lean principles. Int. J. Prod. Res. **58**(6), 1644–1661 (2020)

# Multicriteria Decision Making Methods in Engineering Design: An Application on Medical Devices

Cristina Renzi[✉] and Francesco Leali

Department of Engineering Enzo Ferrari, University of
Modena and Reggio Emilia, Modena, MO, Italy
cristina.renzi@unimore.it

**Abstract.** The design of medical devices is challenging, due to strict geometry specifications and criteria belonging to several disciplines. The aim of this paper is to provide a design methodology which seems to lack in the literature of design of biomedical devices. In particular, the case study proposed in this paper concerns with the design of Bone-Biopsy (BB) needle devices. Following the design process of Pahl and Beitz, a functional analysis is carried out, to point out the interfaces between parts of existing BB needles. A morphology matrix is used to generate alternatives for the cannula holder, the core part of the product. Generated alternatives have been ranked and scored by means of the Pugh's Controlled Convergence method according to the decision makers' opinions. In this work, the design peculiarities related to the BB needles which have been pointed out could be extended for guiding the design of other biomedical devices.

**Keywords:** Bone-Biopsy needle · Multicriteria decision making · Design of biomedical devices · Conceptual design · Pugh's Controlled Convergence

## 1 Introduction

The design of bone biopsy needles is challenging, due to several criteria which need to be considered concurrently, as well as strict geometry specifications requested. Hence, design methods should be adopted in order to satisfy constraints and select most suitable alternatives, considering all the multiple criteria and actors involved. The case study of this paper concerns the design of bone biopsy (BB) needle devices. Bone biopsy is a diagnostic procedure consisting in taking a sample of bone tissue to be analysed in the laboratory, by means of a device with a hollow cannula surrounding a stylet (Fig. 1). The cannula and stylet are coaxially assembled and are used to penetrate through the outer layer of the bone, (cortex), which is harder than the trabecular bone layer and the marrow. Reliable devices are characterized by optimized shapes of the handle, needle and cannula.

In the literature, papers related to the design of biopsy needles are mostly aimed at enhancing the biopsy efficiency in relation to the geometry of the cannula and needle. This is for enhancing the sample geometry [1–3], or the efficiency in taking it from the

© The Author(s) 2021
L. Roucoules et al. (Eds.): JCM 2020, LNME, pp. 406–411, 2021.
https://doi.org/10.1007/978-3-030-70566-4_64

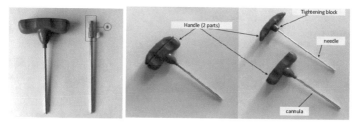

**Fig. 1.** Bone-Biopsy needles (Jamshidi type) and relative components

bone [4–6]. More in general, the use of MCDM methods in the design of medical devices has been proposed in the literature, as in [7–9]. At the authors' best knowledge, no works regard the application of design methods for generating and selecting new conceptual designs. Hence, this work is aimed at describing the requirements related to the design of bone biopsy needles, as well as the steps required for the generation and selection of concepts. This is to enhance the state of the art on the design of medical devices, and guide designers in the assessment of design of novel medical devices, by providing a design methodology which seems to lack in the design of biomedical devices.

## 2 Method

The engineering design process adopted for the generation and selection of biomedical devices is based on the systematic approach proposed by Pahl and Beitz [10]. In Fig. 2, the method followed is schematically depicted. Product specifications are identified within the involved disciplines, (medicine, engineering design, and manufacturing). In the planning phase, a functional analysis is carried out, to identify interfaces between product parts, and analyze them in terms of response to product requirements. To this aim, a benchmark of the existing needles has been performed first, in order to find out the common features among the existing BB-needles. Hence, a functional analysis is carried out, for evaluating the core components of the needle assembly. Related concept designs are generated considering the functions performed by each part of the device. The parts have been collected within a morphology table, screened and scored by means of the Pugh's Controlled Convergence (PuCC) method, which is included in multicriteria decision making (MCDM) methods as described in [11]. PuCC method is used to classify and rank design alternatives candidate solutions, considering the judgments of the decision makers belonging to different technical backgrounds.

## 3 Results

The functional analysis of the BB-needle is schematically reported in (Fig. 3). The handle 1 is made of components A1 (cap) and B1 (handle body) and is the interface with the operator. In the handle, the interface between parts A1 and B1 requires for a reliable tightening and ease of connection. Existing devices use locks as the bayonet lock, snap fits, and cam-clamping. The interface between B1 and 2 is fundamental, as

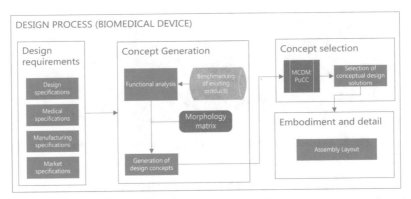

**Fig. 2.** Design process for a biomedical device

the central body is used to couple handle and cannula. The "2-to-3" interface regards the connection of the cannula in the central body. The "3-to-4" interface between the cannula and the needle requires strict coaxial tolerances, hence the dimensions of these parts are constrained by regulations. The "5-to-patient" interface is fundamental as it limits the depth penetration of the needle. The height is regulated by means of a thread ("5 to 2" interface), to guarantee the sliding of the regulator. The interface between B1 and the cannula 3 is the cannula holder 6, which is the core of the product, as it holds the cannula 3 during its penetration into the bone. This is also responsible for the stresses and forces on the operator and patient during biopsy operations. Hence, the generation and selection of novel design are dedicated to the cannula holder component 6. Functions related to the cannula holder are *connection with the syringe (6-1)*, *connection with the handle body (6-2)*, and *connection with the depth regulator 5 (6-3)*. Conceptual designs for the cannula holder are generated by means of a morphology table, taking into account the interfaces with the coupled components as in the functional analysis (Fig. 4).

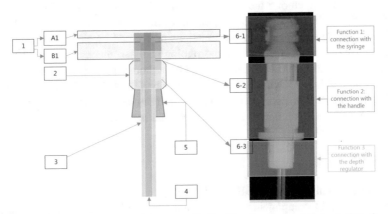

**Fig. 3.** Functional scheme of a bone biopsy needle (1 – Handle. (cover A1 – body B1); 2 – central body; 3 – cannula; 4 – needle (stylet); 5 – depth regulator (guard), 6 – cannula holder: functions 6-1, 6-2, 6-3).

In particular: the connection with a syringe is via a Luer Lock thread (1A) or one contour thread (1B). The tightening on the handle is carried out by means of snap fits of different geometries (2A, 2B and 2C, Fig. 4). The lower side clamping with the regulator (3A, 3B and 3C in Fig. 4) is based on snap-fits. To avoid pairwise comparing and analyzing all 18 generated combinations, a modified PuCC method is used. In particular, concepts are screened according to the functions covered, namely function 1: connection to the syringe (concepts 1A, 1B); function 2: connection to the handle body (concepts 2A, 2B, 2C); function 3: connection to the depth regulator (concepts 3A, 3B, 3C). Concept designs are screened by means of the PuCC method, (Table 1) according to four criteria, namely: ease of manufacturing, easy geometry (presence of undercuts), easy connection with the coupled component, reliable connection with the coupled component. The reference product is the component 6 of Fig. 3. The judgments given to the alternatives in the PuCC matrix derive from a former discussion among the decision makers, namely the designer, the manufacturer, the surgeon and the business manager of a SME in the northern of Italy, which is specialized in biopsy needles. They provided judgments for each design alternative, depending on the specific function (1, 2, 3) and criterion that each part of the device must perform. According to the PuCC method, the most suitable combination for the cannula holder geometry seems to be the 1B-2C-3C Fig. 5. Unlike the AHP method, the PuCC method does not involve analytic calculations for ranking the alternatives. Nevertheless, it maintains the objectivity of the judgments by means of a discussion among the decision makers [11].

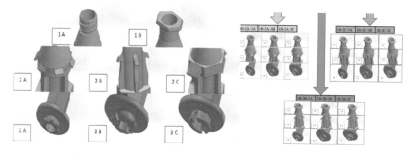

**Fig. 4.** Morphologic table for the inner connection device

**Table 1.** Modified Pugh's Controlled Convergence table for screening concepts according to the functional analysis.

| | Reference product | Function 1: connection to the syringe | | Function 2: connection to the handle body | | | Function 3: connection to the depth regulator | | |
|---|---|---|---|---|---|---|---|---|---|
| Criteria | 6 | 1A | 1B | 2A | 2B | 2C | 3A | 3B | 3C |
| Ease of manufacturing | 0 | − | + | 0 | − | 0 | 0 | − | + |
| Easy geometry (no undercuts) | 0 | 0 | + | − | − | 0 | + | − | + |
| Easy connection with the coupled component | 0 | 0 | + | 0 | + | + | + | − | + |
| Reliable connection with the coupled component | 0 | 0 | + | 0 | − | + | − | + | + |
| Sum of + | | 0 | 4 | 0 | 1 | 2 | 2 | 1 | 4 |
| Sum of 0 | | 3 | 0 | 4 | 0 | 2 | 1 | 0 | 0 |
| Sum of − | | 1 | 0 | 1 | 3 | 2 | 1 | 3 | 0 |
| SUM | | 1 | 4 | −1 | −2 | 0 | 1 | −2 | 4 |

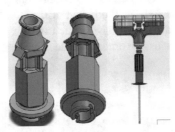

**Fig. 5.** The selected design for the cannula holder (left) and the layout of the entire Bone biopsy needle (right).

# 4 Discussion and Conclusions

The conceptual design of medical devices is challenging, due to strict regulations and criteria to be satisfied. In this paper, the systematic method of Pahl and Beitz [10] is used for the functional analysis of the components in the assembly. Hence a modified PuCC method [11] is adopted for screening and scoring the alternatives for the cannula holder in a BB-needle. The proposed cannula holder seems to provide a better connection to the coupled components, with respect to existing products. This work is aimed at enhancing the state of the art on the design of medical devices, guiding designers in the design of novel medical devices.

# References

1. Keulers, A., Cunha-Cruz, V.C., Bruners, P., Penzkofer, T., Braunschweig, T., Schmitz-Rode, T., Mahnken, A.: Bone biopsy needles: mehanical properties, needle design and specimen

quality [Knochen-biopsiesysteme: Mechanische eigenschaften, nadeldesign und probenqualität]. RoFo Fortschritte auf dem Gebiet der Rontgenstrahlen und der Bildgebenden Verfahren **183**(3), 274–281 (2011)

2. Misiewicz, J., Nguyen, B., Poudyal, A., Wachrathit, K., Chen, Y., Vossoughi, J.: A fiber optic enhanced bone biopsy needle. In: 2012 38th Annual Northeast Bioengineering Conference, NEBEC 2012, pp. 382–383. (2012) Art. no. 6207124
3. Giovannini, M., Ren, H., Cao, J., Ehmann, K.: Study on design and cutting parameters of rotating needles for core biopsy. J. Mech. Behav. Biomed. Mater. **86**, 43–54 (2018)
4. Roberts, C.C., Morrison, W.B., Leslie, K.O., Carrino, J.A., Lozevski, J.L., Liu, P.T.: Assessment of bone biopsy needles for sample size, specimen quality and ease of use. Skeletal Radiol. **34**(6), 329–335 (2005)
5. Uhl, M., Theves, C., Geiger, J., Kersten, A., Strohm, P.C.: The percutaneous bone biopsy: in vitro study for comparison of bone biopsy needles [Die perkutane Knochenbiopsie: Experimentelle Studie zum Vergleich von Knochenbiopsiesystemen]. Zeitschrift für Orthopädie und Unfallchirurgie **147**(3), 327–333 (2009)
6. Islam, A.: Bone marrow solid core biopsy needle: a critical assessment of the utility, benefits and limitations of the instruments employed in current day haematology and oncology. J. Clin. Pathol. **71**(6), 475–482 (2018)
7. Aitchison, G.A., Hukins, D.W.L., Parry, J.J., Shepherd, D.E.T., Trotman, S.G.: A review of the design process for implantable orthopedic medical devices. Open Biomed. Eng. J. **3**, 21 (2009)
8. Ivlev, I., Vacek, J., Kneppo, P.: Multi-criteria decision analysis for supporting the selection of medical devices under uncertainty. Eur. J. Oper. Res. **247**(1), 216–228 (2015)
9. Rogalewicz, V., Jurickova, I.: Multiple-criteria decision making: application to medical devices. In: IWBBIO. pp. 1359–1372 (2014)
10. Pahl, G., Beitz, W.: Engineering Design: A Systematic Approach, 2nd edn. Springer, London (1996)
11. Renzi, C., Leali, F., Pellicciari, M., Andrisano, A.O., Berselli, G.: Selecting alternatives in the conceptual design phase: an application of Fuzzy-AHP and Pugh's Controlled Convergence. Int. J. Interact. Des. Manuf. **9**(1), 1–17 (2013)

# Industry 4.0 and BIM: Do They Share the Same Objectives?

Laurent Joblot[1]([✉]), Christophe Danjou[2], Robert Pellerin[2], and Samir Lamouri[1]

[1] LAMIH-UMR CNRS 8201, Arts et Métiers, Paris, France
Laurent.joblot@ensam.eu
[2] Polytechnique Montréal, Montreal, Canada

**Abstract.** In recent years, many changes have been taking place within the construction sector which is much more prone to innovate than in the past. New forms of processes are emerging with the introduction of digital technologies. This article aims to shed light on recent scientific advances that link Industry 4.0 to this sector. To this end, a review of articles published over the past 10 years reporting experiences and gains from 4.0 technologies applied to construction was conducted. It turns out that recent technological developments have brought new functionalities and new perspectives to companies. Some of these were not initially claimed with the sole deployment of a BIM approach. These new opportunities have ultimately raised questions as to whether and how they could impact the speed at which a digital transformation of the sector could take place.

**Keywords:** Building Information Modeling (BIM) · Industry 4.0 (Ind4.0) · Digital transformation · Construction industry

## 1 Context of the Study

The construction industry plays a key role in the global economy. Despite its importance, it is still described as less productive and innovative than other sectors. However, this situation seems to be changing and several actors have recognized the need to innovate to ensure their competitiveness. New forms of work are emerging thanks to the introduction of technologies linked to Industry 4.0 (Ind4.0). In France and Quebec, several large companies have already adopted such technologies (BIM, virtual and augmented reality, prefabrication, the Internet of Things (IoT), additive manufacturing, robotization, etc.), but their implementation is still in its infancy, particularly among small businesses. These solutions must be encouraged in order to create a true digital transformation of construction. However, companies are still reluctant, not always knowing how to proceed or prioritize and what these technologies can actually bring to them.

In order to help various partners (French and Canadian companies), we felt it necessary to shed light on recent scientific advances that link Industry 4.0 and the construction sector. Several questions emerged at the beginning of our study in June 2019. The one we will try to answer here is, *"What is the relationship between Industry 4.0 and BIM? Do these concepts share the same objectives when applied to the construction sector?"*

L. Roucoules et al. (Eds.): JCM 2020, LNME, pp. 412–418, 2021.
https://doi.org/10.1007/978-3-030-70566-4_65

To answer this question, in the following section we will first present the strategy of the systematic literature review conducted on these questions. A synthesis of the 58 selected articles is then presented in part 3. Part 4 proposes then a deeper analysis of the content, and finally the conclusion sheds light on these recent functionalities and industrial opportunities. Some of these have not been previously when it came to exploiting the BIM model alone.

## 2 Research Methodology

To characterize the contributions of Ind4.0 in the construc tion sector, we first provide a review of the studies on this subject published prior to June 2019. To do so, we have analyzed the scientific publications using the SCOPUS database[1] and post-2010 studies. The query focuses on title, abstract and author keywords on terms related to the construction sector and Ind4.0 in the broadest sense. This initial work made it possible to filter the available publications and to highlight 347 documents. In order to respond to our issue, we chose to study in detail only the articles using *"allow or target or objective or aim or goal"* in the title, abstract or keywords. This additional filter made it possible to keep only articles that openly claimed an intention or purpose. The 68 articles that responded to this new query and that could provide clarification to our partners are detailed and analyzed in the following section.

## 3 Content of the Selected Papers

After an in-depth study of the selected articles, 10 turned out to be irrelevant. The technologies tested and the experiments carried out ultimately led to many advances presented in these articles and listed in Table 1 below[2]:

Numerous overlaps and convergences were thus revealed. For example, **items in bold in the table had an environmental intent**. The following section therefore seeks to define what might be the "major families" of advances brought about by Industry 4.0 technologies in the construction sector.

---

[1] Query used: ((TITLE-ABS("construct* indus*" OR "indus* construct*" OR "build* indus*" OR "indus* build*" OR "innovat* construct*" OR "innovat* build*" OR "construct* site*" OR "building system*" OR "construction sector*") OR AUTHKEY("construct* indus*" OR "indus* construct*" OR "build* indus*" OR "indus* build*" OR "innovat* construct*" OR "innovat* build*" OR "construct* site*" OR "building system*" OR "construction sector*") OR (TITLE("BIM" OR "building information model*") OR ABS ("BIM" OR "building information model*") OR AUTHKEY("BIM" OR "building information model*"))) AND ((TITLE("indus* 4.0" OR "build* 4.0" OR "techno* 4.0" OR "revolution* 4.0" OR "construct* 4.0") OR ABS("indus* 4.0" OR "build* 4.0" OR "techno* 4.0" OR "revolution* 4.0" OR "construct* 4.0") OR AUTHKEY ("indus* 4.0" OR "build* 4.0" OR "techno* 4.0" OR "revolution* 4.0" OR "construct* 4.0") OR TITLE ("digiti*" OR "digitali*" OR "digit* construct*" OR "digit* build*")).

[2] Upon request, the corresponding author can send all of the references that were selected for this study.

**Table 1.**  List of recognized purposes and gains.

| | |
|---|---|
| Improve management of resources and reduce project duration and payment disputes | Monitoring of planning/creation of a "knowledge/experience" database |
| Improve performance/increase collaboration | Increase interoperability |
| Continuous and integrated information exchange and sharing | Ease the transfer of architectural shapes into reality |
| Improve real-time monitoring/discover mismatch between the real discharging place and the target/collect data for construction organizing optimization/quantify logistic density | **Increase sustainable development**/ease management of production/production optimization (3D printers)/optimize production to survive in the global market |
| Deliver greater value/**addressing sustainability/zero carbon and enhanced resilience objectives** | High performance mobile mapping enables a paradigm shift in the way buildings are designed, tested, built, maintained and **refurbished** |
| (Precast concrete and 3D printing) increase creativity and thermal characteristics | Enhance communication/increase ability to manage communication |
| Accelerate (more accurate) data acquisition (real time)/reduce time for data analysis | Reduce complexity and uncertainty/enhance information exchange |
| Transform and facilitate construction operations | **More sustainable approach** |
| Increase competitiveness thanks to BIM implantation | Organize information, processes, people, and/or firms |
| Mass-customisation/Modularisation | Increase productivity |
| Reduction of conflicts/improve project collaboration | Improve the management of complexity of information flow |
| Achieve goals of construction projects | Real-time construction progression monitoring |
| Provide simulation and optimization to the current processes | Ensure the efficient operation of the workflow, project and financial management systems |
| ICT creation of business value (not only productivity enhancements but also, competitive advantage, inventory reduction, new organizational capability among other factors) | Improve decision-making based/on visualizations and simulations/improved information sharing and transparency/improve efficiently and effectively |
| Ensures faster delivery of projects and services to clients | More feedback/mass customized product creation |
| Improve planning/creation of a "knowledge/experience" database | Reduces information losses/helping to reduce waste of time and resources |

(*continued*)

**Table 1.** (*continued*)

| Reducing wastes of resources | Drives internal and external innovation |
|---|---|
| Receive/send data in almost real-time/Automated Construction/increase ability to simulate, analyze, and optimize production | Reduction of construction time and production costs/**development of sustainable built environment** |
| Reduce the amount of construction rework | **Reduction carbon emissions** |
| Rapid data acquisition/progress control | reduction of the direct or indirect costs |
| Improved performance | Increase safety |
| Track concrete trucks with RFID technologies and Building information modelling to reduce quantity of cement and **wastes** | Support the **improvement sustainability** of the construction sector, namely, in environmental terms |
| Enhance the level of incorporation of 4,0 concepts to achieve **sustainable development** | Improvements and automatize design choices in terms of construction issues |
| Increase cooperation | Monitoring the ambient vibration |
| **Energy use tracking** | Improve training |

## 4   Analysis

All of the studies selected aimed to highlight the contributions of digital technologies or approaches to the construction sector. They can be grouped and summarized in the following table. The first one, Table 2, presents the main gains noted on at least six occasions during our review and generally credited to the BIM approaches (as we had noticed in a completely different study relative to the BIM Maturity Model [2]).

**Table 2.** First set of observed gains.

| Productivity or efficiency | Sustainability approach | Cost profit | Cooperation, collaboration or interoperability | Information | Quality |
|---|---|---|---|---|---|
| 26 times | **13 times** | 11 times | 8 times | 8 times | 6 times |

However, the summary in Table 3 reveals six of the most frequently achieved "original" improvements to be credited to technologies such as IoT [3], drones [4], and 3D printing [5], the latter facilitating, for example, prefabrication. Artificial Intelligence and Cloud Computing are presented as solutions of interest in the improvement of feedback or in the automation of choices [6, 7].

This final synthesis makes it possible to highlight new functions to open, in detail, new perspectives for construction companies. Indeed, more and more real-time monitoring is being studied and used. These follow-ups of objects and participants now make real-time simulations and optimization possible (for example, for flows, time or resources) [8, 9].

**Table 3.** New improvements and prospects.

| Real-time track monitoring data acquisition | Optimization, simulation (real time) | Customisation, Pre-fabrication | Resource management and optimization | Feedback | Production control |
|---|---|---|---|---|---|
| 12 times | 7 times | 5 times | 4 times | 4 times | 3 times |

The monitoring of planning and the creation of a knowledge and experience database are also being tested and could be generalized [10]. These elements are therefore likely to provide new sources of interest or questioning of companies… Whereas BIM is only occasionally used by less than one third of French companies [11].

Finally, this observation opens up new research perspectives, which are described in more detail in the last part of this article.

## 5    Conclusions and Perspectives

This literature review first confirmed that 4.0 technologies, generally associated with the manufacturing sector, are already benefiting from some precursors of the construction industry. Among the hundreds of studies conducted by academics and companies in this sector, the sixty or so analyzed here have confirmed that the reasons for the existence of BIM (i.e., to improve productivity, communication, quality, or to reduce costs) is reinforced by all of these complementary technologies. The latter, which is increasingly robust and accessible due to their massive diffusion, also make it possible to obtain new and promising functionalities or gains, such as real-time data acquisition to increase reactivity or to optimize process and simulations. As we recall, this exceeded the expectations that had initially been set during the deployment of BIM. However, the companies we met and worked with are still hesitant and limited in terms of financial resources. This analysis now needs to be continued, and for us to become even more proactive. This is why, as with ongoing studies concerning the complementarity between LEAN and Ind4.0 [12], we will attempt to answer the following nested research questions in future collaborative work:

i.    Do the skills acquired by companies in deploying and operating BIM provide easier access to new "4.0" functionalities and gains summarized in Table 3?
ii.   Conversely, will the acquisition and mastery of new "4.0" technologies facilitate BIM implementation?
iii.  Will the two approaches soon be inseparable, BIM becoming the equivalent of the digital twin as defined by Grieves et al. for the construction sector (as already suggested by many companies today [13]) [14]?
iv.   …or will they remain without real mutual effects?

In order to answer these questions, in early 2020, we have begun major studies and support work with one of France's leading construction companies involved in numerous rehabilitation and new construction programs in the area around Paris.

# References

1. Kagermann, H., Lukas, W.-D., Wahlster, W.: Industrie 4.0: Mit dem Internet der Dinge auf dem Weg zur 4. industriellen Revolution, VDI nachrichten, vol. 13, p. 2011 (2011)
2. Joblot, L., Paviot, T., Deneux, D., Lamouri, S.: Building information maturity model specific to the renovation sector. Autom. Constr. **101**, 140–159 (2019). https://doi.org/10.1016/j.autcon.2019.01.019
3. Correa, F.R.: Cyber-physical systems for construction industry. In: IEEE Industrial Cyber-Physical Systems, ICPS 2018, pp. 392–397 (2018). https://doi.org/10.1109/ICPHYS.2018.8387690.
4. Tezel, B.A., Aziz, Z.U.H.: From conventional to IT based visual management: a conceptual discussion for lean construction. ITcon **22**, 220–246 (2017)
5. Cangelli, E., Conteduca, M.: Architecture on demand. New scenarios for the design project and the construction industry. TECHNE 16, 96–104 (2018). https://doi.org/10.13128/Techne-23036.
6. Khan, K.I.A., Flanagan, R., Lu, S.L.: Managing information complexity using system dynamics on construction projects. CME 34(3), 192–204 (2016). https://doi.org/10.1080/01446193.2016.1190026.
7. Trani, M.L., Cassano, M., Todaro, D.: Standard information for construction equipment in a relational database. In: présenté à ISEC 2017 - 9th International Structural Engineering and Construction Conference: Resilient Structures and Sustainable Construction (2017)
8. Laurini, E., Rotilio, M., Lucarelli, M., De Berardinis, P.: Technology 4.0 for buildings management: from building site to the interactive building book. In: ISPRS Annals of the Photogrammetry, Remote Sensing and Spatial Information Sciences, vol. 42, pp. 707–714 (2019). https://doi.org/10.5194/isprs-Archives-XLII-2-W11-707-2019
9. Favier, A., Scrivener, K., Habert, G.: Decarbonizing the cement and concrete sector: integration of the full value chain to reach net zero emissions in Europe. In: présenté à IOP Conference Series: Earth and Environmental Science, vol. 225 (2019). https://doi.org/10.1088/1755-1315/225/1/012009
10. Tibaut, A., Zazula, D.: Sustainable management of construction site big visual data. Sustain. Sci. **13**(5), 1311–1322 (2018). https://doi.org/10.1007/s11625-018-0595-9
11. AxeoBIM - Plateforme collaborative BIM : Observatoire du BIM 2018 [En ligne]. Disponible sur (2018). https://www.axeobim.fr. [Consulté le: 23-janv-2020]
12. Rosin, F., Forget, P., Lamouri, S., Pellerin, R.: Impacts of Industry 4.0 technologies on Lean principles. IJPR, 1–18 (2019). https://doi.org/10.1080/00207543.2019.1672902
13. Le jumeau numérique|SNCF Réseau (2019). https://www.sncf-reseau.com/fr/entreprise/newsroom/sujet/innovation-jumeau-numerique. [visited le: 1-janv-2020]
14. Grieves, M., Vickers, J.: Digital twin: Mitigating unpredictable, undesirable emergent behavior in complex systems. In: Transdisciplinary Perspectives on Complex Systems, pp. 85–113 (2017). Springer

# Smart Tightening Development for Aeronautical Bolted Assemblies in an Industry 4.0

Charly Foissac[1]([⊠]), Alain Daidie[1], Stephane Segonds[1], and Clément Chirol[1,2]

[1] Université de Toulouse, Institut Clément Ader, UMR CNRS 5312, INSA/UPS/ISAE/Mines, Toulouse, France
charly.foissac@insa-toulouse.fr
[2] Airbus Operations S.A.S., Toulouse, France

**Abstract.** Smart tightening development is part of the Industry 4.0 transformation with the introduction of smart tools, and preload in bolted assemblies is of major interest in today's aircraft manufacturing process. So far, it has been difficult to estimate the importance of each parameter for tightening process quality, mainly because of the large number of combinations and configurations that exist.

The present work aims at evaluating the effects and the interactions between different parameters that have to be taken in consideration in future torquing strategy. Many experimental tests have been conducted on an Automatica test bench using a Taguchi strategy and an analysis of the first main results is now presented, highlighting the complexity of the phenomena studied.

All these points will help us to better understand tightening, so as to improve performance during installation, maintenance and repair.

**Keywords:** Smart tightening · Design of experiments · Taguchi · Bolted assemblies

## 1 Introduction

With millions of fasteners needed to complete the A350, the challenge of mastering the tightening of bolted connections cannot be overlooked.

The aim of this research work was to develop a predictive tool to better control tightening behaviour through the torque/tension relationship, integrating strongly non-linear local phenomena. Controlling the preload within the bolted assembly can be beneficial for the integrity of the aircraft [1, 2] but this tool developed for an industry 4.0 which is geolocated and able to feed a dynamic database can also be an advantage for quality control and maintenance [3]. However, the preload is a very dispersive variable [4] and is difficult to control in production with current clamping tools [5]. One of the objectives is to optimize the tightening instructions that meet aeronautical requirements, in an uncertain environment [6]. This initiative should minimize the cases of over-torque and under-torque and lead to an assessment of the risks implied for the tightening efficiency. It should also provide an understanding of the mechanical behaviour of fastening elements impacted by faults. However, if we consider all the parameters of a bolted connection,

© The Author(s) 2021
L. Roucoules et al. (Eds.): JCM 2020, LNME, pp. 419–425, 2021.
https://doi.org/10.1007/978-3-030-70566-4_66

we obtain a total of approximately 7,050 combinations per diameter. Table 1 lists the number of possible parameter solutions, without taking compatibility into account, when we use a bolted assembly.

**Table 1.** Number of solutions for each parameter.

| Fastener | Technology | Material | Surface finish | Lubrication | Diameter |
|----------|-----------|----------|----------------|-------------|----------|
| Bolt | 64 | 4 | 6 | 4 | 26 |
| Nut | 65 | 5 | 7 | 6 | |

Given the large number of possible combinations, the behaviour of bolted connections can only be understood by creating a large database, in particular using design of experiments (DOE). Many research works have aimed to develop methods of constructing designs of experiments that make it possible to obtain a maximum of information with a minimum of known points, whilst remaining robust with respect to random and approximation errors [7]. These include:

- Mixed level DOEs, with as many factors as desired, each factor taking the number of levels necessary for proper execution of the study;
- Box-Behnken DOEs [8], which respect the rotational isovariance criterion;
- Rechtschaffner DOEs [9], simplified fractional DOEs which allow the effects of factors and second order interactions to be determined;
- Taguchi DOEs, based on the most famous theory for quality, which allow the effects of factors and interaction to be determined.

The most common way to implement an orthogonal multifactor DOE is to use Taguchi tables [10] because results are simple to obtain and the effects of each factor can be better isolated. However, the DOE should be usable in our study case, which means that none of the parameters intervening in the behaviour of the bolted assemblies is allowed to be chaotic. That is why the Taguchi fractional DOE is compared with a complete DOE here. Moreover, this will enable us to highlight the interactions between the parameters, which is not always possible with a multifactor design of experiments.

## 2 Experimental Campaign

To evaluate the behaviour of bolts, an Automatica® test bench [11] was used in the experiments described here. It allows the tightening torque, the preload and the tightening angle to be acquired in real time. The measurement protocol is to tighten the bolt at the maximum speed of Airbus internal standards until the threads of the nut break. Thus, the approach phase of the nut, the linear phase of tensioning the bolt and the non-linear phase of failure can be observed.

This first experimental campaign was limited to EN6115 titanium screws, ASNA aluminium nuts and diameter codes 3 (4.76 mm), 3A (5.56 mm) and 4 (6.35 mm). There are three different surface finishes for the screws and three types of nut, so the campaign comprised a total of 27 combinations. Table 2 gives more details on these combinations. In addition, for reasons of repeatability and reliability, 20 tests were carried out per combination, giving a total of 540 tests.

**Table 2.** Summary table of screw and nut types.

EN6115 – Screws

| Code | Material | Surface finish | Lubrication | Diameter range |
|---|---|---|---|---|
| T | Titanium alloy 6AL – 4 V $R_cmin = 650\,MPa$ | Sulfuric-acid anodizing | Cethyl alcohol | 3(4.76 mm) 3A (5.56 mm) 4 (6.35 mm) |
| V | | Ion Vapour Deposition of aluminium | | |
| K | | Resin based Aluminium coating | | |

ASNA – Nuts

| Code | Material | Surface finish | Lubrication | Diameter range |
|---|---|---|---|---|
| 2528 | Aluminium alloy | Anodizing as per MIL-A-8625 | Cethyl alcohol | 3(4.76 mm) 3A (5.56 mm) 4 (6.35 mm) |
| 2529 | | Anodizing as per MIL-A-8625, Type II, class 2 | | |
| 2537 | Nut: Aluminium alloy | Anodizing as per MIL-A-8625 | Cethyl alcohol | |
| | Washer: Corrosion Resistance steel | Cadmium plating | None | |

For each combination, an average maximum breaking torque is extracted, which corresponds to the average of the torques for which there is loss of the preload in each repeatability test. Also, using the same principle, the average maximum preload is extracted and, as the torque/preload relationship is known to be very dispersive, the standard deviation of the two previous data is also noted.

## 3 Taguchi Factorial Experiment Plan

A fractional DOE enables the number of trials to be reduced to what is strictly necessary to conclude on the same trends as a complete design of experiments. To respect the experimental campaign, three factors (coating of the screw, type of nut and diameter) at three levels will be used. A Taguchi Table 3 is chosen, and this reduces the campaign to 180 tests, a third of the number needed for the complete experiment plan.

**Table 3.** $L_9(3^3)$ Taguchi factorial experiment plan.

| Tests | Screw surface finish | Type of nut | Diameter |
|---|---|---|---|
| 1 | 1 | 1 | 1 |
| 2 | 1 | 2 | 3 |
| 3 | 1 | 3 | 2 |
| 4 | 2 | 1 | 3 |
| 5 | 2 | 2 | 2 |
| 6 | 2 | 3 | 1 |
| 7 | 3 | 1 | 2 |
| 8 | 3 | 2 | 1 |
| 9 | 3 | 3 | 3 |

The factor columns are balanced and orthogonal. This means that, in each pair of columns, all combinations of factors occur the same number of times. The orthogonal plans will allow us to estimate the effect of each factor on the response, independently of all the other factors. For the moment, no interaction is taken into account even though the type of nut and the coating of the screw could be considered as one, since the coatings of the nuts are not similar and thus can locally modify the coefficient of friction between the thread of the screw and the thread of the nut.

## 4  Results

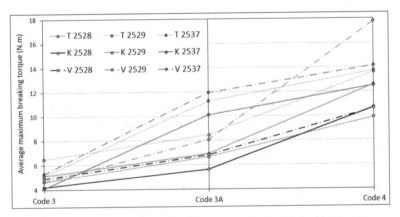

**Fig. 1.** Average maximum breaking torque for the 27 combinations

The first results that can be extracted are directly linked to the data from the test bench: tightening torque and preload. For example, thanks to the complete plan, the main trends in the behaviour of the bolted connection is observable in Fig. 1. The coating of the screw has a small impact on the maximum average breaking torque; a classification is even possible: V > T > K. In contrast, the type of nut has a marked influence on the response and the diameter undeniably has an impact that is almost proportional to the tightening torque.

Next, the results of Fig. 2 correspond to the differences between the fractional DOE and the complete design of experiments. To detect this, we use an effect graph corresponding to each factor. There is no difference for the diameter, possibly because of the continuity of this parameter, and the effect of the parameter supports the previous observation. However, there is a big trend difference for the type of nut. It is likely that this comes from the non-linearity of the parameter and the fact that the screw/nut interaction is disregarded. However, when we look at the scale, we see that this effect is globally negligible (<1.5%).

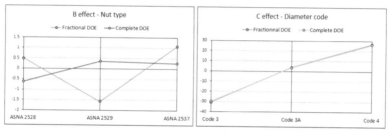

**Fig. 2.** Differences between fractional and complete DOE on the average maximal preload for the nut and diameter factors expressed in percentages.

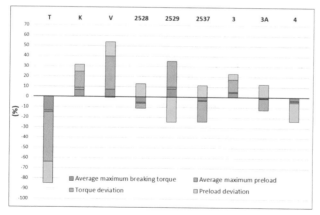

**Fig. 3.** Differences between multifactor experiments and complete design of experiments expressed in percentage depending on parameters and responses.

Figure 3 summarizes all the differences between the two designs of experiments as a percentage. The average maximum breaking torque and the average maximum preload are rather stable, unlike the standard deviations. This is not surprising: the greater the number of tests completed, the greater the dispersion, in particular because of interactions. The most unstable parameters preventing us from using a fractional DOE are the T and V coatings and the ASNA2529 nut. The differences for the coating of screw T and the diameter code 4 are exclusively negative. In other words, our multifactor experiments are the "worst case" for these factors. Conversely, K and V coatings and diameter code 3 show exclusively positive differences.

## 5  Conclusion

Finally, this study shows us that, on all points, the complexity of the behaviour of a bolted connection does not allow us to approximate it with a fractional design of experiments without taking the interactions of the parameters into account, and these interactions are obtained only through analysis of a complete design of experiments. Now, the complete design of experiments must be analysed in greater depth to highlight predominant points of the behavior of the bolted assemblies. This will allow us to better target our objectives for future experimental campaigns and for future digital models.

## References

1. Benhaddou, T., Chirol, C., Daidie, A., Guillot, J., Pierre, S., Tuery, J.-B.: Pre-tensioning effect on fatigue life of bolted shear joints. Aerosp. Sc. and Techn. **36**, 36–43 (2014)
2. Benhaddou, T., Stephan, P., Daidie, A., Alkatan, F., Chirol, C., Tuery, J.-B.: Effect of axial preload on durability of aerospace fastened joints. Int. J. Mech. Sci. **137**, 214–223 (2018)
3. Stock, T., Seliger, G.: Opportunities of sustainable manufacturing in industry 4.0. Procedia Cirp. 40, 536–541 (2016)
4. Grzejda, R.: Determination of bolt forces and normal contact pressure between elements in the system with many bolts for its assembly conditions. Adv. Sci. Technol. Res. J. **9**(13), 116–121 (2019)
5. Rafik, V. Etude du devissage d'un assemblage aéronautique sous sollicitations transverse. Ph.D. thesis, University of Toulouse (2019)
6. Dols, S. Développement d'une nouvelle méthode de serrage intelligente pour le contrôle des assemblages boulonnés. Ph.D. thesis, University of Toulouse (2016)
7. Goupy, J.: Les plans d'experiences. Rev. MODULAD **34**, 74–116 (2006)
8. Box, G., Behnken, D.: Some new three level designs for the study of quantitative variables. Technometrics **2**(4), 455–475 (1960)
9. Rechtschaffner, R.L.: Saturated fractions of 2n and 3 factorial designs. Technometrics **9**(4), 569–575 (1967)
10. Pillet, M.: Introduction aux Plans d'expériences par la méthode Taguchi. Les Editions d'organisation, Paris (1992)
11. Automaticas. CB2 – automatica (2020). https://www.automatica.info/en/products/testing-machines/tightening-bench-torsion-machine/cb2/

# Optimization of Reconfigurable Manufacturing Systems Configuration: A Literature Review

Rachel Campos Sabioni[(✉)], Joanna Daaboul, and Julien Le Duigou

Université de Technologie de Compiègne, CNRS, Roberval (Mechanics Energy and Electricity), Centre de recherche Royallieu - CS 60 319 - 60 203, Compiègne Cedex, France
rachel.campos-sabioni@utc.fr

**Abstract.** Reconfigurable Manufacturing Systems (RMS) have gained importance in the current context of increasing high variety demand, Mass Customization (MC) and market instability, due to their ability of being quickly modified to adjust their production capacity to attain sudden fluctuations in market demands as well as to accommodate operations of new products. RMS can be configured at system and machine levels. Many papers have described the RMS configuration as combinatorial optimization problems and proposed several techniques to optimize them in terms of different responses of interest. This paper presents a literature review that seeks to understand how RMS configuration has been addressed in terms of configuration level, optimization problem modelling and techniques applied to solve it. This work aims to assist researchers working on RMS configuration to identify trends and new research opportunities.

**Keywords:** Reconfigurable Manufacturing Systems · System configuration · Optimization · Layout design · Machine selection · Machine configuration

## 1 Introduction

Increased demand for high variety and mass-customized products forces companies to increase their flexibility throughout their value chain, which can be achieved in manufacturing via Reconfigurable Manufacturing Systems (RMS). RMS are designed to be rapidly modified to adjust their production capacity and functionality, for responding to abrupt variations on market demands [1]. To achieve that, RMS count with Computer Numerically Controlled (CNC) machines, and Reconfigurable Machine Tools (RMT), which are modular machines capable to be configured in different ways for performing specific operations [2, 3].

RMS configuration can be divided into system and machine levels [4]. While system-level configuration is related to the layout design, the machine-level corresponds to change or adjust auxiliary machine-modules [5]. Both configuration levels require supplementary costs during the production process that cannot be neglected [6, 7]. Many researchers have proposed different methods to optimize the RMS configuration, considering system or machine-level independently or jointly.

© The Author(s) 2021
L. Roucoules et al. (Eds.): JCM 2020, LNME, pp. 426–435, 2021.
https://doi.org/10.1007/978-3-030-70566-4_67

However, as far as we know, there is still no paper focused on understanding optimization issues of RMS configuration. To fill this gap, this paper presents a literature review to investigate (1) How RMS configuration optimization problems are modelled, (2) Which are the main methods used to solve them and (3) Which are the main focus of papers when optimizing the RMS configuration in terms of: (a) configuration-level, (b) objective function and (c) process/production planning.

This review was conducted in four databases: SCOPUS, Science Direct, ISI Web of Science and Taylor & Francis with the keywords "Reconfigurable Manufacturing System" AND "configuration". The 924 papers found were screened in order to filter only journal papers from 2009 to 2020, and to remove all duplicates and papers not written in English or without the full text available. Finally, 54 papers were selected. This paper is organized as follows. Section 2 brings a broad overview of papers found in literature and Sect. 3 presents the conclusion.

## 2  Overview of RMS Configuration Optimization Papers

### 2.1  Main Focus of Papers Optimizing the RMS Configuration

Table 1 presents the configuration levels addressed by papers optimizing the RMS configuration, which are divided in three main types: (1) *system and machine levels* jointly (68.5%), (2) *system-level* (26%) and (3) *machine-level* (5.5%).

**Table 1.** Configuration-level addressed by papers optimizing the RMS configuration.

| Config. level | | Characteristics | Authors |
|---|---|---|---|
| System and machine level | | System configuration is related to machine addition or removal. Further, each machine can be reconfigured by changing its auxiliary-modules or tools | [2–7, 22–52] |
| System level | Layout design | System configuration is based on machines selection and their position in layout. Machine-configuration is not addressed | [15–21] |
| | - | System configuration is based on the operations path or on adding/removing machines from the system. There is no reference to layout design issues | [8–14] |
| Machine configuration | | Configuration of a machine singly to produce a specific product or product family | [53–55] |

However, machine configuration as well as its layout placement could highly affect manufacturing costs. Therefore, layout design and machine configuration must be considered together in the RMS configuration for process planning (PP) or production planning (ProP). Few papers worked on the RMS configuration to improve their PP (21.1%). PP is the translation of a *single* product design data into a method to manufacture it, including machine and configuration selection and operations sequencing [29].

This is why PP papers mainly worked with a single product [11, 16, 19, 43]. ProP was the mainly focus of the most part of papers addressing the RMS configuration (78.9%). ProP assigns operations defined by PP while respecting some constraints such as due time, machine capacity etc. and generally includes the plan of multiple parts/products simultaneously [3, 10]. In fact, ProP papers from this review mostly optimized the RMS configuration for one or many product families and multiple product/parts simultaneously, by respecting machine capacity and specific demand rates in a time period [15, 24, 29, 40].

Since RMS is a relatively new type of production system, it is still hard to find completely reconfigurable systems in industries; hampering the implementation of real case studies. Papers mainly conducted case studies through numerical illustration or simulation (83%). They mainly used simple or generic parts (few features) and hypothetical values for required parameters (e.g. machine allocation and configuration cost) [3, 6, 7, 44]. This not exactly represents the real life, since modular products can be much more complex, especially in high variety context. Hence, more research with complex products (many features) is still required.

Papers applying real case studies (17%) mostly used a work-piece provided by an industrial partner to map required operations and, based on that, identify all machine-configurations capable to execute these operations. Although their propositions were focused on RMS, their case studies were usually applied in process composed by CNC or dedicated machines, without including RMTs [17, 23, 24, 53]. Some researchers conducted their case study in a reconfigurable assembly line of an automotive industry [10, 25]; while others did not clearly present the machine types that made up their case study [8, 11, 15].

## 2.2 Modelling and Optimizing the RMS Configuration

Optimization problems of RMS configuration found in literature were mainly multi-variate and multi-objective. This confirms the complexity of optimizing RMS configuration, since researchers must consider many decision variables whilst optimizing various responses of interest. Table 2 summarizes the techniques used to model and solve these problems. They were mainly modelled with integer variables varying between nonlinear and linear models, with the predominance of the latter.

Nonlinear problems were mostly solved by approximate or hybrid approaches using GA singly [13, 17, 50, 52] or coupled with other methods, like Monte Carlo [30] and dynamic programming [42]. Linear problems were also solved by approximate methods, but papers mostly tried to validate a new heuristic [43] or metaheuristic method [44, 54] by comparing their solution with those obtained by the well-known NSGA-II. Since commercial software, like LINGO, are capable to find a global optimum for ILP and MILP problems, some papers compared their results with those obtained by approximate [5, 53, 54] or hybrid methods [51] to verify the reliability of those methods. Others just used these solvers singly [28, 47, 48]. Linear problems were also solved by different exact approaches, like enumerative [2, 24] and iterative techniques [11, 41, 44].

Some papers modelled their problems as Multi-Criteria Decision-Making (MCDM) ones, by comparing divergent criteria of multiple alternatives and ranking them according to its suitability. MCDM problems mainly compared different system configurations [21, 55], but comparison of resource [16] and scheduling [25] alternatives were also found. They mainly addressed qualitative attributes (e.g. system reconfigurability, convertibility) being mostly solved by heuristic [16, 21, 25] or enumerative methods [34, 55].

Many papers from literature partially detailed their optimization problems, without including all information about decision variables, constraints etc., while others presented the whole model without classifying their problem [23, 38, 39, 45]. Not surprisingly, they mostly used metaheuristics (50%), like GA and NSGA-II, which have proven their effectiveness to solve optimization problems related to RMS configuration. Metaheuristics are known to not being problem-specific, meaning that they can solve several problems with few modifications in the algorithm [56]. Therefore, it allows people solving complex problems, like RMS configuration, even if they do not totally know how to model their optimization problems. Further, papers dealing with multi-objective problems also hybridized multi-objective metaheuristics, like NSGA-II and AMOSA, with TOPSIS, which attributes weights to each objective for ranking solutions in the Pareto front [3, 6, 22].

Researchers tried to optimize many objectives, but cost stood out as the most addressed, being minimized by 74.1% of works. The three main types of costs were: (1) *Capital cost*: to attain new market demands [17, 23], to satisfy pre-fixed demand scenarios [2, 27, 33], or to deal with stochastic demands [13, 14]. (2) *Reconfiguration cost*: mainly related to machine allocation or configuration for reducing costs of changing product's production within the same family [32, 36, 37]. (3) *Production cost*: of single/multiple parts [3, 44, 53]. Most of the time, papers addressed the minimization of these costs simultaneously [3, 5–7,36].

Due to the RMS ability to rapidly change their production capacity or to accommodate new operations required by new product launches, they are known as key enablers of MC. Nevertheless, few papers (16,7%) have addressed MC, and those who have considered it mostly focused on optimizing the RMS for responding to given demand scenarios and due times [5, 10, 21, 49, 50, 52]. One paper focused on increasing system modularity to accommodate high variety in MC [6], while others cited MC without clearly explain which were their scientific contribution to enable MC throughout the RMS configuration optimization [19, 44].

The increasing attention to the worldwide environmental sustainability have reflected the challenges faced by works optimizing the RMS configuration. Recent publications have showed their interest in minimizing the energy consumption of RMS [18, 24, 44, 54]. However, these works represent only 7.4% of papers found, meaning that there are opportunities to do more investigations in this domain.

**Table 2.** Approaches used to model and solve optimization problems of RMS configuration.

| Optimization models | | | Linear | | | Nonlinear | | MCDM | Undefined model |
|---|---|---|---|---|---|---|---|---|---|
| Optimization methods | | | LP | ILP | MILP | MINLP | NIP | | |
| Exact approaches | Enumerative | CKSP or e-constraint | | [2] | [49] | | | | |
| | | Fuzzy logic or ELECTRE | | | | | | [34, 55] | |
| | | Topological sort or weighted sum | [24] | | | | | | [7, 40] |
| | Iterative | Negotiation algorithm | | | | | [44] | | |
| | | Others | [11] | [41] | [44] | | | | |
| Approximate approaches | Metaheuristic | NSGA-II | | [41] | [5, 44, 54] [18] | | | | [6, 26, 29] |
| | | GA | | | [18] | [13, 17] | [50, 52] | | [9, 23, 31, 38, 39] |
| | | MOPSO | [4] | | | | | | [27] |
| | | SA or AMOSA | | | [44, 54] | | | | [6, 19, 45] |
| | | TS | | | | [14] | | | |
| | Heuristic | Shannon entropy; Intelligent search | | | | | | [16, 21, 25] | |
| | | Other heuristics | | [43] | [53] | [20] | | | [10, 46] |

*(continued)*

**Table 2.** (*continued*)

| Optimization methods | Linear | | | Nonlinear | | MCDM | Undefined model |
|---|---|---|---|---|---|---|---|
| | LP | ILP | MILP | MINLP | NIP | | |
| Hybrid approaches — Stochastic programming | | | | | | | [8] |
| (NSGA-II or AMOSA) + TOPSIS | | | | | | | [3, 6, 22] |
| GA + other algorithms | | [51] | | | [30, 42] | | |
| Decision tree + Markov analysis | | | | | | | [15] |
| Software and Solvers (CPLEX/Gurob /LINGO/GAM) | | [41, 47, 48, 51] | [5, 18, 28, 49, 53, 54] | [13] | | | |
| Undefined method | | | [30, 52] | | | | [30, 52] |

**Optimization models:** LP: Linear programming; ILP: Integer Linear Programming; MILP: Mixed integer linear programming; MINLP: Mixed Integer Non-Linear Programming; NIP: Nonlinear integer programming; MCDM: Multi-Criteria Decision-Making problems.
**Optimization methods.** SA: Simulated annealing; AMOSA: Archived Multi-Objective SA; GA: Genetic Algorithm; NSGA-II: Non-dominated sorting GA; MOPSO: Multiple Objective Particle Swarm Optimization; ELECTRE: Elimination and Choice-Translating; CKSP: Constrained K-shortest path. TOPSIS: Technique for Order of Preference by Similarity to Ideal Solution.

## 3  Conclusion

This paper presents a literature review of RMS configuration in terms of (1) configuration level and (2) optimization techniques for modelling and solving problems related to the RMS configuration as well as (3) the main focus of RMS configuration optimization. This review highlights that although papers mostly work with system and machine level configurations, there is still no work dealing with layout design and machine-configuration issues simultaneously. Indeed, few papers were dedicated to the layout design optimization. Very few papers conducted real case studies and those who have done it did not address RMTs, evidencing that it is still very hard to find laboratories or industries equipped with RMS or their prototypes. RMS configuration optimization problems are mainly modelled with integer variables or no optimization model is available. In any case, metaheuristic methods highlight as the most used ones due to their ability of solving complex optimization problems by providing optimal acceptable solutions in relatively little time. Cost minimization is the main focus of papers optimizing the RMS configuration; however, recent publications also show a tendency to focus on sustainability and MC issues.

**Acknowledgments.** This work was supported by the French National Research Agency under Grant ANR-17-CE10-0010-01.

## References

1. Koren, Y., et al.: Reconfigurable manufacturing systems. CIRP Annal. Manufact. Technol. **48**(2), 527–540 (1999)
2. Dou, J., Dai, X., Meng, Z.: Graph theory-based approach to optimize single-product flow-line configurations of RMS. Int. J. Adv. Manufact. Technol. **41**(9–10), 916–931 (2009)
3. Goyal, K.K., Jain, P.K., Jain, M.: Optimal configuration selection for reconfigurable manufacturing system using NSGA II and TOPSIS. Int. J. Prod. Res. **50**(15), 4175–4191 (2012)
4. Goyal, K.K., Jain, P.K.: Design of reconfigurable flow lines using MOPSO and maximum deviation theory. Int. J. Adv. Manufact. Technol. **84**(5–8), 1587–1600 (2016)
5. Dou, J., Li, J., Su, C.: Bi-objective optimization of integrating configuration generation and scheduling for reconfigurable flow lines using NSGA-II. Int. J. Adv. Manufact. Technol. **86**(5–8), 1945–1962 (2016)
6. Benderbal, H.H., Dahane, M., Benyoucef, L.: Modularity assessment in reconfigurable manufacturing system (RMS) design: an archived multi-objective simulated annealing-based approach. Int. J. Adv. Manufact. Technol. **94**, 729–749 (2018)
7. Saxena, L.K., Jain, P.K.: A model and optimisation approach for reconfigurable manufacturing system configuration design. Int. J. Prod. Res. **50**(12), 3359–3381 (2012)
8. Copani, G., Urgo, M.: Innovative flexibility-oriented business models and system configuration approaches: an industrial application. CIRP J. Manufact. Sci. Technol. **11**, 10–17 (2015)
9. Xiaowen, X., Beirong, Z., Wei, X.: Configuration optimization method of reconfigurable manufacturing systems. Res. J. Appl. Sci. Eng. Technol. **6**(8), 1389–1393 (2013)
10. Pattanaik, L.N., Jena, A.: Tri-objective optimisation of mixed model reconfigurable assembly system for modular products. Int. J. Comput. Integr. Manuf. **32**(1), 72–82 (2018)

11. Lv, C., Li, A.P., Xu, L.Y.: Research and optimization of reconfigurable manufacturing system configuration based on system reliability. Kybernetes **39**(6), 1058–1065 (2010)
12. Hasan, F., Jain, P.K., Kumar, D.: Optimum configuration selection in reconfigurable manufacturing system involving multiple part families. Opsearch **51**(2), 297–311 (2014)
13. Abbasi, M., Houshmand, M.: Production planning and performance optimization of reconfigurable manufacturing systems using genetic algorithm. Int. J. Adv. Manufact. Technol. **54**(1–4), 373–392 (2011)
14. Abbasi, M., Houshmand, M.: Production planning of reconfigurable manufacturing systems with stochastic demands using Tabu search. Int. J. Manufact. Technol. Manage. **17**(1–2), 125–148 (2009)
15. Abdi, M.R., Labib, A.: RMS capacity utilisation: product family and supply chain. Int. J. Prod. Res. **55**(7), 1930–1956 (2016)
16. Michalos, G., Makris, S., Mourtzis, D.: An intelligent search algorithm-based method to derive assembly line design alternatives. Int. J. Comput. Integr. Manufact. **25**(3), 211–229 (2012)
17. Wang, W., Koren, Y.: Scalability planning for reconfigurable manufacturing systems. J. Manufact. Syst. **31**(2), 83–91 (2012)
18. Ghanei, S., Algeddawy, T.: An integrated multi-period layout planning and scheduling model for sustainable reconfigurable manufacturing systems. J. Adv. Manufact. Syst. **19**(1), 31–64 (2020)
19. Haddou Benderbal, H., Benyoucef, L.: Machine layout design problem under product family evolution in reconfigurable manufacturing environment: a two-phase-based AMOSA approach. Int. J. Adv. Manufact. Technol. **104**(1–4), 375–389 (2019)
20. Guan, X., Dai, X., Qiu, B., Li, J.: A revised electromagnetism-like mechanism for layout design of reconfigurable manufacturing system. Comput. Ind. Eng. **63**(1), 98–108 (2012)
21. Gupta, A., Jain, P.K., Kumar, D.: Configuration selection of reconfigurable manufacturing system based on performance. Int. J. Ind. Syst. Eng. **20**(2), 209–229 (2015)
22. Benderbal, H.H., Dahane, M., Benyoucef, L.: Flexibility-based multi-objective approach for machines selection in reconfigurable manufacturing system (RMS) design under unavailability constraints. Int. J. Prod. Res. **55**(20), 6033–6051 (2017)
23. Koren, Y., Wang, W., Gu, X.: Value creation through design for scalability of reconfigurable manufacturing systems. Int. J. Prod. Res. **55**(5), 1227–1242 (2016)
24. Choi, Y.-C., Xirouchakis, P.: A holistic production planning approach in a reconfigurable manufacturing system with energy consumption and environmental effects. Int. J. Comput. Integr. Manuf. **28**(4), 379–394 (2015)
25. Prasad, D., Jayswal, S.C.: Reconfigurability consideration and scheduling of products in a manufacturing industry. Int. J. Prod. Res. **56**(19), 6430–6449 (2017)
26. Bensmaine, A., Dahane, M., Benyoucef, L.: A non-dominated sorting genetic algorithm based approach for optimal machines selection in reconfigurable manufacturing environment. Comput. Ind. Eng. **66**(3), 519–524 (2013)
27. Goyal, K.K., Jain, P.K., Jain, M.: Applying Swarm intelligence to design the reconfigurable flow lines. Int. J. Simul. Model. **12**(1), 17–26 (2013)
28. Eguia, I., Molina, J.C., Lozano, S., Racero, J.: Cell design and multi-period machine loading in cellular reconfigurable manufacturing systems with alternative routing. Int. J. Prod. Res. **55**(10), 2775–2790 (2017)
29. Mohapatra, P., Benyoucef, L., Tiwari, M.K.: Integration of process planning and scheduling through adaptive setup planning: a multi-objective approach. Int. J. Prod. Res. **51**(23–24), 7190–7280 (2013)
30. Renna, P.: A decision investment model to design manufacturing systems based on a genetic algorithm and Monte-Carlo simulation. Int. J. Comput. Integr. Manuf. **30**(6), 590–605 (2016)

31. Asghar, E., Zaman, U.K., Baqai, A.A., Homri, L.: Optimum machine capabilities for reconfigurable manufacturing systems. Int. J. Adv. Manufact. Technol. **95**(9–12), 4397–4417 (2018)

32. Mittal, K.K., Kumar, D., Jain, P.K.: a systematic approach for optimum configuration selection in reconfigurable manufacturing system. J. Inst. Eng. India Ser. C **99**(6), 629–635 (2018)

33. Gupta, A., Jain, P.K., Kumar, D.: A novel approach for part family formation for reconfiguration manufacturing system. Opsearch **51**(1), 76–97 (2014)

34. Rehman, A.U., Babu, A.S.: The evaluation of manufacturing systems using concordance and disconcordance properties. Int. J. Serv. Oper. Manage. **5**(3), 326–349 (2009)

35. Li, X., Bayrak, A.E., Epureanu, B.I., Koren, Y.: Real-time teaming of multiple reconfigurable manufacturing systems. CIRP Annal. Manufact. Technol. **67**(1), 437–440 (2018)

36. Ashraf, M., Hasan, F.: Configuration selection for a reconfigurable manufacturing flow line involving part production with operation constraints. Int. J. Adv. Manufact. Technol. **98**(5–8), 2137–2156 (2018)

37. Moghaddam, S.K., Houshmand, M., Valilai, O.F.: Configuration design in scalable reconfigurable manufacturing systems (RMS); a case of single-product flow line (SPFL). Int. J. Prod. Res. **56**(11), 3932–3954 (2018)

38. Vafadar, A., Hayward, K., Tolouei-Rad, M.: Drilling reconfigurable machine tool selection and process parameters optimization as a function of product demand. J. Manufact. Syst. **45**, 58–69 (2017)

39. Xie, N., Li, A., Xue, W.: Cooperative optimization of reconfigurable machine tool configurations and production process plan. Chin. J. Mech. Eng. (English Edn.) **25**(5), 982–989 (2012)

40. Mittal, K.K., Jain, P.K., Kumar, D.: Configuration selection in reconfigurable manufacturing system based on reconfigurability. Int. J. Logist. Syst. Manage. **27**(3), 363 (2017)

41. Yu, J.M., Doh, H.H., Kim, H.W., Kim, J.S., Lee, D.H., Nam, S.H.: Iterative algorithms for part grouping and loading in cellular reconfigurable manufacturing systems. J. Oper. Res. Soc. **63**(12), 1635–1644 (2012)

42. Bryan, A., Hu, S.J., Koren, Y.: Assembly system reconfiguration planning. J. Manuf. Sci. Eng. **135**(4), 041005 (2013)

43. Touzout, F.A., Benyoucef, L.: Multi-objective multi-unit process plan generation in a reconfigurable manufacturing environment: a comparative study of three hybrid metaheuristics. Int. J. Prod. Res. **57**(24), 7520–7535 (2019)

44. Touzout, F.A., Benyoucef, L.: Multi-objective sustainable process plan generation in a reconfigurable manufacturing environment: exact and adapted evolutionary approaches. Int. J. Prod. Res. **57**(8), 2531–2547 (2019)

45. Musharavati, F., Hamouda, A.S.M.: Enhanced simulated-annealing-based algorithms and their applications to process planning in reconfigurable manufacturing systems. Adv. Eng. Softw. **45**(1), 80–90 (2012)

46. Zhang, L., Rodrigues, B.: Modelling reconfigurable manufacturing systems with coloured timed Petri nets. Int. J. Prod. Res. **47**(16), 4569–4591 (2009)

47. Moghaddam, S.K., Houshmand, M., Saitou, K., Fatahi Valilai, O.: Configuration design of scalable reconfigurable manufacturing systems for part family. Int. J. Prod. Res. **58**, 2974–2996 (2019)

48. Bortolini, M., Galizia, F.G., Mora, C., Pilati, F.: Reconfigurability in cellular manufacturing systems: a design model and multi-scenario analysis. Int. J. Adv. Manufact. Technol. **104**(9–12), 4387–4397 (2019)

49. Dou, J., Su, C., Zhao, X.: Mixed integer programming models for concurrent configuration design and scheduling in a reconfigurable manufacturing system. Concurr. Eng. Res. Appl. **28**(1), 32–46 (2020)

50. Dou, J., Dai, X., Meng, Z.: Optimisation for multi-part flow-line configuration of reconfig-urable manufacturing system using GA. Int. J. Prod. Res. **48**(14), 4071–4100 (2010)
51. Dou, J., Dai, X., Meng, Z.: Precedence graph-oriented approach to optimise single-product flow-line configurations of reconfigurable manufacturing system. Int. J. Comput. Integr. Manuf. **22**(10), 923–940 (2009)
52. Dou, J., Dai, X., Meng, Z.: A GA-based approach for optimizing single-part flow-line configurations of RMS. J. Intell. Manufact. **22**(2), 301–317 (2011)
53. Battaïa, O., Dolgui, A., Guschinsky, N.: Decision support for design of reconfigurable rotary machining systems for family part production. Int. J. Prod. Res. **55**(5), 1368–1385 (2017)
54. Liu, M., An, L., Zhang, J., Chu, F., Chu, C.: Energy-oriented bi-objective optimisation for a multi-module reconfigurable manufacturing system. Int. J. Prod. Res. **57**(19), 5974–5995 (2019)
55. Mpofu, K., Tlale, N.S.: Multi-level decision making in reconfigurable machining systems using fuzzy logic. J. Manufact. Syst. **31**(2), 103–112 (2012)
56. Blum, C., Roli, A.: Metaheuristics in combinatorial optimization: overview and conceptual comparison. ACM Comput. Surv. **35**(3), 268–308 (2003)

# Author Index

L. Roucoules et al. (Eds.): JCM 2020, LNME, pp. 437–439, 2021.
https://doi.org/10.1007/978-3-030-70566-4